Springer Series in
Surface Sciences
27

Editor: G. Ertl

Springer Series in Surface Sciences

Editors: G. Ertl, R. Gomer and D. L. Mills   Managing Editor: H. K. V. Lotsch

E. Hulpke (Ed.)

Helium Atom Scattering from Surfaces

With Contributions by
G. Benedek   V. Celli   G. Comsa
R. B. Doak   J. W. M. Frenken   B. J. Hinch
H. Hoinkes   E. Hulpke   K. Kern
A. M. Lahee   J. Lapujoulade   J. R. Manson
L. Miglio   D. Neuhaus   B. Poelsema
K.-H. Rieder   H. Wilsch   Ch. Wöll

With 130 Figures

Springer-Verlag
Berlin Heidelberg New York
London Paris Tokyo
Hong Kong Barcelona
Budapest

Dr. E. Hulpke

MPI für Strömungsforschung, Bunsenstrasse 10
W-3400 Stuttgart 80, Fed. Rep. of Germany

Series Editors

Professor Dr. Gerhard Ertl

Fritz-Haber-Institut der Max-Planck-Gesellschaft, Faradayweg 4-6,
1000 Berlin 33, Fed. Rep. of Germany

Professor Robert Gomer, Ph. D.

The James Franck Institute, The University of Chicago, 5640 Ellis Avenue,
Chicago, IL 60637, USA

Managing Editor: Dr. Helmut K. V. Lotsch

Springer-Verlag, Tiergartenstrasse 17,
W-6900 Heidelberg, Fed. Rep. of Germany

ISBN 3-540-54605-7 Springer-Verlag Berlin Heidelberg New York
ISBN 0-387-54605-7 Springer-Verlag New York Berlin Heidelberg

Library of Congress Cataloging-in-Publication Data. Helium atom scattering from surfaces / E. Hulpke
(ed.) ; with contributions by G. Benedek ... [et al.]. (Springer series in surface sciences ; 27)
ISBN 0-387-54605-7 (U. S.) 1. Surfaces (Physics) 2. Scattering (Physics) 3. Helium. I. Hulpke, E. II. Series.
QC173.4.S94H45 1992 530.4'17--dc20   92-4029

Typesetting: Data conversion by Danny L. Lewis, Buchproduktion Berlin with TeX
Printing: Mercedes-Druck, Berlin; Binding: Lüderitz & Bauer, Berlin
54/3020-5 4 3 2 1 0 – Printed on acid-free paper

Preface

The first results of experiments in which helium atoms were scattered from crystal surfaces were published more than 60 years ago. However, it was more than four more decades before these experiments, originally intended to confirm basic quantum mechanical principles, had matured into an established tool in modern surface science. The close relationship to neutron scattering with its well known applications to questions of structure, dynamics, phase transitions and diffusion in bulk solids and liquids, alongside the extreme surface sensitivity and non-destructive nature of the technique, distinguish helium scattering as a unique method for investigations of the analogous properties and processes on solid surfaces.

An important breakthrough was achieved about 10 years ago with the introduction of He nozzle beams and energy analysis of the backscattered atoms increasing the field of applications significantly. Those developments and the availability of a sound theoretical understanding are responsible for the fact that the number of laboratories devoted to such studies is steadily increasing. At present about 15 groups in Europe and North America are involved in a great variety of atom–surface scattering experiments.

This volume presents a comprehensive account of the state of the art and of the contributions made by He scattering experiments to our present knowledge about solid surfaces. The 12 chapters, dealing with the different aspects and present applications of He–surface scattering, are written by distinguished experts in these fields. One intention is to enable fellow surface scientists to make an informed judgement on the benefits and shortcomings of He–surface scattering. A further aim is to assist those who wish to get involved in the exciting experiments using this gentle and surface-sensitive tool.

This book is dedicated to Professor J.P. Toennies on the occasion of his sixtieth birthday. He and his collaborators have pioneered phonon inelastic helium atom scattering and have not only contributed to the experimental sophistication that is now the state of the art in such experiments, but have also stimulated theorists to further improve an understanding of the scattering process, of helium–surface interactions, and of the dynamics of the solid surface itself. I would like to thank all the authors who have contributed their expertise by writing the individual chapters of this book and I am particularly grateful to them for their timely submission of the manuscripts. Special thanks go to Heidi Witte for her accurate and untiring work in preparing the TeX files.

Göttingen, April 1991

E.Hulpke

Contents

VIII

Contributors

Benedek, G.
  Dipartimento di Fisica dell'Università
  Via Celoria 16, I–20133 Milano, Italy

Celli, V.
  Department of Physics, University of Virginia
  McCormick Rd., Charlottesville, VA 22901, USA

Comsa, G.
  Institut für Grenzflächenforschung, KFA Jülich
  Postfach 1913, W–5170 Jülich 1, Germany

Doak, R.B.
  AT&T Bell Laboratories
  Room IE-350, 600 Mountain Ave., Murray Hill, NJ 07974, USA

Frenken, J.W.M.
  FOM Institute for Atomic and Molecular Physics
  Kruislaan 407, 1098 SJ Amsterdam, Netherlands

Hinch, B.J.
  Dept. of Chemistry, State University of New Jersey-Rutgers
  P.O. Box 939, Piscataway, NJ 08855-0939, USA

Hoinkes, H.
  Physikalisches Institut, Universität Erlangen-Nürnberg
  Erwin-Rommel-Strae, W–8520 Erlangen, Germany

Hulpke, E.
  MPI für Strömungsforschung
  Bunsenstr. 10, W–3400 Göttingen, Germany

Kern, K.
  École Polytechnique FED de Lausanne, Institute de Physique
  Expérimentale, PH-Ecublens
  CH–1015 Lausanne, Switzerland

Lahee, A.M.
  Springer-Verlag GmbH
  Postfach 105280, Tiergartenstr. 17, W–6900 Heidelberg, Germany

Lapujoulade, J.
Service de Physique des Atomes et de Surface
CEN Sarclay, Bat. 62, F–91191 Gif sur Yvette, Cedex, France

Manson, J.R.
Department of Physics and Astronomy, Clemson University
Clemson, SC 29633-1911, USA

Miglio, L.
Dipartimento di Fisica dell'Università
Via Celoria 16, I–20133 Milano, Italy

Neuhaus, D.
Institut für Raumsimulation, DLR
Postfach 906058, W–5000 Köln 90, Germany

Polsema, B.
Institut für Grenzflächenforschung, KFA Jülich
Postfach 1913, W–5170 Jülich 1, Germany

Rieder, K.-H.
Institut für Experimentalphysik, Freie Universität Berlin
Arnimallee 14, D–1000 Berlin 33, Germany

Seriani, G.
Osservatorio Geofisico Sperimentale
CP 2011 Opicinia, I–34016 Trieste, Italy

Wöll, Ch.
Institut für Angewandte Physik. Chemie, Universität Heidelberg
Im Neuenheimer Feld 253, W–6900 Heidelberg, Germany

Wilsch, H.
Physikalisches Institut, Universität Erlangen-Nürnberg
Erwin-Rommel-Strae, W–8520 Erlangen, Germany

1. Introduction

E. Hulpke

Prior to the advent of methods that provide real space images of solid surfaces, and in particular prior to the scanning tunneling microscope, diffraction techniques were virtually the only source of information on surface structure, ordered adsorbate phases, deviations from translational symmetry (surface defects) and two-dimensional structural phase transitions. This book is devoted to the only genuinely non-destructive surface sensitive diffraction method, helium atom scattering. The broad spectrum of applications of this technique to surface science problems is discussed in the next eleven chapters, each of which is written by one or more prominent experts in the field. A look at the wealth of information gained from helium atom scattering reveals that the exploration of reciprocal space and that of real space are quite complementary. Diffraction methods that are sufficiently well understood theoretically will continue to be irreplacable tools in surface science.

The Fourier transformation that is at the root of diffraction emphasizes effects related to translational symmetry, i.e., periodicity. Therefore a diffraction experiment is sensitive to changes in surface atom position which are so small that they are obscured by the noise of a real-space imaging instrument. Information on the density and the actual shape of defects such as voids, adatoms and steps can also be simply and accurately extracted from the diffraction data. A realm in which helium atom scattering is intrinsically and uniquely superior to real space imaging techniques is the investigation of the lattice dynamical properties of the surface. Coherent inelastic scattering of monochromatic de Broglie waves from surfaces is governed by the coupling of these waves to surface waves in accordance with the conservation laws for the frequencies and the components of the wave vectors parallel to the surface. Thus in inelastic scattering single phonon modes are projected out of the multimode excitations which dominate the picture in real space.

In recent years the theory of elastic and inelastic helium scattering from surfaces has undergone a remarkable development, and this progress, in conjunction with the refinement of the experiments, has been responsible for the impressive success of the method. Theoretical work has provided sophisticated methods as well as more convenient, but nonetheless well understood approximative schemes to evaluate data of elastic and single phonon inelastic helium

1

Springer Series in Surface Sciences, Vol. 27   **Helium Atom Scattering from Surfaces**
Editor: E. Hulpke   © Springer-Verlag Berlin, Heidelberg 1992

scattering. Both structural and dynamical studies have benefited immensely from the introduction of highly expanded nozzle beam sources and high resolution time-of-flight techniques. This very fortunate interaction between experiment and theory has, within the last ten years, made possible the elucidation of the surface phonon dispersion on a large number of metal and some semiconductor and insulator surfaces [1.1]. In many cases, unexpected anomalies have been found in the surface lattice dynamics. In some cases it has been possible to demonstrate surface phonon softening which is directly related to structural instability and reconstruction.

In addition to its intrinsic surface sensitivity, helium atom scattering is, in contrast to most other surface science tools, applicable also to insulating surfaces. Even though insulator surfaces were historically the first systems to be investigated by the helium beam method, these interesting surfaces have subsequently been treated as the poor relative in surface science. It is impossible to predict future developments in helium atom–surface scattering. There may indeed be a stronger emphasis on such insulating surfaces. It is also possible that it will be more frequently applied to the problems of surface diffusion or simply to accurately titrate the concentrations of defects and adatoms. The importance of this method in the field of structural and dynamical investigation will certainly remain.

It should be stated at the outset that the objective of this book is to present He–surface scattering as a tool in surface science. Thus general aspects of gas–surface interaction will not be treated. The experiments and theory descibed here are confined to elastic and single phonon inelastic scattering. The realm of multiphonon inelastic interactions which dominate surface collisions of heavier or more energetic projectiles is deliberately avoided.

The next chapter, "Experimental Limitations and Opportunities", introduces the reader to the modern experimental machinery of elastic and inelastic He–surface scattering. Its author, R.B. Doak, was the first experimentalist to successfully apply He nozzle beams in conjunction with time-of-flight analysis of the scattered atoms to the determination of surface phonon dispersion curves on LiF surfaces. The chapter contains a discussion of energy and angular resolution and their interrelation as well as a discussion of signal versus noise problems and possible signal improvement schemes achievable by He atom focussing or by the exploitation of scattering resonances. In addition, ways are presented of evaluating experimental results in terms of the collision kinematics. Kinematical considerations also enter into the accuracy to which phonon energies and momenta can be measured.

The following five chapters are devoted to different aspects of elastic He scattering from surfaces. This section is headed in Chap. 3 by a theoretical treatise on the He–surface interaction potential. V. Celli introduces the reader to more or less complex model potentials as well as ways of calculating attractive and repulsive contributions to these potential hypersurfaces. In addition to static considerations, the influence of surface vibrations on the interaction is discussed in terms of a dynamical potential.

2

Elastic He scattering provides information on the geometrical structure of surfaces. Chapter 4, by K.-H. Rieder, deals with the determination of structures on well–ordered surfaces. It contains a presentation of approximative methods of extracting single crystal surface structures from measured He atom diffraction data. Data inversion schemes based on the hard corrugated wall approximation are discussed as well as the limitations of such procedures. Examples of reconstructed and unreconstructed surfaces with both small and large corrugations illustrate the achievements of atom–surface scattering experiments in the field of structure determination.

The He–surface interaction is particularly sensitive to deviations from a perfectly periodic structure on the investigated surface in the form of point defects or steps. The authors of Chap. 5, A.M. Lahee and Ch. Wöll, were the first to investigate the diffuse elastically scattered He intensity in order to obtain information on the shape of step edges and adsorbed molecules. This chapter also shows what one can learn about defect concentrations and step–height or terrace–width distributions from such experiments. Chapter 6 by J. Lapujoulade concentrates on the investigation of steps and, in particular, on the interesting phenomenon of roughening, a structural phase transition specific to surfaces with a finite step density.

Elastic scattering resonances due to bound states in the attractive atom–surface potential well were among the interesting features observed in the early He scattering experiments of O. Stern and collaborators. The determination of scattering conditions for these "selective adsorption" resonances represents the most accurate way of measuring the actual form of the He–surface interaction potential. The last contribution to the section on elastic scattering, Chap.7 by H. Hoinkes and H. Wilsch, is devoted to modern studies of such scattering resonances.

The second part of this work contains five chapters on single phonon inelastic He–scattering experiments. The connection to the first part on elastic scattering is provided by by J.R. Manson in Chap. 8 via his comprehensive description of elastic and single phonon inelastic scattering theory and associated approximation methods.

The theory of surface lattice dynamics had reached a high level of sophistication long before He–scattering experiments could provide surface phonon dispersion data. The subject of Chap. 9 by G. Benedek and L. Miglio is the development of these theoretical efforts and the influence that the availability of experimental results had on the subsequent improvements of this theory. Two chapters follow which illustrate the type of information on surface lattice dynamics that can be gained from He scattering experiments. The first of these (Chap. 10) is written by three authors from the He scattering group in Jülich, the other (Chap. 11) originates from the involvement of B. Feuerbacher in this field. The final contribution to this book is by J.M.W. Frenken and B.J. Hinch and deals with yet another fascinating aspect. In a fashion analogous to bulk diffusion studies, which use quasielastic neutron scattering, quasielastic He scattering allows investigations of surface diffusion provided that the diffusion coefficients are sufficiently large and the energy resolution of such experiments is on the order of 100 μeV.

Quite recently another facet has been added to the applications of He scattering in surface science, namely the study of thin film growth on crystalline substrates [1.2–6]. It has been customary to monitor film growth that proceeds layer by layer by measuring the oscillating behavior of the specular reflection of electrons with deposition time. He scattering in the antiphase condition produces similar oscillations and has been successfully applied in the fast growing field of thin epitaxial film studies.

References

1.1   W. Kress, F.de Wette (eds.): *Surface Phonons*, Springer Ser. Surf. Sci. Vol. 21 (Springer, Berlin, Heidelberg 1991)

1.2   L.J. Gomez, S. Bourgeal, J. Ibanez, M.Salmeron: Phys. Rev. **B31**,2551 (1985)

1.3   J. Ferron, J.M. Gallego, A. Cellabolla, J.J. Miguel, S. Ferrer: Surf. Sci. **211/212**, 797 (1989)

1.4   B.J. Hinch, C. Kozioł, J.P. Toennies, G. Zhang: Vacuum **42** 309 (1991); Europhys. Lett. **10** 341 (1989)

1.5   R. Kunkel, B. Poelsema, L.K. Verheij, G. Comsa: Phys. Rev. Lett. **65**, 733 (1990)

1.6   Wei Li, J.S. Liu, M. Karimi, C. Moses, G. Vidali: Appl. Surf. Sci. **48/49**, 160 (1991)

2. Experimental Limitations and Opportunities in Single-Phonon Inelastic Helium Scattering

R.B. Doak

Somewhat over a decade ago, researchers in Göttingen [2.1] and Saclay [2.2] discovered that it was possible to produce helium atomic beams with a velocity spread of well under 1%. Attainment of this unprecedented energy resolution (a fraction of 1 meV) made it feasible to explicitly resolve the small energy losses or gains which ensue through inelastic interaction of a helium atom with the vibrational modes of a solid surface. This immediately stimulated tremendous interest in the use of inelastic helium scattering to experimentally probe surface lattice dynamics [2.3]. The first measurement of a complete surface phonon dispersion curve was reported in 1981 [2.4], heralding a period of dramatic growth in inelastic helium scattering. Today there are about twenty high resolution helium beam machines in existence, distributed among a dozen groups worldwide. Single-phonon inelastic helium scattering has clearly found a niche.

The reasons for this are fairly obvious. Apart from having excellent energy resolution and high intensity, a helium atom beam offers the intrinsic advantages of being strictly surface sensitive, completely inert, and well-matched in energy and momentum to the surface phonons under study. With helium scattering it is possible to measure large wave vector phonons — which are inaccessible to optical spectroscopies — and to resolve closely-spaced modes — which may be beyond the resolution of inelastic electron scattering. Helium scattering therefore both complements and augments other surface vibrational spectroscopies.

Given this unique niche and the still somewhat nascent nature of inelastic helium scattering, there have been only sporadic attempts [2.5–7], to delineate the experimental limitations of the technique. With helium scattering, as with any experimental method, there exists a hierarchy of constraints which determines the range and resolution of possible measurements. The fundamental constraints are those imposed by the use of a neutral atom as the probe species. But there are technological constraints as well, having to do with the cost and complexity of experimental apparatus required to accommodate these intrinsic restrictions. This chapter will attempt to outline in a coherent fashion these various limitations of inelastic helium scattering, drawing upon the wealth of

5

Springer Series in Surface Sciences, Vol. 27   **Helium Atom Scattering from Surfaces**
Editor: E. Hulpke   © Springer-Verlag Berlin, Heidelberg 1992

experimental data which has accrued over the past decade. An effort will also be made to show where and by what means these boundaries might be further expanded.

2.1 Experimental Limitations

To introduce the subject of experimental limitations, we first discuss briefly the experimental techniques employed to probe surface phonon characteristics with helium scattering. A detailed discussion of the experimental methods is given elsewhere [2.8]. A typical phonon measurement is depicted schematically in Fig.2.1. The intense, nearly monoenergetic helium beam is directed towards a target surface at a particular angle of incidence and the scattered intensity monitored at a given angle of emergence. Creation or annihilation of surface phonons will produce discrete shifts in the energy of the scattered beam. Experimental conditions are chosen to ensure that single phonon scattering dominates, in which case energy analysis of the scattered beam yields directly the energy of the surface phonons. From the measured energy exchange and knowing the angles of scattering, the momentum of the phonon is also easily computed to provide one point $\omega(Q)$ on the phonon dispersion curve. By varying kinematic parameters (beam energy, scattering angles) it is possible to sample other $\omega(Q)$ points and to map out the full dispersion relation.

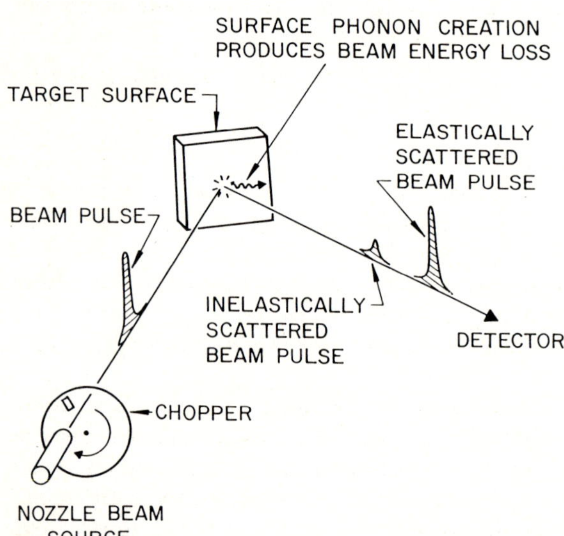

Fig.2.1. Schematic depiction of an inelastic helium scattering measurement using time-of-flight (TOF) spectroscopy to energy-analyze the scattered beam. A phonon creation (Stokes) event is illustrated. Phonon annihilation (anti-Stokes) scattering is also possible

Virtually all inelastic helium scattering experiments make use of time-of-flight energy analysis [2.9,10]. As illustrated in Fig.2.1, this entails pulsing the beam mechanically and measuring the transit time of the pulse over a known flight path, e.g. the transit time t_{cd} of an elastically scattered atom over the flight path x_{cd} from chopper to detector yields the incident beam velocity $v_i = x_{cd}/t_{cd}$. Inelastic scattering at the target surface transfers energy to or from the atoms, changing their velocity and thereby their flight time from target to detector. A measurement of detector output as a function of time — a TOF spectrum — displays peaks which are shifted relative to the elastic flight time. From these shifts the velocity v_f of the scattered beam, and thereby the amount of energy gained or lost at the target surface, can be computed. A set of typical TOF spectra is shown in Fig.2.2.

In making surface phonon measurements, there are two matters of primary concern. The first is the resolution with which a phonon frequency and wave vector can be measured. Instrumental broadening will determine the minimum separations in phonon frequency and wave vector which can be explicitly resolved. For both energy and momentum measurements this broadening will depend upon:

1. the characteristics of the incident beam, namely its velocity spread and angular spread;
2. the kinematics, which determine range of accessible phonon energy and momentum given this non-zero spread of velocity and angle in the incident beam; and
3. the TOF resolution, since the acts of pulsing and detecting the beam introduce limitations of their own.

The second concern is that of signal intensity, which will be a function of:

1. the source intensity,
2. the beam collimation and scattering geometry,
3. the TOF parameters,
4. the inelastic scattering cross-sections and
5. the detector sensitivity.

As usual there will be trade-offs between resolution and intensity although it is interesting to note that, as far as *beam generation* is concerned, no such trade-off is necessary: increasing the resolution of a supersonic nozzle beam also improves its intensity!

We restrict the present discussion to questions of resolution and intensity alone. Helium scattering is presently pushing into new and different areas, sampling the envelope of other experimental limitations, e.g. in generating ultra-cold beams for atom optics applications or in using excited metastable helium to yield a probe with a magnetic moment and allow scattering from magnetic surfaces. To limit this discussion to a reasonable length, however, we concentrate on single-phonon inelastic scattering, for which intensity and resolution are the overriding concerns.

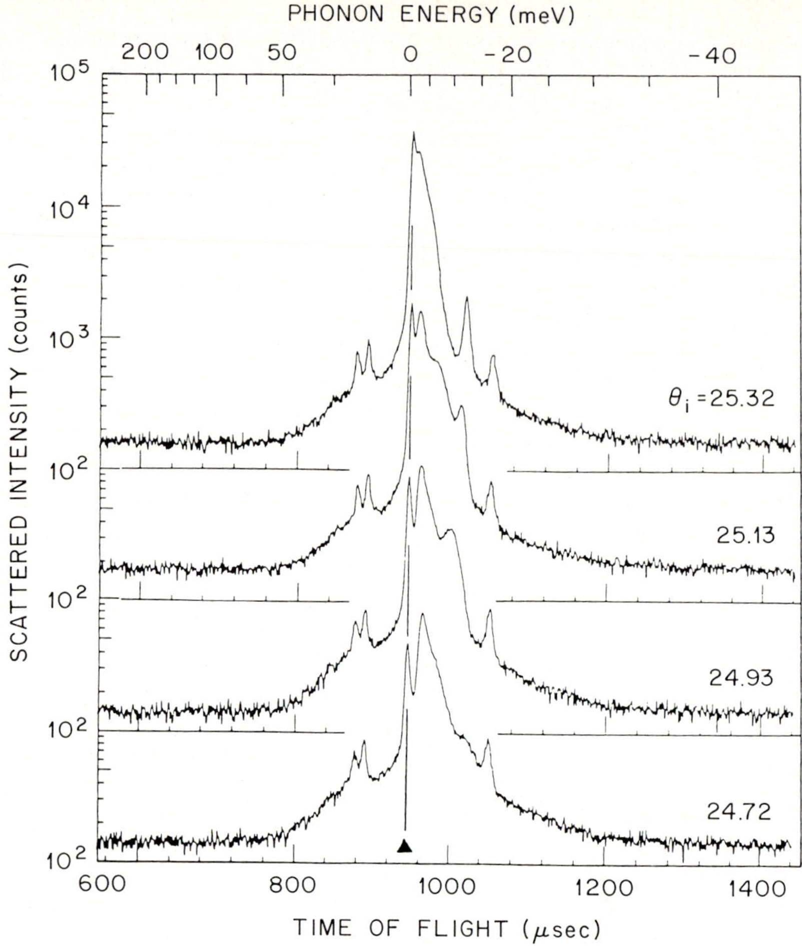

Fig.2.2. Typical TOF spectra, here for scattering from (1×1) arsenic-terminated Si(111) surface along the $\langle 112 \rangle$ azimuth with a room temperature beam [2.8]. The elastic flight time is marked by the small triangle. This elastic peak arises through incoherent elastic scattering from surface defects. Peaks at longer flight times than the elastic are due to inelastic scattering involving creation of a single surface phonon. Peaks at shorter flight times arise through phonon annihilation. Note that the ordinate is logarithmic. The amplitude of the phonon peaks decreases dramatically with increasing phonon frequency and wave vector, i.e. for flight times shifted further away from the elastic flight time

2.2 Beam Resolution

Thanks to some fortuitous quantum mechanical enhancements [2.11–13] the intrinsic energy resolution of a ^{4}He beam is exceptionally and uniquely excellent. A beam velocity distribution of $\Delta v/v = 6 \times 10^{-3}$, FWHM, is routinely attainable, giving energy widths of about 250 neV and 800 neV, respectively, for a liquid-nitrogen-cooled and a room temperature nozzle [2.6]. This narrow velocity distribution is achieved by operating with as high a nozzle stagnation pressure p_0 and as large a nozzle diameter D^* as possible, conditions which are obviously incompatible with maintaining vacuum conditions in the nozzle chamber to avoid Beer's law attenuation of the beam. One solution is to simply mount tremendous vacuum pumps ($20\,000\ell/s$) on the source chamber, the configuration perfected by *Toennies* and colleagues in Göttingen [2.1,10]. This approach is facilitated by using very high nozzle pressures (up to 1000 atm) and very small nozzle diameters (5–20 μm) to yield high values of $p_0 D^*$ and a corresponding narrow velocity spread while simultaneously maintaining reasonable values of gas throughput, which scales roughly as $p_0 D^{*2}$. An alternative approach is to abandon any attempt to maintain vacuum conditions in the source chamber and to operate instead in a continuum flow regime, extracting the beam from the "zone of silence" [2.13] bounded by the barrel shock and Mach disk of the free-jet expansion. This is the configuration developed in Saclay.

Both of these configurations are in use for inelastic scattering measurements, although the Göttingen configuration seems to be the prevalent choice. Both can provide the sub-meV energy spreads quoted above. Although this energy resolution is more than adequate for most elastic and inelastic scattering experiments, attempts are being made to improve beam performance still further. This can be done by pushing to yet higher values of $p_0 D^*$, although there is an intrinsic limitation imposed by condensation in the expansion, the presence of which spoils the beam resolution [2.14]. The pressure at which condensation appears decreases with decreasing nozzle temperature [2.14,15]. At 77K and below, clustering is the limiting factor for nozzles in the range of 5–20 μm diameter. Since condensation scales roughly as $p_0^2 D^*$, its avoidance requires higher $p_0 D^*$ values to be obtained by operating at lower source stagnation pressures and larger D^*. This, in turn, vastly increases $p_0 D^{*2}$ and thereby the nozzle throughput, probably beyond the capability of any vacuum pump of reasonable size for cw source operation. Higher Mach number beams will therefore almost certainly be based on pulsed nozzle technology [2.16]. Pulsed operation introduces complexities of its own, including the fact that a pulsed nozzle has a finite turn-on time [2.17]. The beam must still be chopped, therefore, with the concomitant TOF limitations to be discussed below. An alternative route to higher resolution is to simply velocity-select a conventional cw nozzle beam, either mechanically [2.18] or perhaps by diffractive monochromatization [2.9c].

2.3 Kinematic Smearing

For a given spread in beam velocity and angle, it is a straightforward matter to predict the kinematic smearing in energy and momentum of a phonon measurement. An inelastic scattering event can be characterized by the energy ΔE and parallel momentum ΔK transferred from the surface to the atom [2.8]. The kinematics of the inelastic scattering is quickly visualized by superimposing a "scan curve" — that locus of points $\Delta E(\Delta K)$ which satisfy conservation of energy and momentum — upon an extended-zone plot of the phonon dispersion curves. For an "in-plane" scattering geometry the scan curve relation is given by

$$\Delta E = \frac{(\sin\theta_i + \Delta K/k_i)^2}{\sin^2\theta_f} - 1 \quad . \tag{2.1}$$

A typical set of scan curves is shown in Fig.2.3. Each TOF spectrum samples a cut through $\Delta E(\Delta K)$-space along its specific scan curve. Every point of intersection with a phonon dispersion curve represents a kinematically accessible phonon $\omega(\boldsymbol{Q})$ and, provided the scattering cross-section is adequate, one will observe a corresponding single phonon peak in the TOF spectrum. Varying any kinematic parameter, e.g. scattering angles as in Fig.2.3, shifts the intersections to sample different points along the dispersion curve.

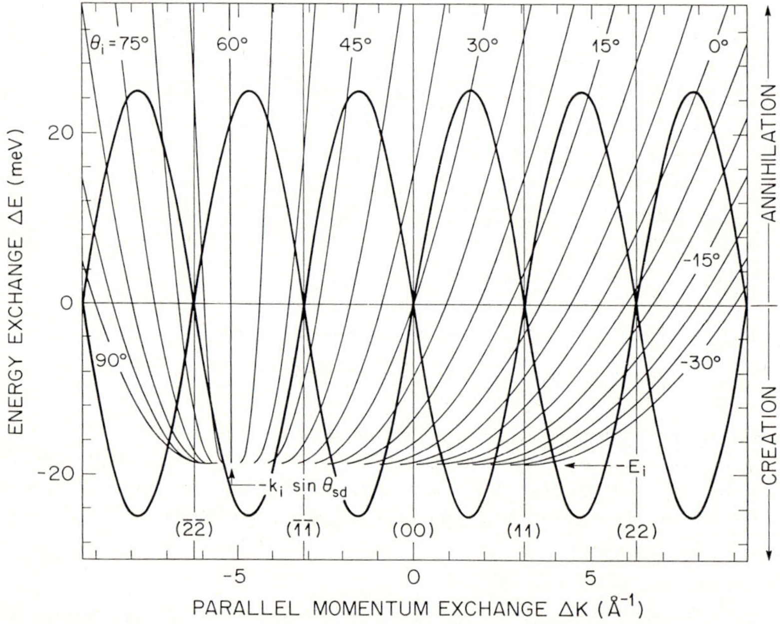

Fig.2.3. Scan curves for a liquid-nitrogen-cooled beam ($k_i = 6.00$Å^{-1}, $E_i = 18.8$ meV) scattering from a surface under a fixed source–target detector angle $\theta_{sd} = \theta_i + \theta_f = 60°$ [2.8]. The lighter, parabolic line curves are the scan curves for the incident angles θ_i shown. The darker sinusoids are representative surface phonon dispersion curves. Intersections between a given scan curve and the various dispersion curves give the set of phonons which may be created ($\Delta E < 0$) or annihilated ($\Delta E > 0$) at any particular value of θ_i. Phonon creation with $|\Delta E| > E_i$ is clearly not possible

The scan curves of Fig.2.3 correspond to those of an ideal beam: zero velocity spread and zero angular spread. To investigate kinematic smearing it is a simple matter to plot the *envelope* of scan curves corresponding to the actual distribution of k_i values. This is done in Fig.2.4 for a full width velocity spread of 6% and in Fig.2.5 for full width angular spread of 3°. These spreads are exaggerated by a factor of *ten* over those of an actual beam in order to yield a tangible width to the scan curve envelopes. The range of ΔE and ΔK given by the intersection of these scan curve envelopes with the dispersion curves constitutes the kinematical smearing.

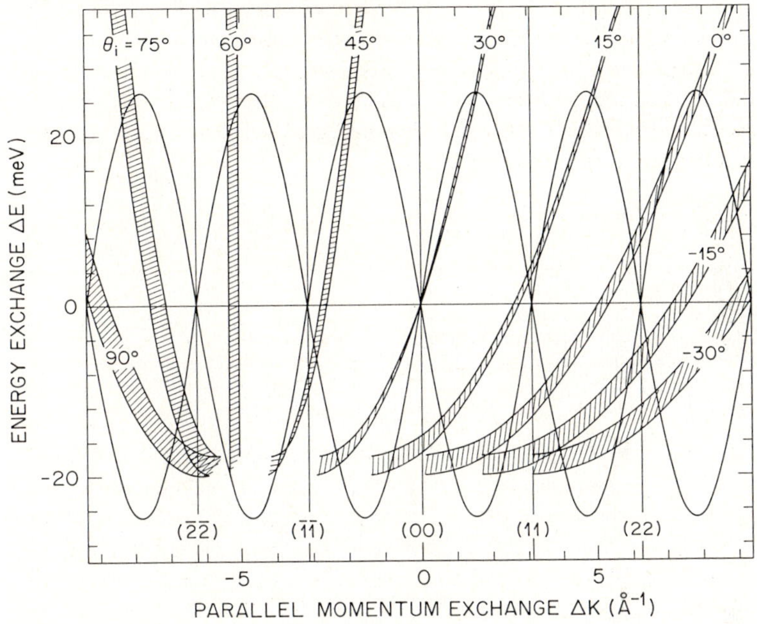

Fig.2.4. Demonstration of kinematic smearing due to the incident beam velocity spread [2.8]. The envelope of scan curves at each θ_i corresponds to $k_i = 5.82$ to 6.18 Å^{-1} or *ten times* an actual velocity spread. Note that near the specular lattice rod ($\Delta K = 0$), the width of the scan curve envelope narrows substantially. The scattering geometry is the same as that of Fig.2.3

It is obvious that the kinematical smearing is a strong function of both the scattering geometry and of the shape of the dispersion curves. For an actual beam (with one-tenth of the spreads shown in Figs.2.4,5) it is generally possible to pick scattering angles such that the kinematic smearing is not a major concern. In calculating the smearing due to the angular spread, Fig.2.5, the finite size of the target was neglected. In fact, at large incident angles, $\theta_i \gtrsim 60°$, the target itself becomes the limiting factor in determining the angular spread of the beam and the scan curve envelope will actually narrow to zero width as θ_i approaches 90° (with a concomitant loss in intensity). Thus, overall, the

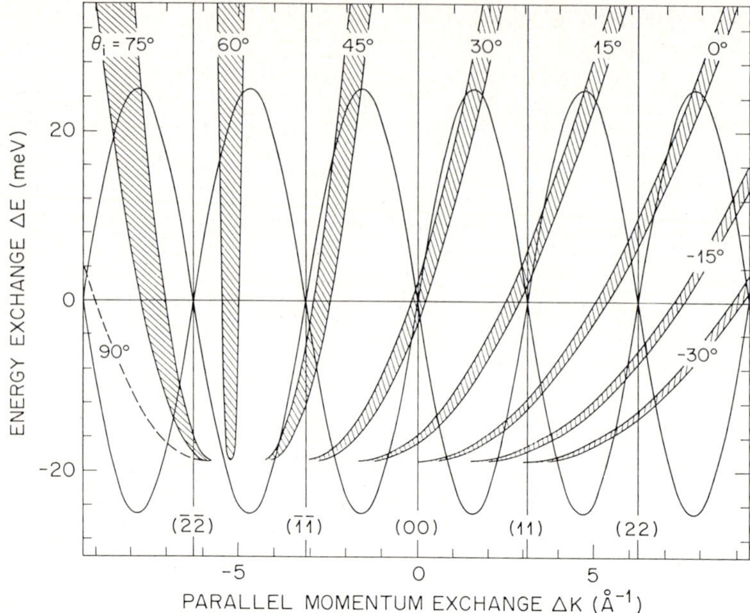

Fig.2.5. As in Fig.2.4, but demonstrating the kinematic screening due to the angular spread of the beam [2.8]. The scan curve envelopes correspond to a full-angle spread of 3.0° or again *ten times* that of an actual beam

smearing due to a 0.6% velocity spread is roughly compatible with that due to a 0.3° angular spread. It can be seen, however, that the velocity smearing virtually disappears at near-specular angles. Hence an attempt to further reduce kinematic smearing would probably concentrate on reducing the angular divergence. This is of interest in conjunction with the beam focussing discussed below which could, in principle, be used to produce a parallel beam. Kinematical smearing is a straightforward geometrical effect. Unless the scattering cross-sections vary dramatically over the width of the scan curve envelope it is thus possible, once the dispersion curves have been "roughed-out", to make a second pass through the data, using the measured dispersion curves to fit the TOF peak shapes and deconvolute the smearing effects. Given the minor contribution of the kinematical smearing to the overall resolution, there is little incentive to do so, however. Kinematical smearing poses no major problems for inelastic helium scattering.

2.4 Time-of-Flight Resolution

To resolve two peaks in a TOF spectrum such as those of Fig.2.2, it is the *time* width of the peaks which is of importance. The *velocity* spread of the beam will contribute to this time width but need not be the dominant factor. Indeed, it turns out that for the high energy phonon annihilation processes it is never the beam velocity distribution which limits energy resolution but rather time widths introduced by the TOF analysis itself. To understand this we consider in a rather simplistic but transparent way [2.6–8] the various contributions to a TOF peak width. First of all there is the chopper shutter function, which produces a beam pulse of time width

$$\Delta_{\mathrm{S}} \; = \; \frac{w_{\mathrm{s}}\, N_{\mathrm{s}}\, \tau_{\mathrm{s}}}{2\,\pi\, r} \quad , \tag{2.2}$$

where w_{s} is the effective chopper slit/beam width [2.8], N_{s} the number of slits on the disk (at radius r), and τ_{s} the chopper slit/pulse frequency. The beam pulse will disperse in time as it travels from chopper to target contributing an amount

$$\Delta_{\mathrm{F}}(\mathrm{ct}) \; = \; \frac{-x_{\mathrm{ct}}\, \Delta v_{\mathrm{i}}}{v_{\mathrm{i}}^{2}} \tag{2.3}$$

to the overall time width, where x_{ct} is the chopper–target flight path and Δv_{i} the velocity spread of the incident beam. The atoms scatter from the surface to emerge with a velocity v_{f}. The TOF spread in traveling from target to detector is then

$$\Delta_{\mathrm{F}}(\mathrm{td}) \; = \; \frac{\partial t_{\mathrm{td}}}{\partial v_{\mathrm{f}}}\, \frac{\partial v_{\mathrm{f}}}{\partial v_{\mathrm{i}}}\, \Delta v_{\mathrm{i}} \; = \; -x_{\mathrm{td}} \left(1 \, + \, \frac{\Delta E}{E_{\mathrm{i}}}\right)^{-3/2} \Delta v_{\mathrm{i}} \quad , \tag{2.4}$$

where x_{td} is the target–detector flight path, ΔE the energy gained by the atom at the surface ($\Delta E > 0 \Leftrightarrow$ phonon annihilation, $\Delta E < 0 \Leftrightarrow$ phonon creation) and E_{i} the incident beam energy. Finally, there is time smearing due to the finite detector length, or

$$\Delta_{\mathrm{D}} \; = \; \frac{L_{\mathrm{D}}}{v_{\mathrm{f}}} \; = \; L_{\mathrm{D}} \left(1 \, + \, \frac{\Delta E}{E_{\mathrm{i}}}\right)^{-1/2} \quad . \tag{2.5}$$

Ignoring the niceties of a proper convolution integral [2.8] we combine these time widths as if they were Gaussian functions

$$\Delta_{\mathrm{T}} \; = \; (\Delta_{\mathrm{S}}^{2} \, + \, \Delta_{\mathrm{F}}^{2}(\mathrm{ct}) \, + \, \Delta_{\mathrm{F}}^{2}(\mathrm{td}) \, + \, \Delta_{\mathrm{D}}^{2})^{1/2} \tag{2.6}$$

and relate the total time spread Δ_{T} to an effective energy width $\delta(\Delta E)$

$$\delta(\Delta E) \; = \; \frac{\partial \Delta E}{\partial t_{\mathrm{cd}}}\, \Delta_{\mathrm{T}} \tag{2.7}$$

or

$$\frac{\delta(\Delta E)}{E_i} = -2\left(1 + \frac{\Delta E}{E_i}\right)^{3/2} \frac{v_i}{x_{td}} \Delta_T \quad . \tag{2.8}$$

These relations apply when the beam is chopped *before* striking the target. Setting $\Delta_F(ct) = 0$ and letting $t_{td} \rightarrow t_{cd}$ gives the relation for a beam chopped *after* striking the target.

An ideal TOF spectrometer would have Δ_S, Δ_D, and $\Delta_F(ct)$ all negligible with respect to $\Delta_F(td)$. In this case

$$\frac{\delta(\Delta E)}{E_i} = \frac{2\,\Delta v_i}{v_i} \quad . \tag{2.9}$$

This value is often quoted as the "energy resolution" of an apparatus but it is obvious from the above equations that this baseline value does *not* suffice to characterize the *inelastic* resolution at arbitrarily large ΔE *except* for $\Delta_S = \Delta_D = \Delta_F(ct) \equiv 0$. To demonstrate the broadening effects of the other contributions in (2.6) it is useful to consider a concrete example. Table 2.1 lists TOF parameters as measured by *Doak* and *Nguyen* [2.6] for an apparatus at Bell Labs. These values are typical of most of the latest generation of inelastic helium scattering machines. Although $\Delta_F = \Delta_F(ct) + \Delta_F(td)$ limits the resolution for these *elastic* scattering measurements, a plot as a function of phonon energy — Fig.2.6, heavy solid curves — tells a somewhat different story. In particular, the baseline resolution of (2.9) is achieved only for large energy ($\Delta E \simeq -E_i$) phonon creation events. For phonon annihilation events the resolution deteriorates rapidly with increasing phonon energy, reflecting the non-linear mapping of a TOF interval onto an energy interval as given by (2.8). Note that this TOF contribution to the energy broadening will far exceed that due to kinematic smearing or that of the beam itself. Since one often must measure phonon annihilation events (to access $|\Delta E| > E_i$) it is clear that future gain in energy resolution will require improvements in TOF resolution. Two other sets of curves in Fig.2.6 illustrate the possible gains. The light solid lines show the improvement upon eliminating the chopper–target TOF dispersion by moving the chopper to a point just downstream of the target. This produces only minimal gains. By in addition reducing the shutter function and detector length contributions it is possible to effect much more dramatic changes. The broken line curves shown the situation if w_s and L_D are reduced to 20% of the above values. This yields sub-meV resolution up to a phonon energy of 50 meV.

The above discussion makes it clear that TOF resolution is a major limitation at present. While it is possible to attain sub-meV energy resolution for phonon creation events, phonon creation is not always a viable option, specifically when sampling $|\Delta E| \gtrsim E_i$. Nor is it possible to increase E_i indefinitely to this avoid impasse as doing so reduces the baseline resolution, as seen in Fig.2.6. Higher E_i can also move the scattering into the multiphonon regime although this can, to some extent, be countered by reducing the surface temperature or increasing the angle of incidence as the beam energy is increased [2.19].

To improve TOF resolution it is necessary to:

Table 2.1. Summary of TOF Resolution Measurements

Specular Scattering, LiF(001) $\langle 100 \rangle$ $\theta_{sd} = 110°$; Chopper slot frequency ν_s=1010 Hz; Chopper-detector flight path $x_{cd} = 1.6916$ m; Chopper-target flight path $x_{ct} = 0.479$ m; Nozzle diameter=10 μ, nominally.

Beam characteristics

Nozzle temp. [K]	300	77
Nozzle pressure [bar]	116	118
Beam velocity [m/s]	1783.3	964.4
Flight time t_{cd} [μs]	948.6	1754.0

Specular TOF width [FWHM]*

Total Δ_T [μs]	7.8	14.4
Shutter function Δ_S [μs]	3.9± .1	4.4± .1
Velocity distribution Δ_F [μs]	5.2± .8	11.4± 1.2
Detector length Δ_D [μs]	4.3± .7	7.6± 1.3

Calculated parameters*

Effective shutter width w_s [FWHM, mm]	0.70± .01	0.62± .01
Detector length L_D [FWHM, mm]	7.7± 1.3	7.4± 1.2

Resolution

TOF, $\Delta t/t$ [FWHM]	0.0082	0.0082
Velocity, $\Delta v/v$ [FWHM]	0.0055	0.0065
Energy, true [FWHM, meV, over x_{cd}]	0.72	0.25
Energy, effective [FWHM, meV, over x_{td}]	1.52	0.44

* Error bars are 95% confidence intervals to least squares fits and *do not* include systematic errors due to modeling!

1. narrow the chopper shutter function, which requires *both* the angular spread of the beam and the chopper slit width to be reduced,
2. reduce the length of the detector, and/or,
3. increase the target–detector flight path.

Moving the chopper to a point downstream of the target will potentially bring some improvement as in Fig.2.6. However, since the beam will be broader there than near the nozzle, this can also worsen the chopper shutter function given the limits on maximum chopper speed [2.8]. The gain in resolution obtained through any of these changes results in a loss of usable beam signal.

15

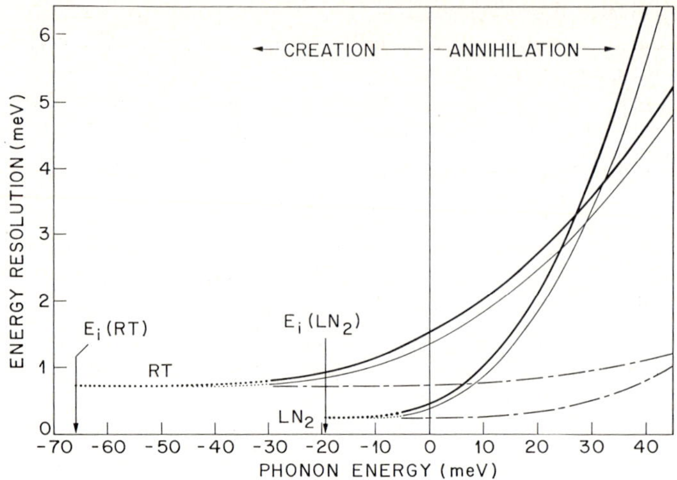

Fig.2.6. The effective energy width (dark solid lines) of a TOF peak as a function of phonon energy, calculated using the chopper shutter function, detector length, and TOF distances of Table 2.1. The lighter solid lines show the improvement obtained by moving the chopper to just downstream of the target to eliminate the time dispersion over the chopper–target flight path (and neglecting changes in beam width). The dot-dashed lines show the additional gain achieved by, in addition, reducing the shutter function and detector length to 20% of the previous values. Dotted lines show the region on the creation side in which TOF dispersion as $v_f \rightarrow 0$ may make measurements difficult

Ultimately, then, the limits on energy resolution couple to limitations on beam intensity and on inelastic scattering cross-sections.

2.5 Intensity Limitations

Signal intensity is every bit as crucial as energy and momentum resolution in determining the success of an inelastic scattering measurement, given the trade-offs between angular resolution and TOF resolution on the one hand and beam intensity on the other. For a selected beam geometry and a given choice of TOF parameters the course to be charted is obvious:

1. maximize beam spectral intensity,
2. minimize background levels at the detector,
3. optimize detector sensitivity, and
4. choose experimental conditions to maximize inelastic scattering cross-sections.

   Gains can still be realized in all of these areas, although in the case of the first three — the "technology" areas — this will probably require development of new methods rather than simple extension or refinement of existing methods. One may therefore prefer to ask: "Given the present state of the art, what are the intensity-imposed limitations on inelastic helium scattering as a surface spectroscopy probe?". In the following we explore this question and offer some suggestions for improving signal intensity.

2.6 Frequency and Momentum "Cut-Off"

Due to small scattering cross-sections and large steric factors, the signal intensities for single phonon scattering are quite low. Pulse counting rates for an unchopped, inelastically scattered beam are typically well under 10^4 counts/s, compared with 10^6–10^7 counts/s for elastic scattering (specular and diffraction) and 2×10^8 counts/s in the incident beam [2.8]. Chopping reduces these levels by the duty factor of the chopper disk, or typically by of 3×10^{-3} for single-slit chopping. Nonetheless, by integrating over many beam pulses it is possible to measure well-defined single phonon loss/gain features, such as those displayed in the TOF spectra of Fig.2.2. The need for long integration times (typically about 1 h) raises problems of its own, since the sample surface must be either maintained clean over the entire integration time or the measurement must be interrupted periodically to re-clean the surface. This is a limitation of a technical nature, which makes a measurement more cumbersome but not impossible. A more intractable problem is that posed by the very nature of the interaction of a neutral atom with a solid surface. Basically, atoms are big and slow, at least in comparison to other possible probe particles such as neutrons, electrons, or photons. Consider a classical picture of a hard sphere atom impacting an isotropic elastic continuum. The atom will produce a deformation of the surface of an extent no smaller than the size of the atom and of an impulse no shorter than the collision time. Since this deformation may be expanded in terms of a Fourier series of the normal modes of vibration of the solid, it is clear that a large, slow atom will be unable to excite high frequency, short wavelength normal modes, i.e. the inelastic scattering will be "cut-off" beyond certain values of phonon energy and wave vector. Helium scattering is, of course, poorly described by classical mechanics, but these "cut-off" effects persist even in a rigorous quantum treatment. Quantum mechanically the "cut-off" is not *absolute*, but rather characterized by a rapid decay of the single phonon intensity at wave vectors beyond that corresponding roughly to the inverse of the hard sphere atomic diameter and/or a frequency corresponding to the inverse of the "turn-around" time of the atom in striking the repulsive wall of the gas-surface potential. These trends are seen in the TOF spectra of Fig.2.2 and are explicitly demonstrated in Figs.2.7 and 2.8. Note that the momentum "cut-off" takes the particularly simple form

$$I(Q) \; = \; I_0 e^{-Q^2/Q_c^2} \; , \tag{2.10}$$

with Q_c the "cut-off" value for phonon momentum. The TOF data of Fig.2.2 demonstrate that phonon peaks may be measured over an intensity range varying by a factor of 10^2 or more.

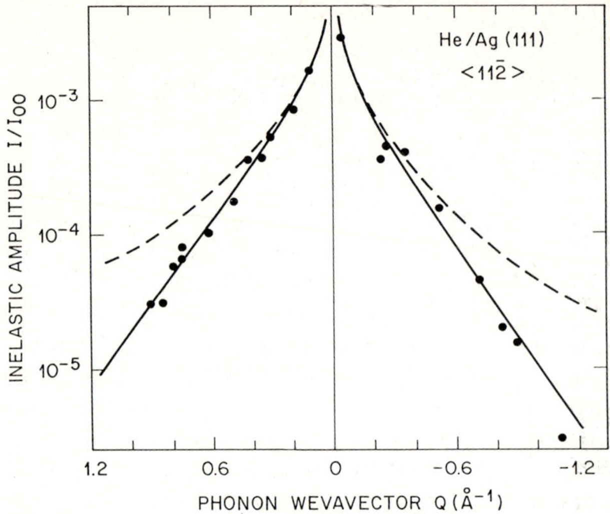

Fig.2.7. Momentum "cut-off" due to the finite size of the helium atom [2.20]. Data points are for helium scattered from Ag(111). The curves show calculations both with a momentum "cut-off" ($Q_c = 0.74$ Å^{-1}, solid line) and without ($Q_c = \infty$, dashed line)

From Figs.2.7 and 2.8 this is seen to correspond to measurements out to about $Q_c = 2$ Å^{-1} and $\hbar\omega_c = 30$–50 meV (depending on the beam energy). Note that the amplitudes of the phonon peaks in Fig.2.2 reflect not just the "cut-off" but also the thermal population of the normal modes. Some enhancement of annihilation peaks may be achieved by increasing the surface temperature, although there are definite limits given by the onset of multi-phonon scattering at high temperatures. *Weare* has discussed the trade-offs among surface temperature, beam energy, and scattering angles in establishing a regime of single phonon interactions [2.19]. It is also worth emphasizing that — with the exception of hydrogen, which is of little use as a beam species due to the tremendous natural background on masses 1 and 2 — helium suffers less from "cut-off" effects than any other atom. It is clear, however, that "cut-off" effects can be as important in establishing the range of measurements as is the TOF resolution.

2.7 Improving Signal Intensity: Bound State Resonances

To increase both the resolution and range of single phonon measurements, it appears that signal intensity is the primary concern to be addressed. Moreover, since *both* the TOF resolution and the "cut-off" effects become limiting at phonon energies of a few tens of millielectron volts, it would seem that *dramatic* increases in signal intensity are required. Is this feasible? The answer is "yes", but with the qualification that a much more painstaking and detailed approach

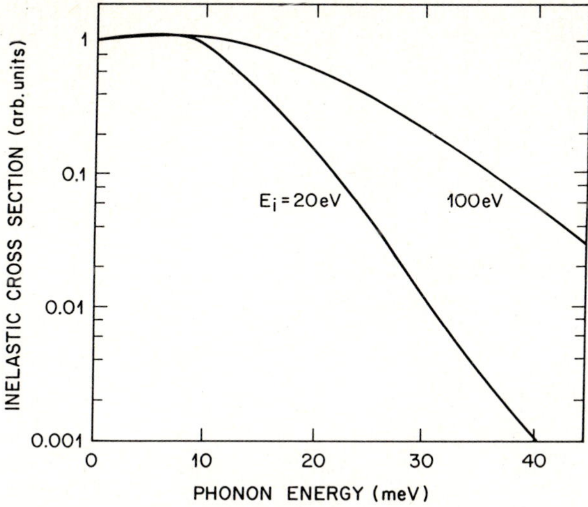

Fig.2.8. Predictions of the energy "cut-off" due to the low speed and therefore protracted collision time of the helium atom [2.21]. Calculations are for a "jellium" surface with potential parameters corresponding to copper. Curves are shown for two different incident helium energies, 20 meV and 100 meV corresponding to nozzle temperatures of 90K and 460K, respectively

to data-taking and data interpretation may be required than has heretofore been generally evident in the field.

Figure 2.9 illustrates the case in point. These TOF spectra [2.22] demonstrate the use of a bound state resonance to enhance inelastic scattering cross-sections, making the S_2 mode at 45 meV easily visible despite its frequency far up into the "cut-off" regime. This resonance involves coupling of the incident beam into a bound state of the surface potential well. Resonance effects have been investigated extensively for elastic scattering [2.23] where they are known to produce dramatic increases or decreases in specular or diffraction intensities as kinematic parameters are varied to move into or out of resonance.

This "selective adsorption" may also take place via inelastic coupling. For phonon coupling in the incident channel, as in the case of Fig.2.9, the effect is to enhance phonons at specific frequencies and wave vectors [2.24]. Effectively, the conditions of bound state resonance generate an additional family of curves similar to the scan curves. By adjusting kinematic parameters to superimpose scan curve and resonance curve on top of a phonon dispersion curve, it is possible to enhance the TOF amplitude for that particular phonon. The use of resonant enhancement places additional demands on both the apparatus and the process of data-taking. The apparatus must provide sufficient kinematic variability — in the form of a continuously variable total scattering angle and/or a continuously variable beam energy — to track the resonances through $\Delta E(\Delta K)$-space. The bound state energies must either be known or they must be measured prior to undertaking the inelastic scattering measurements. Data-

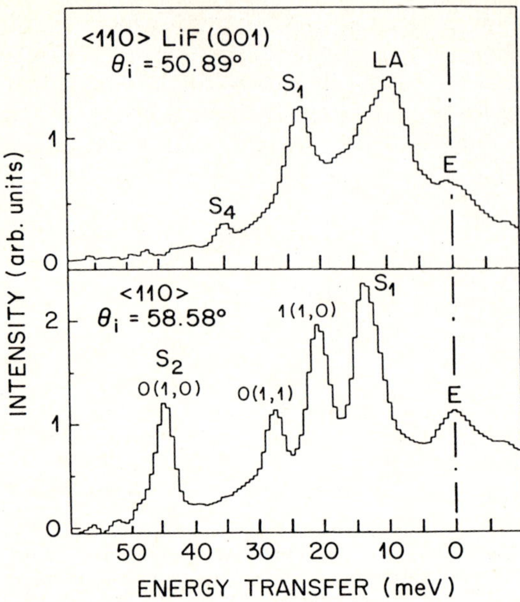

Fig.2.9. Enhancement of high frequency phonon signatures through inelastic coupling into a bound state of the helium surface potential well [2.22]. The top spectrum was recorded under non-resonant, the bottom under resonant conditions. By placing the S_2 mode into resonance with the lowest ($\epsilon_0 = -5.9$ meV) bound state via the (10) reciprocal lattice vector, it becomes easily observable in spite of its relatively high 45 meV energy. Note that some care is needed in interpreting the data: resonances can produce sharp peaks in regions of surface-projected bulk phonon bands, e.g. the middle two peaks in the bottom spectrum, which mimic surface phonon signatures

taking thus becomes a much more comprehensive, demanding process, probably requiring extensive measurement of elastic scattering to extract the bound state energies in addition to the inelastic scattering itself. A higher level of theoretical modeling would be extremely useful since, while the scattering geometry of a given resonant enhancement can be determined from kinematic relations, the kinematics give no information on the degree of enhancement (or lack thereof). Note the similarities to inelastic electron scattering in the impact scattering regime, where accurate calculations of the scattering cross-sections allow kinematic settings to be found at which interactions with one particular mode will dominate the scattering. Given the poorer energy resolution (5–10 meV) for electron scattering, this theoretical analysis is crucial for fitting electron energy loss peaks. Inelastic helium scattering has had no such compelling motivation to date and, given the complexity of calculations involving bound state resonances, only very few such calculations [2.25] have been made.

It is of interest to note that there is a second form of inelastic selective adsorption which involves phonon coupling in the exit channel rather than the entrance channel. This form of bound state resonance has the effect of enhancing *all* of the peaks in a TOF spectrum [2.26]. Many inelastic helium scattering measurements have taken advantage of this enhancement unwittingly, simply because that is where the intensity is! This enhancement occurs at specific values of θ_i (for a given k_i) so that it is a simple matter to enhance specific phonon signatures by adjusting scattering angles. Note that since resonant enhancement requires elastic coupling (diffraction) in one of either the entrance or exit channel, it is possible only for highly corrugated surfaces such as insulators or semiconductors.

2.8 Improving Signal Intensity: Beam Focussing

In striking contrast to charged particle and photon beams, atomic beams are completely unfocussed. The atomic beam diverges radially outward from an effective source point and its solid angle is determined solely by either the cross-sectional area of the detector or the collimator geometry, whichever is limiting. As mentioned above, the full-angle beam divergence is typically about 20 minutes of arc, i.e. only a very small portion of the free-jet expansion is utilized. There are thus potentially enormous gains to be realized by focussing the beam. Helium beam focussing has recently been demonstrated using a bent-crystal mirror formed from epitaxial gold overlayers deposited on mica substrates [2.27]. For the experimental arrangement of Fig.2.10 results are shown in Fig.2.11 for focussing of the specular reflection.

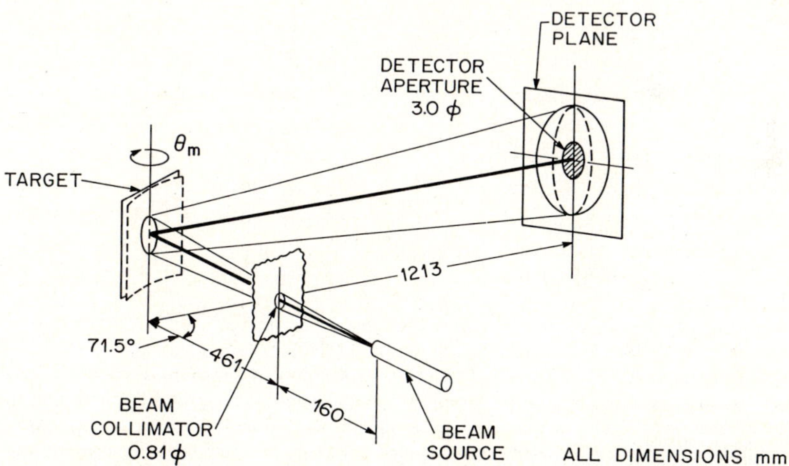

Fig.2.10. Scattering geometry employed to demonstrate focussing of a helium atomic beam [2.27]. The target/mirror, a gold epilayer on a mica substrate, could be bent *in situ* to form a cylindrical mirror, narrowing the width of the specular reflection (dashed lines). To characterize the focussing, the target was rotated (θ_m) to sweep the specular reflection across the detector aperture

In this particular instance, focussing was limited by a 0.12° polar angle mosaic spread of the gold epilayer. Nonetheless a narrowing of the angular spread and an increase in beam intensity is clearly demonstrated. In actual use for beam enhancement, the mirror would be bent to a spherical rather than a cylindrical shape. Moreover a much larger solid angle of the beam would be captured. A beam *converging* on the target with the same angular spread at which the unfocussed beam *diverges* would not alter the kinematic smearing depicted in Fig.2.5 while improving the beam intensity by a factor of more than 10^3. Some of this gain would be sacrificed to non-unity mirror reflectivity, of course, although with careful surface preparation and a cryogenically cooled mirror surface, reflectivities of 80–90% could probably be attained. The mirror surface must be prepared and maintained clean in UHV — effectively one has *two* target surfaces to worry about — and this would add to the experimental complexity. Nonetheless the potential gains in beam intensity make focussing an attractive possibility.

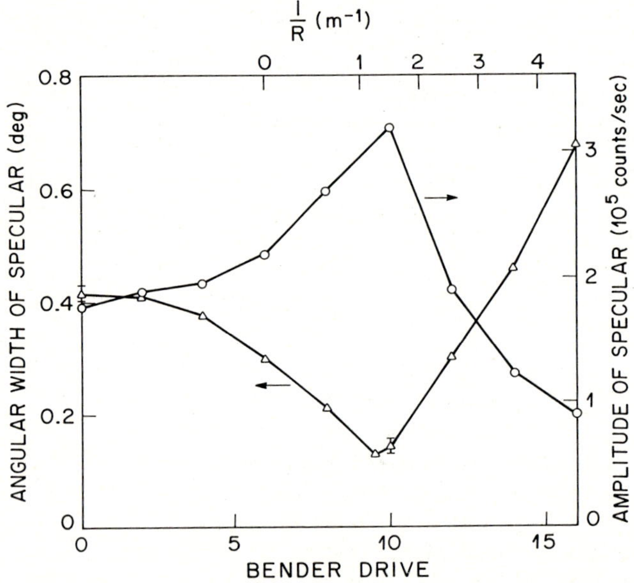

Fig.2.11. Focussing measurements made using the apparatus of Fig.2.10. The amplitude of the specular reflection (circles) and its angular width (FWHM, triangles) are plotted as a function of the inverse radius of curvature of the mirror. Optimum focussing gives an angular width of 0.12°. That this was not due to the detector aperture was verified by reducing the detector from 3 mm dia. to 50 μm dia. Hence we ascribe this limiting width to the mosaic spread of the gold crystallites comprising the surface epilayer, in reasonable agreement with X-ray diffraction measurements of other gold films

2.9 Improving Signal Intensity: Pseudorandom Chopping

A survey of methods for improving signal levels must include pseudorandom chopping [2.10], even though this is an established, mature technology and is in fairly extensive use. Pseudorandom chopping provides a technique for pulsing the beam at time intervals which are much shorter than the TOF spread of inelastic peaks at the detector. This normally leads to a prohibitive overlap of TOF spectra from different chopper pulses. The superposition can be deconvoluted, however, provided the beam is chopped in specific sequence of pulses. The Hadamard-like sequences used for this purpose have the property of a constant off-phase autocorrelation function: for one particular phase of the sequence shifted with respect to itself, the autocorrelation gives a unique value. For every other phase the autocorrelation is constant but differs from the "in-phase" value. This property allows one to deconvolute a "single-slit" spectrum from the pseudorandom spectrum of overlapping pulses. The gains in signal intensity can be quite large, as one can employ a slit pattern of roughly 50% duty factor (as opposed to perhaps 0.3% for single-slit chopping).

Pseudorandom chopping adds complexity to the experimental apparatus: The chopper must be strictly phase-locked to the multichannel analyzer (MCA) sweep and the chopper period must exactly equal the MCA sweep time to within a small fraction of one MCA time channel. This places great demands on the stability of the chopper and chopper electronics. Magnetically supported choppers are often used. All of this technology, while complicated, is well in hand.

On the other hand, pseudorandom chopping works poorly if the TOF spectra — as in Fig.2.2 — contain very small peaks in the presence of very large peaks. Superposition in the pseudorandom spectrum distributes the statistical noise of the large peaks across the entire spectrum. This persists in deconvoluting to the single slit spectrum so that the actual signal-to-noise ratio on the small peaks can in fact be *worse* than with single slit chopping. The key to successful use of pseudorandom chopping is thus to arrange the scattering geometry to sample only phonon peaks of comparable intensities in any given TOF spectrum. Use of a flow-through detector and beam dump is also mandatory since with the higher beam intensity of pseudorandom chopping the beam load would otherwise produce unacceptably high background levels of helium in the detector chamber [2.8].

2.10 Summary

Inelastic helium scattering is clearly best suited to the study of low energy phonons. Up to energies of a few tens of millielectron volts, helium scattering is the undisputed method of choice by virtue of its high energy resolution and strict surface sensitivity. While limitations due to TOF smearing and to "cut-off' in scattering cross-sections make measurements of higher energy phonons more difficult, these limitations have never been critically probed. A concerted effort incorporating selective adsorption enhancement and pseudorandom chopping would undoubtedly push phonon measurements to energies well above 50 meV. With focussing of the atomic beam in addition, this could probably be done with an energy resolution of under one millielectron-volt over the entire range. The additional complexity of the experimental apparatus could be substantial but then, even in its present realization, helium scattering is not a pursuit for the faint-hearted!

References

2.1   G. Brusdeylins, H.-D. Meyer, J. P. Toennies, K. Winkelmann: In *Rarefield Gas Dynamics*, Vol. 51, Part II, ed. by J. L. Potter, AIAA, New York (1977) p. 1047

2.2   R. Campargue, A. Lebéhot, J. C. Lemonnier: ibid., p. 1033

2.3   For a discussion of the early history of inelastic helium scattering see J.P.Toennies: In *Dynamics of Gas-Surface Interaction*, ed. by G.Benedek, U.Valbusa, (Springer, Berlin, Heidelberg 1982), p. 208

2.4   G. Brusdeylins, R. B. Doak, J. P. Toennies: Phys. Rev. Lett. **46**, 437 (1981)

2.5   G. Brusdeylins, R. B. Doak, J. P. Toennies: Phys. Rev. B **27**, 3662 (1983)

2.6   R. B. Doak, D. B. Nguyen: In *Rarefied Gas Dynamics: Physical Phenomena*, Vol. 117, ed. by E. P. Muntz, D. P. Weaver, D. H. Campbell, AIAA, Washington, DC (1989) p. 187

2.7   D.-M. Smilgies, J. P. Toennies: Rev. Sci. Instrum. **59**, 2185 (1988)

2.8   For a discussion of the experimental aspects of inelastic helium scattering see R. B. Doak: In *Atomic and Molecular Beam Methods*, Vol. II, ed. by G. Scoles (Oxford University Press, Oxford 1990)

2.9   It is also possible to use diffraction from a crystal surface to disperse the beam in angle: see B. F. Mason, B. R. Williams: Rev. Sci. Instrum. **49**, 897 (1978)

2.10   For a discussion of TOF analysis see D. J. Auerbach: In *Atomic and Molecular Beam Methods*, Vol. I, ed. by G. Scoles (Oxford University Press, Oxford 1988) p. 362

2.11   J. P. Toennies, K. Winkelmann: J. Chem. Phys. **66**, 3965 (1977)

2.12   For a comprehensive discussion of helium nozzle beams see D. R. Miller: In *Atomic and Molecular Beam Methods*, Vol. I, ed. by G. Scoles (Oxford University Press, Oxford 1988) p. 14

2.13   R. Campargue: J. Phys. Chem. **88**, 4466 (1984)

2.14   K. Kern, R. David, G. Comsa: Rev. Sci. Instrum. **56**, 369 (1985)

2.15   O. F. Hagena, W. Obert: J. Chem. Phys. **56**, 1793 (1972)

2.16   W. R. Gentry: In *Atomic and Molecular Beam Methods*, Vol. I, ed. by G. Scoles (Oxford University Press, Oxford 1988) p. 54

2.17   B. D. Kay, T. D. Raymond, J. K. Rice: Rev. Sci. Instrum. 57, 2266 (1986)

2.18   C. J. N. van der Meijdenberg: In *Atomic and Molecular Beam Methods*, Vol. I, ed. by G. Scoles (Oxford University Press, Oxford 1988) p. 345

2.19   J. H. Weare: J. Chem. Phys. **61**, 2000 (1974)

2.20   V. Celli, G. Benedek, U. Harten, J. P. Toennies, R. B. Doak, V. Bortolani: Surf. Sci. Lett. **143**, L376 (1984)

2.21   H. Ibach, T. S. Rahman: In *Chemistry and Physics of Solid Surfaces V*, ed. by R. Vanselow, R. Howe (Springer, Berlin, Heidelberg 1984) p. 455

2.22   G. Bracco, M. D'avanzo, C. Salvo, R. Tatarek, S. Terreni, F. Tommasini: Surf. Sci. **189/190**, 684 (1987)

2.23   H. Hoinkes: Rev. Mod. Phys. **52**, 933 (1980)

2.24   G. Benedek, J. P. Toennies, R. B. Doak: Phys. Rev. B **28**, 7277 (1983)

2.25   D. Eichenauer, J. P. Toennies: J. Chem. Phys. **85**, 532 (1986)

2.26   G. Brusdeylins, R. B. Doak, J. P. Toennies: J. Chem. Phys. **74**, 1784 (1981)

2.27   R. B. Doak: To be published

3. Interaction Potentials

V. Celli

Much of the usefulness of He scattering to surface science stems from data analysis by simple kinematics: the two-dimensional diffraction pattern from an ordered surface directly gives the reciprocal lattice vectors G of the surface; the broadening of the diffraction peaks indicates the type and amount of disorder; dips or peaks in the intensity of a diffracted beam as a function of the incidence angles (θ_i, ϕ_i), or of the incident energy E_i, give the surface bound state energies, ε_b ; and sharp peaks in the time-of-flight (TOF) spectra directly give the dispersion curves $\omega_j(Q)$ of the surface phonon branches. Nevertheless, for a full analysis of all these experiments a knowledge of the He–surface interaction is useful, and often essential.

Already implicit in the simple kinematic analysis is the assumption that the He–surface interaction consists of a short-range repulsion $V_{\rm rep}$ and a long range attraction $V_{\rm att}$, with $V_{\rm rep}$ coming predominantly from the top atom layer on the surface and $V_{\rm att}$ almost entirely from the van der Waals dispersion force.

In the simplest approximation $V_{\rm rep}$ is replaced by a corrugated hard wall and $V_{\rm att}$ is neglected, except of course when dealing with resonant scattering, i.e. with the dips and peaks caused by resonance with a surface-bound state. We briefly review the predictions of the hard wall models in Sect. 3.1.

Hard wall models are still widely used, especially for the analysis of angular distributions at fixed incidence parameters (E_i, θ_i, ϕ_i). Deviations from the hard wall predictions are often called "softness corrections". They can be substantial enough to make it worth abandoning the hard wall picture, except for comparison. In particular, the effective hard wall itself varies with incidence parameters [3.1].

Empirical "soft" potentials, with adjustable parameters are used with success to correlate experimental data [3.2]. A general expansion of the static potential, which comes naturally for coupled channels calculations, is

$$V(r) = \sum_{G} V_G(z) \exp(iG \cdot R) \quad . \tag{3.1}$$

(We follow the standard notation where the z axis is perpendicular to the mean surface, pointing outwards, and the "lateral" vector R is in the xy plane). The laterally averaged potential $V_0(z)$ has often been approximated by a Morse potential

25

Springer Series in Surface Sciences, Vol. 27   **Helium Atom Scattering from Surfaces**
Editor: E. Hulpke   © Springer-Verlag Berlin, Heidelberg 1992

$$V_0(z) = D(e^{-2\beta z} - 2e^{-\beta z}) \quad , \tag{3.2}$$

although this does not reproduce the correct asymptotic behavior of $V_0(z)$ for large z, which is z^{-3}. D is the surface well depth and β is the "softness parameter". If the attractive well can be neglected, $V_0(z)$ is simply approximated by an exponential. For $G \neq 0$ there is good justification, as we discuss below, for taking

$$V_G(z) = V_G e^{-\beta_G z} \quad , \tag{3.3}$$

where $\beta_G \cong (\beta^2 + G^2)^{1/2}$. (Often however one simply takes $\beta_G = \beta$). The parameters D, β, and V_G are adjusted to reproduce the experimental He diffraction data, with good success for weakly corrugated simple surfaces.

Generally, the attractive well in the averaged potential $V_0(z)$ is characterized by the depth D and by one or more shape parameters, such as β. For a given model potential, these parameters can be accurately determined from observations of resonant scattering, supplemented in some cases by thermodynamic data on adsorbed He films. Surface resonant scattering, also called selective adsorption, is reviewed in Chaps. 7 and 8 by H. Hoinkes and H. Wilsch and by J.R. Manson; it is also described in Sect.(3.3.2) for processes involving one-phonon exchange. In general, all the scattered intensities, elastic and inelastic, vary rapidly when the incident energy E_i passes through one of the resonant conditions given by (3.35). A similar variation can be observed more conveniently as a function of the angles θ_i and ϕ_i. The bound state energies ϵ_b, hence D and β, are determined by matching the observed resonances to (3.35).

The Fourier coefficients V_G for $G \neq 0$, or some equivalent corrugation parameters, are usually inferred from diffraction data taken far away from resonance; however, these V_G also determine the shape of the resonances. Hard wall models are often used for a first estimate of the surface corrugation.

Hunting for parameters in empirical potentials becomes expensive for complicated surfaces. A priori calculations of the He–surface interaction have then come increasingly into play, for practical as well as for aesthetic and conceptual reasons. If nothing else, they have suggested more realistic empirical models, with fewer adjustable parameters. We discuss these theories in Sect.3.2 for the static potential and in Sect.3.3 for the dynamic interaction. Further empirical potentials are also reviewed.

3.1 Hard Wall Models

For elastic scattering the total potential $V = V_{\mathrm{att}} + V_{\mathrm{rep}}$ is simply a function of the He coordinates $(\boldsymbol{R}, z)$. The effective position of the hard wall, $\zeta(\boldsymbol{R})$, is given approximately by the locus of classical turning points

$$V(\boldsymbol{R}, \zeta(\boldsymbol{R})) = E_{iz} \quad , \tag{3.4}$$

where $E_{iz} = E_i \cos^2 \theta_i$ is the incident perpendicular energy. It is clearly interesting to know the shape of the hard-wall corrugation profile $\zeta(\boldsymbol{R})$, which gives a physical picture of the surface.

Similarly, the simplest picture of inelastic scattering is in terms of the vibrating hard wall model. If $\boldsymbol{u}_\ell(t)$ is the displacement of the ℓ-th atom in the solid from equilibrium, the time-dependent generalization of (3.4)

$$V(\boldsymbol{R}, \zeta(\boldsymbol{R}, t), \boldsymbol{u}_\ell(t)) = E_{iz} \quad , \tag{3.5}$$

defines the vibrating hard-wall corrugation profile $\zeta(\boldsymbol{R}, t)$. Here the usual assumption is that, to a first approximation, only the displacements $\boldsymbol{u}_\ell(t)$ of the top surface layer determine the time dependence of ζ.

Historically, the hard wall model [3.3,4] was initially quite successful in predicting the dependence of the diffracted intensities on θ_i, ϕ_i and E_i with a very simple expression for $\zeta(\boldsymbol{R})$: often just the first term in a Fourier series expansion is sufficient. For instance, in the classic case of the alkali halide (001) surfaces one can use

$$\zeta(\boldsymbol{R}) = h[\cos(2\pi x/a) + \cos(2\pi y/a)] \quad , \tag{3.6}$$

where a is the nearest neighbor distance and $4h$ is the peak-to-valley height of the corrugation. Similarly for the basal plane of graphite [3.5]

$$\zeta(\boldsymbol{R}) = h[\cos(2\pi x/a) + \cos(2\pi y/a) + \cos(2\pi(x-y)/a)] \quad . \tag{3.7}$$

In such cases it is clearly more practical to treat h as an adjustable parameter than to try to calculate it from (nearly) first principles.

The fitting procedure to determine the parameters in $\zeta(\boldsymbol{R})$ is very efficient [3.6] if one can use the eikonal approximation for the scattering amplitude S_G. The complete eikonal formula [3.7,8] is $S_G = P_G I_G$, where

$$P_G = \frac{1 + \cos\theta_i \cos\theta_G - \sin\theta_i \sin\theta_G \cos(\phi_i - \phi_G)}{\cos\theta_G(\cos\theta_i + \cos\theta_G)} \tag{3.8}$$

is a kinematic prefactor of order unity that depends on the initial angles (θ_i, ϕ_i) and on the final angles (θ_G, ϕ_G), and

$$I_G = -\int \frac{\mathrm{d}^2\boldsymbol{R}}{A_c} \exp\left\{-\mathrm{i}[\Delta p\zeta(\boldsymbol{R}) + \boldsymbol{G} \cdot \boldsymbol{R}]\right\} \quad . \tag{3.9}$$

Here A_c is the area of the surface unit cell and Δp is the perpendicular momentum transfer in the hard wall collision. The effect of V_{att} is fairly well simulated by changing E_{iz} to $E_{iz} + D$, where D is the surface well depth. This is known as the Beeby correction; physically, it means that the He atom is accelerated by V_{att} on its way to the hard-wall collision with the surface, and is decelerated on its way out. Then $\Delta p = |p_i| + |p_G|$, where $\hbar^2 p_i^2 = 2m(E_{iz} + D)$, and similarly for p_G . The explicit eikonal result for the simple corrugation (3.6) is expressed in terms of Bessel functions as

$$I_G = -\mathrm{i}^{-|m|-|n|} J_{|m|}(h\Delta p) J_{|n|}(h\Delta p) \tag{3.10}$$

27

for $G = (2\pi/a)(m, n)$.

A feature of surface scattering that is naturally explained by the corrugated hard wall model is the surface rainbow. Classically, the rainbow angles $\theta_{\max}$ and $\theta_{\min}$ correspond to the maximum possible deflections and the scattered intensity vanishes for polar angles greater than $\theta_{\max}$ or smaller than $\theta_{\min}$. The rainbow deflections $\theta_{\max} - \theta_i$ and $\theta_i - \theta_{\min}$ are simply twice the maximum inclination angles of the effective corrugation, which is the intersection of $\zeta(\boldsymbol{R})$ with the scattering plane. A saddle-point evaluation of the eikonal integral (3.9) reproduces this classical result; a remnant of the rainbow is often clearly discernible in the full eikonal result. The positions of the rainbow peaks offer the simplest rough estimates of the parameters appearing in $\zeta(\boldsymbol{R})$, such as h in Eqs.(3.6,7).

At first sight it appears that a detailed knowledge of $V(\boldsymbol{r})$ is required to predict the bound state energies ϵ_{b} and the resonance shapes [3.9], but in fact the simple model [3.10,11] of a structureless well in front of a corrugated hard wall produces good predictions. A resonance with a surface-bound state appears as a dip or a peak in the scattered intensity, or, more generally, as an asymmetric Wigner–Fano shape, depending on the relative phase of the direct and resonant contributions to the total scattering amplitude. If the two are in phase a maximum results, if they are in antiphase there will be a minimum, unless the resonant amplitude is greater than twice the direct amplitude, in which case a maximum results. What determines the shape of the resonance are the phases of the hard-wall collision amplitudes for the transition into a bound state and out of it, and for the direct transition. For the corrugation (3.6) these phases are given explicitly by (3.10); the resulting predictions of maxima and minima have been well verified.

The vibrating hard wall model has been used to treat inelastic scattering, including one-phonon and multiphonon processes [3.4]. Inelastic resonances have been successfully discussed using the model of a vibrating hard wall with an attractive well [3.12,13]. Further discussion of inelastic scattering can be found in Sect.(3.3).

3.2 The Static Atom–Surface Potential

A compilation of data on the laterally averaged He-surface potential $V_0(z)$ is contained in a recent review by *Vidali* et al. [3.14] that focuses on the shape of the surface attractive well [3.15]. For He scattering we are more interested in the corrugation of $V(\boldsymbol{r})$ for positive energies.

It is common practice to divide the atom–surface interaction into an attractive part V_{att} due to van der Waals forces and a repulsive part V_{rep} due to the overlap of the surface wave functions with the closed-shell He orbitals. The theory has been developed for the static potential, but the results can be taken over directly to the dynamic interaction, within the usual Born-Oppenheimer approximation.

3.2.1 The Static Repulsive Potential

It has been shown by several methods that to a good approximation

$$V_{\rm rep}(\boldsymbol{r}) = A\rho(\boldsymbol{r}) \tag{3.11}$$

where $\rho(\boldsymbol{r})$ is the unperturbed electronic density of the surface and the constant A depends only weakly on the nature of the surface. Unfortunately in the important case of metal surfaces different approaches yield different values of A.

Esbjerg and *Nørskov* [3.16] and *Stott* and *Zaremba* [3.17] were the first to obtain (3.11) for a He atom immersed in homogeneous electron gas with a positive uniform charge background (the jellium model). The values of A calculated from this effective medium approach range from 305 to 329 eV $\cdot a_{\rm Bohr}^3$, with the latter value presumed to be the most accurate [3.18]. However, the He atom lies outside the jellium background in a region of varying $\rho(\boldsymbol{r})$. When the appropriate corrections are made, (3.11) still holds, but the effective A changes from 329 to 255 eV $\cdot a_{\rm Bohr}^3$. Hartree-Fock calculations for He interacting with a metal atom cluster can also be fitted to (3.11) with A=373 eV $\cdot a_{\rm Bohr}^3$ [3.19].

A different approach starts with the He atom well outside the surface and treats it as a perturbation on the surface electrons. This approach was successfully used by *Zaremba* and *Kohn* [3.20] to predict the laterally averaged interaction with metal surfaces and was perfected by *Harris* and *Liebsch* [3.21]. By carrying out the calculation to second order in the He–surface overlap integral, *Nordlander* and *Harris* [3.22,23] arrive at (3.11) again, but with $A \cong 500$ eV $\cdot a_{\rm Bohr}^3$. When the He is well outside the surface, it is actually possible to infer $V_{\rm rep}$ from electron–He scattering data. In the simplest approximation, that goes back to *Fermi* [3.24], the interaction between the He atom at $\boldsymbol{r}$ and the electron at $\boldsymbol{r}_{\rm e}$ is represented by the pseudopotential $(2\pi\hbar^2 a_{\rm s}/m)\delta(\boldsymbol{r}_{\rm e} - \boldsymbol{r})$, where $a_{\rm s}$ is the s-wave scattering length. This gives (3.11) with $A = 2\pi\hbar^2 a_{\rm s}/m \cong 200$ eV $\cdot a_{\rm Bohr}^3$. However, one should exclude the contribution to $a_{\rm s}$ coming from the long-range polarization force, because this force cannot be represented by a δ-like pseudopotential and in fact gives rise to the van der Waals attraction [3.25]. Further, the electron–He scattering data must be continued to negative energies of a few eV, because the surface electrons are bound. The resulting best estimate of A [3.26,27] is about 500 eV $\cdot a_{\rm Bohr}^3$, in agreement with the *Nordlander* and *Harris* [3.23] calculations, but about twice the value given by the effective medium theory [3.17]. These refinements of the pseudopotential method show that it is subject to considerable uncertainty, so that values of A ranging from 250 to 600 eV $\cdot a_{\rm Bohr}^3$ are obtained, depending also on the surface work function [3.27].

For insulators, one can start directly from pairwise repulsive interactions between the He atom and the surface constituents, which in a first approximation can be taken to be undistorted atoms in the case of graphite and of noble gas overlayers [3.28–32] , and undistorted ions in the case of alkali halides [3.33–36]. In these cases $\rho(\boldsymbol{r})$ is by construction the sum of the densities of the constituents and (3.11) turns out to be valid within the uncertainties of the calculation. The fact that one can take undistorted ions for LiF(100), however, results from accidental cancellation of several corrections[3.34]; in the other ionic crystal surface that has been studied in detail, NaCl(100), ion distortion is important, but good agreement with He diffraction data is obtained also in this case when all the corrections to the simple model are carefully evaluated [3.35].

The main trouble with the attractively simple formula (3.11) is that, while it works well for the laterally averaged atom–surface potential, it generally overestimates the effective corrugation for the metal surfaces that have been carefully studied, such as Cu, Ni and Ag (110). Actually, the correct corrugation is obtained if V_{att} is neglected altogether [3.37], but this is of course inconsistent. The effect of V_{att} is to move the classical turning point of the collision closer to the surface, where the variation of the density is greater [3.38]. *Takada* and *Kohn* [3.26] have pointed out that in the pseudo-potential approach the discrepancy could in principle be removed by the contribution of p-wave scattering; however, this correction turns out to be small for Cu, Ni and Ag [3.39]. It is possible that the error comes from uncertainties in the separation of V_{rep} from the attractive part of the potential, to be discussed below. It could also come from a failure of the methods used to compute $\rho(\boldsymbol{r})$ in the region of interest for the scattering of 20–60 meV He atoms ($z=3$–4 Å). Surprisingly, the full calculation of $\rho(\boldsymbol{r})$ gives a result that, in the region of interest, is practically indistinguishable [3.44] from a simple superposition of atomic densities, as tabulated [3.45] by *Herman* and *Skillman* and by *Clementi* and *Roetti*.

An additional term in the short-range potential arises from the hybridization of the He (1s) orbital with the unoccupied states above the Fermi level. *Annett* and *Haydock* [3.40-42] have shown that this term is approximately of the form $V_{\mathrm{hyb}}(\boldsymbol{r}) = -\gamma\rho_{\mathrm{u}}(\boldsymbol{r})$, where $\rho_{\mathrm{u}}(\boldsymbol{r})$ is the density of unoccupied states in the conduction band and γ is a constant. V_{hyb} has the desired effect of reducing the corrugations predicted by Eq. (3.11), but the magnitude of γ has been a subject of controversy: *Annett* finds that it is sufficient to remove the discrepancy with the experiment, while *Harris* and *Zaremba* [3.43] find that it is negligible.

As a practical matter the laterally averaged He–surface repulsion is generally well approximated by

$$V_{\mathrm{rep}}(z) = U_{\mathrm{rep}}\mathrm{e}^{-\beta z} \tag{3.12}$$

which is consistent with a pairwise sum of Yukawa potentials $U_\kappa \mathrm{e}^{-\beta r}/r$ over a monatomic top layer, with $U_{\mathrm{rep}} = 2\pi U_\kappa/\beta A_{\mathrm{c}}$. However it is more customary to approximate the pairwise repulsion by the Born-Mayer form $U_\kappa \mathrm{e}^{-\beta r}$, which then gives $(2\pi U_\kappa/\beta^2 A_{\mathrm{c}})\,(1+\beta z)\mathrm{e}^{-\beta z}$ instead of (3.12). If V_{att} is neglected, the potential (3.12) defines the simple soft wall model, and for any exponential- type potential β is called the softness parameter. The hard-wall model is recovered in the limit $\beta \to \infty$.

3.2.2 The Static Attractive Potential

V_{att} is usually dominated by the van der Waals dispersion force. At large distances z from the surface (but not large enough that retardation effects come into play), $V_{\text{att}} = -C_3/z^3$. The exact value of C_3 is given by the Lifshitz formula

$$C_3 = \frac{\hbar}{4\pi} \int_0^\infty du \frac{\varepsilon(iu) - 1}{\varepsilon(iu) + 1} \alpha_{\text{He}}(iu) \quad , \tag{3.13}$$

where ε is the dielectric function of the surface material and α_{He} is the He atom polarizability. The extension of ε and α_{He} to imaginary frequencies iu is easily performed if these functions are known for real frequencies; further, the integral is numerically convenient because $\varepsilon(iu)$ and $\alpha(iu)$ are real, non-negative, and monotonically decreasing functions of u. Thus C_3 can often be accurately evaluated, but the corrections to $-C_3/z^3$ are much less certain.

For a crystal consisting of localized electronic units (atoms, molecules, or ions) having polarizability α, the crudest way to compute V_{att} is to sum pairwise interactions $-c_6/\mid \boldsymbol{r} - \boldsymbol{r}_\ell \mid^6$, where

$$c_6 = \frac{3\hbar}{\pi} \int_0^\infty \alpha(iu)\alpha_{\text{He}}(iu)du \quad . \tag{3.14}$$

The sum of $-c_6/\mid \boldsymbol{r} - \boldsymbol{r}_\ell \mid^6$ over a monatomic plane gives a laterally averaged interaction $-C_4/(z - z_\ell)^4$, with $C_4 = (\pi/2A_{\text{c}})c_6 \cdot A_{\text{c}}$ is the surface area per atom. The sum over equally spaced planes of $-C_4/(z-z_\ell)^4$ gives asymptotically $-C_3/(z - d/2)^3$, where the origin of z is on the top atom plane and d is the interplanar spacing. The value of C_3 is $C_4/3d$, or $(\pi/6A_{\text{c}}d)c_6$, in agreement with (3.13) in the low-density limit where $\varepsilon = 1+4\pi\alpha/A_{\text{c}}d$. However, $-C_3/(z-d/2)^3$ gives a poor approximation to the sum over planes in the region of interest ($z \cong 3$ Å); a better approximation is

$$V_{\text{att}}(z) = -C_3/(z + d/2)^3 - C_4/z^4 \tag{3.15}$$

which treats the top atom layer separately.

A systematic improvement of the pairwise sum formula can be obtained by including n-body forces, beginning with the Axilrod-Teller three body interaction [3.46–48]. The result is approximately of the form (3.15), with C_3 now given by (3.13) and C_4 in general different from $3dC_3$.

A very general expression for V_{att} can be written down, assuming only that the He charge density does not overlap the surface charge density and that the response to electrical fluctuations is linear. In the limit when retardation effects can be neglected, V_{att} involves only the surface density response function $\chi_G(z_{\text{e}}, z'_{\text{e}}, \boldsymbol{Q}, \omega)$ and the 2^ℓ-pole He polarizabilities $\alpha_{\text{He}}^{(\ell)}(\omega)$. The surface density response to an external electric potential $\Phi(z'_{\text{e}}, \boldsymbol{Q}, \omega)$ is in general given by

$$e\rho(z_{\text{e}}, \boldsymbol{Q}, \omega) = \sum_G \int dz'_{\text{e}} \, \chi_G(z_{\text{e}}, z'_{\text{e}}, \boldsymbol{Q}, \omega) \, \Phi(z'_{\text{e}}, \boldsymbol{Q} - \boldsymbol{G}, \omega) \quad . \tag{3.16}$$

Here z_e and z_e' are electron coordinates. When the surface corrugation is negligible, one finds [3.49]

$$V_{\text{att}}(z) =$$

$$-\sum_{\ell=1}^{\infty} \frac{2^\ell}{2\pi\ell!(2\ell-1)!!} \int_0^\infty dQ \int_0^\infty du\, Q^{2\ell}\, \chi(Q,iu)\, \alpha_{\text{He}}^{(\ell)}(iu)\, e^{-2Qz} \quad,$$

$$(3.17)$$

where

$$\chi(Q,\omega) = \frac{2\pi}{Q} \int_{-\infty}^\infty dz_e \int_{-\infty}^\infty dz_e'\, \chi_0(z_e,z_e',Q,\omega)\, \exp\left[Q(z_e+z_e')\right] \quad.$$

$$(3.18)$$

(Recall that the vacuum side is for $z > 0$, so that χ_0 vanishes for z_e or $z_e' \to \infty$). The difficulty of course is in the evaluation of $\chi(Q,\omega)$. For large z, V_{att} is obtained by expanding

$$\chi(Q,\omega) = \chi(0,\omega) + Q\chi'(0,\omega) + \frac{1}{2}Q^2\chi''(0,\omega) \quad. \qquad (3.19)$$

Since $\chi(0,\omega) = [\varepsilon(\omega)-1]/[\varepsilon(\omega)+1]$, the leading term in (3.17) (from $\ell = 1$) is $-C_3/z^3$, with C_3 given by (3.13). The next term, coming from $Q\chi'(0,\omega)$, is proportional to z^{-4}. The result can then be written in the form (3.15) with an appropriate value of C_4. However, it is possible to make $\chi'(0,\omega)$ vanish by shifting the origin of z_e and z_e' by an appropriate amount z_{vW}; the origin of the He coordinate z must of course be shifted by z_{vW} too. One can then write, up to terms of order z^{-5}

$$V_{\text{att}}(z) = -C_3/(z-z_{\text{vW}})^3 \quad. \qquad (3.20)$$

The value of z_{vW} for a variety of surfaces (jellium model and noble metals) was first given by *Zaremba* and *Kohn* [3.50]. A later paper by *Liebsch* [3.51] suggested that the exact values of z_{vW} are uniformly less — typically about 20% closer to the surface — than calculated by Zaremba and Kohn. Based on these values, the currently accepted z_{vW} for the noble metals is close to $d/2$, which, in the jellium model, is the jellium edge.

The form (3.20) has become customary for metal surfaces, although the pairwise sum approach suggests that (3.15) is a better approximation. Comparison of (3.15) and (3.20) shows that z_{vW} differs from $d/2$ if C_4 for the topmost atom plane differs from $3dC_3$.

Corrections to (3.15) or (3.20) are of three kinds: (i) multipolar interactions, (ii) charge overlap, and (iii) lateral variation.

(i) Multipolar corrections from (3.17) begin with $-C_5/z^5$ and consist of two terms [3.49], one coming from $\alpha_{\text{He}}^{(2)}$ and $\chi(0,\omega)$ (He quadrupole–surface dipole), the other from α_{He} and $\frac{1}{2}Q^2\chi''(0,\omega)$ (He dipole–surface quadrupole). The latter is generally more important and hard to estimate. Here again, the pairwise summation model gives explicit results. Pairwise multipolar terms c_8, c_{10}, and c_{12} are routinely included in interatomic potentials: they are of the form (3.14) with α and/or α_{He} replaced by $\alpha^{(\ell)}$ and $\alpha_{\mathrm{He}}^{(\ell)}$. The pairwise sum of $-c_{2n}/\mid \boldsymbol{r} - \boldsymbol{r}_\ell \mid^{2n}$ gives asymptotically $-C_{2n-3}/z^{2n-3}$ with $C_{2n-3} = [\pi/2n(2n-1)](N_a/d)c_{2n}$.

(ii) The effect of charge overlap is to reduce the attraction at close range. In pairwise interactions it is found semiempirically that the reduction of $-c_n/\mid \boldsymbol{r} - \boldsymbol{r}_\ell \mid^{2n}$ is often well described by the Tang-Toennies "damping factor" [3.52]

$$f_n(x) = 1 - e^{-x} \sum_{k=0}^{n} \frac{x^k}{k!} \quad , \tag{3.21}$$

where $x = \beta \mid \boldsymbol{r} - \boldsymbol{r}_\ell \mid$ and β is approximately the same parameter that describes the fall-off of V_{rep}. Other "damping factors" have been proposed, but (3.21) works as well as any and does not introduce new parameters. A pairwise sum of Tang-Toennies interactions gives the following modification of (3.15)

$$V_{\mathrm{att}}(z) = -C_3 F_3[\beta(z + d/2)]/(z + d/2)^3 - C_4 F_4(\beta z)/z^4 \tag{3.22}$$

with

$$F_3(x) = 1 - e^{-x}[1 + x + \frac{x^2}{2} + \frac{2x^3}{45} + \frac{x^4}{180}] \tag{3.23}$$

$$F_4(x) = 1 - e^{-x}[1 + x + \frac{x^2}{2} + \frac{x^3}{6} + \frac{7x^4}{180} + \frac{x^5}{180}] \quad . \tag{3.24}$$

Another popular form of "damping" is obtained [3.23] by cutting off the integral in (3.17) at $Q = \beta$. To leading order, this prescription gives

$$V_{\mathrm{att}}(z) = -C_3 f_2(\beta \mid z - z_{\mathrm{vW}} \mid)/\mid z - z_{\mathrm{vW}} \mid^3 \quad . \tag{3.25}$$

For the same β (3.25) gives a stronger attraction than (3.22).

(iii) The lateral variation of V_{att} for a sum of pairwise interactions is caused almost entirely by the atoms in the top surface layer and falls off like e^{-Gz}. The dipole–dipole term takes the form [3.53]

$$- (C_4/2) \sum_{\boldsymbol{G}\neq 0} (G/2z)^2 \exp(i\boldsymbol{G} \cdot \boldsymbol{R}) K_2(Gz) \quad , \tag{3.26}$$

where K_2 is a modified Bessel function and $C_4 = \pi c_6 / A_c$. A formula of the same type, with $C_4/2$ replaced by an appropriate C_G, is obtained from the generalization of (3.27) to a corrugated surface [3.54,55]. For a material consisting of point polarizabilities, $C_G = C_4/2 = 3dC_3/2$, with C_3 given by the Lifshitz formula (3.13); this is an upper limit for the true C_G [3.54].

3.2.3 The Total Static Potential

In order to obtain relatively small effects such as the corrugation of metal surfaces, it may not be a good idea to separate V_{static} into V_{att} and V_{rep}. Around the classical turning point of a typical He–metal collision there is a large cancellation of V_{att} and V_{rep} , which tends to amplify the uncertainties.

A direct calculation of the total V_{static} is feasible, using density functional theory in the local density approximation, for a closed shell atom interacting with a jellium surface. *Lang* and *Nørskov* [3.56] find by this method very reasonable shapes of V_{static} for $z \leq z_0$, where z_0 is the bottom of the attractive well. The values of the well depth D are also in fair agreement with the values obtained by *Zaremba* and *Kohn* [3.19], and by others [3.21], using the separate evaluation of V_{att} and V_{rep}, as discussed above. Of course for $z > z_0$ the Lang-Nørskov potential decays exponentially, rather than as z^{-3}; thus it will not give the correct energies of the surface-bound states. However, it is presumably accurate around the classical turning point; hence a calculation of He interacting with a corrugated metal surface would be very useful.

It seems prudent at present to regard the corrugations of metal surfaces as adjustable parameters to be determined experimentally. If one wants to preserve the relation $V_{\text{rep}}(\boldsymbol{r}) = A\rho(\boldsymbol{r})$ with $\rho(\boldsymbol{r})$ a superposition of atomic charge densities, one must assume non-spherical atoms without theoretical justification. Thus *Dondi* et al. [3.57] are able to fit the corrugation of Ag(110) by superimposing atomic potentials of the form $U_\kappa \exp[-(\beta_x^2 x^2 + \beta_y^2 y^2 + \beta_z^2 z^2)^{1/2}]$ with $\beta_x = 2.03 \text{ Å}^{-1}, \beta_y = 2.17 \text{ Å}^{-1}, \beta_z = 2.78 \text{ Å}^{-1}$. Here the x direction is along the close-packed (110) atom rows. For comparison, the Herman-Skillman tables [3.45] give $\beta = 2.42 \text{ Å}^{-1}$.

A possible way to account for the metal corrugations is to include a large hybridization correction. For example, *Annett* [3.42] has constructed potentials that fit remarkably well the diffraction intensities for the whole set of Cu surfaces: (100), (110), (113), (115), (117). He obtains the total $V(\boldsymbol{r})$ by adding the Nørskov repulsion $A\rho(\boldsymbol{r})$, the hybridization term $-\gamma\rho(\boldsymbol{r})$, the laterally averaged $V_{\text{att}}(z)$ of Eq. (3.25), and the corrugated V_{att} of Eq. (3.26). He gets $\rho(\boldsymbol{r})$ and $\rho_{\text{u}}(\boldsymbol{r})$ from a simple tight binding calculation and adjusts the controversial constant γ so that the laterally averaged potential agrees with that calculated from a jellium model of the Cu surface. *Goldberg* et al. [3.58] have developed an alternative a priori method based entirely on tight binding and have reported a calculation giving good agreement with the experimental corrugations of the Cu(110), Ag(110) and Ni(110), when the hybridization term $-\gamma\rho_{\text{u}}(r)$ is included in a way similar to *Annett*.

An outcome of all the theory discussed so far are various semi-empirical forms of the atom-surface potential. For the laterally averaged $V_0(z)$, the simplest form is

$$V_0(z) = U e^{-\beta z} - C_3 z^{-3} F_3(\beta z) \tag{3.27}$$

with F_3 given by Eq. (3.23), or alternatively replaced by f_2 from Eq. (3.21). Compared to the Morse potential, Eq. (3.27) has the disadvantage of introducing three adjustable parameters instead of two, although C_3 can be obtained from other experiments according to Eq. (3.13). Thus the Morse potential, scaled by the well depth D, is a universal function of βz, while (3.27) cannot be put in a universal form using scaled variables. However, for the purpose of describing the bound states of $V_0(z)$ (but not usually to describe scattering) one can set $F_3 = 1$ and rewrite (3.27) in the scaled one-parameter form suggested by *Vidali* and *Cole* [3.15]

$$\frac{V_0(z)}{D} = \frac{3}{u-3} e^{-ux/a} - \frac{1}{(x+a)^3} \tag{3.28}$$

with $x = \beta z$ and $a = (1 - 3/u)^{1/3}$. In most cases $u \approx 5.25$, so that the form (3.28) is nearly universal.

The simplest way to describe the surface corrugation is to replace $e^{-\beta z}$ with $\exp[-\beta(z - \zeta(\boldsymbol{R})]$ in Eq. (3.27). Further, $\zeta(\boldsymbol{R})$ is Fourier-expanded as in Eq. (3.6) or (3.7). However, this simple replacement does not account for the fact that the Fourier components $V_{\boldsymbol{G}}(z)$ decay more rapidly with z than $V_0(z)$, as displayed in Eqs. (3.29-31) below for a sum of pairwise potentials. More accurately, the $\boldsymbol{G}$ Fourier component of $\zeta(\boldsymbol{R})$ should be multiplied by $\exp[(\beta - \beta_{\boldsymbol{G}})z]$. A potential of this type has been constructed, for instance, by *Kirsten* et al. [3.59] for the reconstructed surface of Pt(111)(1 × 2).

For insulators, a better semi-empirical form of the potential may be given by a pairwise sum, with most of the parameters determined from a knowledge of the He-atom (or He-ion) pair potentials. This procedure will automatically give a corrugated potential of the correct functional form. Potentials of this type have been systematically constructed by *Eichenauer* and *Toennies* for the alkali halides [3.36].

3.3 The Dynamic Interaction

In most of the inelastic scattering calculations performed to date, V_{rep} has been approximated by a pairwise sum of Born-Mayer or Yukawa potentials, or even more drastically by a vibrating wall, soft or hard. The dynamical effects of V_{att} can be fully included in the calculation if V_{att} is represented by a pairwise sum, but it is usually assumed with some justification (see below) that V_{att} does not contribute appreciably to the inelastic matrix elements, except at small lateral momentum transfer. Since the prime interest is in the study of surface phonon spectra throughout the surface Brillouin zone, the dynamical V_{att} can then be neglected altogether. We therefore proceed by considering at first V_{rep} alone.

3.3.1 The Dynamic Repulsion and the Cutoff Factor

According to the distorted wave Born approximation (DWBA), what is needed are the matrix elements of the lateral Fourier transform of the pairwise force between the He atom and the κ-th atom in the surface unit cell. For a Yukawa pair potential $V_\kappa(r) = U_\kappa\,e^{-\beta r}/r$, the lateral Fourier transform is

$$V_\kappa(\boldsymbol{P}|z) = \frac{2\pi}{\beta(P)}U_\kappa\,e^{-\beta(P)z} \quad , \tag{3.29}$$

where

$$\beta(P) = \sqrt{\beta^2 + P^2} \quad . \tag{3.30}$$

For a Born-Mayer potential $V_\kappa(r) = U_\kappa\,e^{-\beta r}$

$$V_\kappa(\boldsymbol{P}|z) = \frac{2\pi\beta}{\beta(P)^3}\,[1 + \beta(P)z]\,U_\kappa\,e^{-\beta(P)z} \quad . \tag{3.31}$$

The pairwise force $F_\kappa(\boldsymbol{P}|z)$ has components

$$(\mathrm{i}\boldsymbol{P}, \partial/\partial z)V_\kappa(\boldsymbol{P}|z) \quad . \tag{3.32}$$

If $u_{\ell\kappa}$ is the displacement of the κ-th atom in the ℓ-th cell, the He-atom coupling is $\boldsymbol{F}(\boldsymbol{r} - \boldsymbol{r}_{\ell\kappa}) \cdot \boldsymbol{u}_{\ell\kappa}$.

For $P \ll \beta$ one has $\beta(P) \cong \beta + P^2/2\beta$ and from (3.27) or (3.29), neglecting the P dependence of the prefactor, we have

$$V_\kappa(\boldsymbol{P}|z) = V_\kappa(0|z)\,e^{-P^2 z/2\beta} \quad , \tag{3.33}$$

where $V_\kappa(0|z)/A_{\mathrm{c}}$ is in fact the same as the laterally averaged $V_{\mathrm{rep}}(z)$; cf. (3.12). A simple "rippling wall" picture follows from this formula if the static corrugation is negligible and only the z-component of the relevant $u_{\ell\kappa}$ is different from zero. Suppose for simplicity that only one atom per cell (the atom in the surface layer) contributes appreciably to the repulsive interaction. Then the total He-surface interaction, to first order in $u_{\ell z}$, is approximately the same as

$$V_{\mathrm{rep}}(z - (\beta A_{\mathrm{c}}/2\pi z) \sum_\ell \exp[-\beta(\boldsymbol{R} - \boldsymbol{r}_\ell)^2/2z]u_{\ell z}) \tag{3.34}$$

as long as $\beta z \gg 1$. This condition is satisfied if the important values of z are near the classical turning point z_{t} of the He–surface collision . The equivalence of (3.30) and (3.31) hinges on the fact that $\exp(-\beta R^2/2z)$ and $(2\pi z/\beta)\,\exp(-P^2 z/2\beta)$ are the Fourier transforms of each other. If z is replaced by z_{t} in the "influence factor" $\exp(-\beta(\boldsymbol{R} - \boldsymbol{r}_\ell)^2/2z)$, equation (3.31) describes a "rippling soft wall", which in the limit $\beta \to \infty$ becomes a "rippling hard wall". The influence factor describes how the displacement of the turning point is affected by the displacement of the underlying surface atoms. The corresponding factor in momentum space is the "cutoff factor" $\exp(-P^2 z_t/2\beta)$, according to (3.30). The influence factor in coordinate space averages over the displacements of neighboring surface atoms; the cutoff factor in momentum space reduces the contribution to the scattering of short-wavelength phonons. The underlying physical effect was recognized and discussed qualitatively by *Hoinkes* [3.60] and *Armand* [3.61]. The presentation given here follows the treatment by *Bortolani* et al. [62–65]. The fall off of the scattered one-phonon intensity as $\exp(-Q^2/Q_c^2)$, with $Q_c^2 \cong \beta/z_t$, describes rather well the decay of the intensity with momentum transfer for scattering from a surface with negligible static corrugation.

3.3.2 Dynamical Effects of the Attractive Potential

The dynamical effects of the static part of V_{att} are fully included in DWBA calculations using the correct wave functions of V_{static}. It is however common practice to keep only V_{rep} throughout the calculation and to account for V_{att} approximately by applying the "Beeby correction" , i.e. by increasing the perpendicular He momentum from k_z to $\sqrt{k_z^2 + 2mD/\hbar^2}$. Here D is supposed to be the surface well depth, but comparison with full DWBA calculations shows that it can be considerably larger, especially at low incident He energies [3.66]. This is a reasonable approximation for weakly corrugated surfaces, but does not account for the occurrence of inelastic resonant scattering.

Three types of resonant one-phonon scattering processes can occur on a corrugated surface. In the first type the He atom is elastically scattered (diffracted) from the initial state into a surface-bound state, and then inelastically scattered out of it. This process occurs when the initial energy and angles are close to the resonance condition

$$E_i = \varepsilon_b + (\hbar^2/2m)(\boldsymbol{K}_i + \boldsymbol{G})^2 \quad . \tag{3.35}$$

Here ε_b (which is negative) is the binding energy of a surface-bound state and $\boldsymbol{G}$ is a surface reciprocal lattice vector. In a process of the second type the He atom is inelastically scattered from the initial state into a surface-bound state, and then elastically scattered out of it. This process occurs when the final energy and angles are close to the resonance condition

$$E_f = \varepsilon_{b'} + (\hbar^2/2m)(\boldsymbol{K}_f + \boldsymbol{G'})^2 \quad . \tag{3.36}$$

In a process of the third type both (3.35) and (3.36) are approximately satisfied and an inelastic one-phonon transition occurs between the surface-bound states of binding energies ε_b and $\varepsilon_{b'}$. This process usually causes a large increase of the inelastic cross-section called "supernova" by *Toennies* [3.67].

It is possible to use (3.36) to determine the surface phonon dispersion relation without a TOF apparatus, since the resonance condition itself acts as a final energy analyzer [3.68]. The resonance manifests itself as a sharp variation of the total inelastic intensity. With TOF analysis, the effect of inelastic resonances in LiF, for instance, is very sharp [3.67]. The theory of the inelastic resonance line shape and of its magnitude has been worked out by *Cantini* and *Tatarek* [3.12] in the model of a vibrating hard wall with a stationary attractive potential in front of it. Detailed applications of this model have been made to the analysis of inelastic resonances in LiF [3.13,69]. Resonances are useful to enhance the scattering from surface phonon branches that are weakly coupled to the He atom; however, one must be careful not to mistake a bound-state resonance for a peak in the surface phonon density of states.

We have discussed so far dynamical effects that arise from the presence of a static attraction. There are also direct dynamical effects of V_{att}. They can be evaluated in the pairwise summation model: $V_\kappa(Q \mid z - z_\kappa)$ is of the form (3.26) with $G \to Q$ and $C_4 \to \pi c_6/2$. This force is included in the calculations by *Eichenauer* et al. [3.70,71]. However, it is found to be unimportant in determining the phonon cross sections throughout most of the surface Brillouin zone, because it decays rapidly with Q.

In summary then the pairwise interaction model predicts that the dominant coupling, at least for one-phonon processes, is given by (3.32). This equation makes a definite prediction about the relative contribution of the z-polarized and Q-polarized (longitudinal) vibrations to the observed inelastic scattering. *Bortolani* et al. [3.64, 3.72–74] have relied on these predictions in extracting surface phonon spectral densities from helium TOF data on metal surfaces. It is an open question whether the sum of pairwise Born-Mayer potentials gives a valid representation of the dynamical He–surface interaction, especially in view of the fact that it does not seem to give correctly the static corrugations.

Acknowledgement   This review was written in part while visiting the MPISF in Göttingen with support of the Alexander von Humboldt Foundation

References

3.1   K.-H. Rieder: Surf. Sci. **117**, 13 (1982)

3.2   A. Liebsch, J. Harris, B. Salanon, J. Lapujoulade: Surf. Sci. **123**, 338 (1982)

3.3   U. Garibaldi, A.C. Levi, R. Spadacini, G.E. Tommei: Surf. Sci. **48**, 649 (1975)

3.4   A.C. Levi, H. Suhl: Surf. Sci. **88**, 221 (1979) G. Armand, J.R. Manson: Surf. Sci. **80**, 532 (1979) G. Benedek, N. Garcia: Surf. Sci. **80**, 543 (1979)

3.5   G. Boato, P. Cantini, R. Tartarek: Phys. Rev. Lett. **40**, 887 (1978)

3.6   See T. Engel, K. H. Rieder: *Structural Studies of Surfaces with Atomic and Molecular Beam Diffraction*, Springer Tracts in Modern Physics, Vol. 91 (Springer, Berlin, Heidelberg 1981)

3.7   C. Chiroli, A. C. Levi: Surf. Sci. **59**, 3438 (1972)

3.8   N. R. Hill, V. Celli: Surf. Sci. **75**, 577 (1978)

3.9   C. E. Harvie, J.H. Weare: Solid State Commun. **27**, 1297 (1978)

3.10   N. Garcia, F. O. Goodman, V. Celli, M. R. Hill: Phys. Rev. B **19**, 1908, (1979)

3.11   V. Celli, N. Garcia, J. Hutchison: Surf. Sci. **87**, 112 (1979); J. Hutchison, V. Celli: Surf. Sci. **93**, 263 (1980)

3.12   P. Cantini, R. Tatarek: Phys. Rev. B **23**, 3030 (1981)

3.13 D. Evans, V. Celli, G. Benedek, J. P. Toennies, R. B. Doak: Phys. Rev. Lett. **50**, 1854 (1983)

3.14 G. Vidali,G. Ihm, H.-Y. Kim, M.W. Cole: Surf. Sci. Reports **12**, 133 (1991)

3.15 G. Vidali, M.W. Cole, J.R. Klein: Chem. Phys. Lett. **95**, 199 (1981); G. Ihm, E. Cheng, M.W. Cole: J. Low Temp. Phys. **74**, 519 (1989)

3.16 N. Esbjerg, J.K. Nørskov: Phys. Rev. Lett. **45**, 807 (1980)

3.17 M.J. Stott, E. Zaremba: Phys. Rev. B **22**, 1564 (1980)

3.18 M. Manninen, J.K. Nørskov, M.J. Puska, C. Umrigar: Phys. Rev. B **29**, 2314 (1984)

3.19 I.P. Batra: Surf. Sci. **148**, 1 (1984); I.P. Batra, P.S. Bagus, J.A Barker: Phys. Rev. B **31**, 1737 (1985)

3.20 E. Zaremba, W. Kohn: Phys. Rev. B **15**, 1769 (1977)

3.21 J. Harris, A. Liebsch: Phys. Rev. Lett. **49**, 341 (1982); J. Harris, A. Liebsch: J. Phys. C**15**, 2275 (1982)

3.22 P. Nordlander: Surf. Sci. **126**, 675 (1983)

3.23 P. Nordlander, J. Harris: J. Phys. C**17**, 1151 (1984)

3.24 E. Fermi: Nuovo Cimento **11**, 157 (1934)

3.25 M.W. Cole, F. Toigo: Phys. Rev B **31**, 727 (1985)

3.26 Y. Takada, W. Kohn: Phys. Rev. Lett. **54**, 470 (1985)

3.27 Y. Takada, W. Kohn: Phys. Rev. B **37**, 826 (1987)

3.28 W.E. Carlos, M.W. Cole: Surf. Sci. **77**, L173 (1978); Phys. Rev. Lett. **43**, 697 (1979); Surf. Sci. **91**, 339 (1980)

3.29 M.W. Cole, D.R. Frankl, D.L. Goodstein: Rev. Mod. Phys. **53**, 199 (1981)

3.30 G. Vidali, M.W. Cole: Phys. Rev. B **22**, 4661 (1980); Surf. Sci. **110**, 10 (1981); G. Vidali, M.W. Cole, C. Schwartz: Surf. Sci. **87**, L273 (1979)

3.31 H. Jonsson, J. Weare: Phys. Rev. Lett. **57**, 412 (1986)

3.32 H. Jonsson, J. Weare: J. Chem. Phys. **86**, 3711 (1987)

3.33 V. Celli, D. Eichenauer, A. Kaufhold, J.P. Toennies: J. Chem. Phys. **83**, 2504 (1985)

3.34 P.W. Fowler, J.M. Hutson: Phys. Rev. B **33**, 3724 (1986)

3.35 J.M. Hutson, P.W. Fowler: Surf. Sci. **173**, 337 (1986)

3.36 D. Eichenauer, J.P. Toennies: Surf. Sci. **197**, 267 (1988)

3.37 D.R. Hamann: Phys. Rev. Lett. **46**, 1227 (1981)

3.38 N. Garcia, J.A. Barker, K.H. Rieder: Solid State Commun. 45, 567 (1983)

3.39 J. Tersoff: Phys. Rev. Lett. **55**, 140C (1985); Y. Takada, W. Kohn: Phys. Rev. Lett. **55**, 141C (1985)

3.40 J.F. Annett, R. Haydock: Phys. Rev. Lett. **53**, 838 (1984); Phys. Rev. Lett. **57**, 1382 (1986)

3.41 J.F. Annett, R. Haydock: Phys. Rev. B **29**, 3773 (1984); ibid. **34**, 6860 (1986)

3.42 J.F. Annett: Phys. Rev. B **35**, 7826 (1987)

3.43 J. Harris, E. Zaremba: Phys. Rev. Lett. **55**, 1940C (1985)

3.44 J.A. Barker, N. Garcia, I.P. Batra, M. Baumberger: Surf. Sci. **141**, L317 (1984)

3.45 F. Herman, S. Skillman: *Atomic Structure Calculations*, (Prentice-Hall, Englewood Cliffs 1963); E. Clementi, C. Roetti: At. Data Nucl. Data Tables **14**, 177 (1974)

3.46 H.-Y. Kim, M.W. Cole: Phys. Rev. B **35**, 3990 (1987)

3.47 B.R.A. Nijboer, M.J. Renne: Chem. Phys. Lett. **2**, 35 (1969)

3.48 E. Cheng, M.W. Cole: Phys. Rev. B **38**, 987 (1988)

3.49 J.M. Hutson, P.W. Fowler, E. Zaremba: Surf. Sci. **175**, L775 (1986)

3.50 E. Zaremba, W. Kohn: Phys. Rev. B **13**, 2270 (1976)

3.51 A. Liebsch: Phys. Rev. B **33**, 7249 (1986)

3.52 K.T. Tang, J.P. Toennies: J. Chem. Phys. **80**, 3726 (1984)

3.53 W.A. Steele: *The Interaction of Gases with Solid Surfaces* (Pergamon, New York 1974) p. 22

3.54 N.R. Hill, M. Haller, V. Celli: Chem. Phys. **73**, 363 (1982)

3.55 A.M. Lahee, W. Allison, R.F. Willis: Surf. Sci. **147**, L630 (1984)

3.56 N.D. Lang, J.K. Nørskov: Phys. Rev. B **27**, 4612 (1983)

3.57 M.G. Dondi, S. Terreni, F. Tommasini, U. Linke: Phys. Rev. B **37**, 8034 (1988)

3.58 E.C. Goldberg, A. Martin-Rodero, R. Monreal, F. Flores: Phys. Rev. **B38**, 5684 (1989)

3.59 E. Kirsten, K.H. Rieder: Surf. Sci. **222**, L837 (1989); E. Kirsten, G. Parschau, K.H. Rieder: Phys. Rev. **B41**, 5392 (1990)

3.60 H. Hoinkes, H. Nahr, H. Wilsch: Surf. Sci. **33**, 516 (1972)

3.61 G. Armand, J. Lapujoulade, Y. Lejay: Surf. Sci. **63**, 143 (1977)

3.62 V. Bortolani, A. Franchini, F. Nizzoli, G. Santoro, G. Benedek, V. Celli: Surf. Sci. **128**, 249 (1983); V. Bortolani, A. Franchini, F. Nizzoli, G. Santoro, G. Benedek, V. Celli, N. Garcia: Solid State Commun. **48**, 1045 (1983)

3.63 V. Bortolani, A. Franchini, N. Garcia, F. Nizzoli, G. Santoro: Phys. Rev. B **28**, 7358 (1983)

3.64 V. Bortolani, A. Franchini, F. Nizzoli, G. Santoro: Phys. Rev. Lett. **52**, 429 (1984)

3.65 V. Celli, G. Benedek, U. Harten, J.P. Toennies, R.B. Doak, V. Bortolani: Surf. Sci. **143**, L376 (1984)

3.66 G. Armand, J.R. Manson: J. Phys. France **44**, 473 (1983)

3.67 G. Brusdeylins, R.B. Doak, J.P. Toennies: J. Chem. Phys. **75**, 1784 (1981); G. Lilienkamp, J.P. Toennies: Phys. Rev. B **26**, 4752 (1982); J. Chem. Phys. **78**, 5210 (1983)

3.68 P. Cantini, G.P. Felcher, R. Tatarek: Phys. Rev. Lett. **37**, 606 (1976); in *Proceedings of the Seventh International Vacuum Congress and the Third International Conference on Solid Surfaces*, ed. by R. Dobrozemsky et al. (Vienna, 1977) p. 1357

3.69 W.L. Nichols, J.H. Weare: Phys. Rev. Lett. **56**, 753 (1985)

3.70 D. Eichenauer, J.P. Toennies: In *Dynamics at Surfaces*, ed. by B. Pullman et al. (Reidel, Dordrecht 1984) p. 1; D. Eichenauer, J.P. Toennies: J. Chem. Phys. **85**, 532 (1987)

3.71 D. Eichenauer, U. Harten, J.P. Toennies, V. Celli: J. Chem. Phys. **86**, 3693 (1987)

3.72 V. Bortolani, A. Franchini, G. Santoro, U. Harten, J. P. Toennies: Surf. Sci. **148**, 82 (1984)

3.73 V. Bortolani, A. Franchini, F. Nizzoli, G. Santoro: In *Electronic Structure, Dynamics, and Quantum Structural Properties of Condensed Matter*, ed. by J.T. Devreese, P. Van Camp, (Plenum, New York 1985) p. 401

3.74 V. Bortolani, A. Franchini, G. Santoro, J.P. Toennies, Ch. Wöll, G. Zhang: Phys. Rev. **B90**,3524 (1990)

4. Structural Information from Atomic Beam Diffraction

K.-H. Rieder

Since about 1975, several novel surface spectroscopic methods based on the interaction of thermal energy neutral particles [4.1] even with very reactive solids have become possible due to the progress in combining high-resolution molecular beam production systems [4.2] with UHV techniques [4.3]. Because of their low kinetic energies (5–300 meV), beams of (nonreactive) atoms or molecules probe the topmost layer of the surface in an absolutely nondestructive manner. Using light species, especially helium, the scattering is predominantly elastic and, as the de Broglie wavelengths are of the order of several tenths of an Ångstrom, diffraction effects dominate on well-ordered surfaces [4.4]. Measurements of the angular locations of the Bragg-diffraction beams allow determination of the size and orientation of the surface unit cells and analyses of diffraction intensities yield the surface corrugations, which very often provide direct pictures of the geometrical arrangement of the surface atoms [4.4]. In the diffraction regime, accurate determination of the particle–surface physisorption potential becomes possible via resonant scattering into bound state channels [4.4,5]. In this contribution a brief but self-contained survey is given of the experimental methods used to obtain atomic beam diffraction data and the theoretical procedures used to analyze them quantitatively to arrive at the pertinent corrugation functions. Selected examples of structural results serve to illustrate the potential and scope of atomic beam diffraction and allow discussion of its position relative to other surface structure methods like scanning tunneling microscopy (STM) [4.6] and low energy electron diffraction (LEED) [4.7].

Springer Series in Surface Sciences, Vol. 27   **Helium Atom Scattering from Surfaces**
Editor: E. Hulpke   © Springer-Verlag Berlin, Heidelberg 1992

4.1 The Particle–Surface Interaction Potential

Diffraction intensity analyses yield the distribution of scattering centers within the unit cell (i.e. ion cores for X-rays and electrons, nuclei for neutrons [4.8]). Appreciation of the nature of the scattering centers in the case of atomic beam diffraction requires a discussion of the atom–surface interaction potential: At distances not too far away from the surface, the impinging atoms first feel an attraction owing to van der Waals forces. Closer to the surface they will be repelled due to the overlap of their electron clouds with those of the atoms forming the solid surface (Pauli exclusion); this causes the repulsive part of the interaction potential to rise steeply. In general, the classical turning points will be farther away for particles impinging on top of the surface ions than for particles impinging between the ions; this gives rise to a periodic modulation of the repulsive part of the potential parallel to the surface. The locus of classical turning points follows a surface of constant total electron density [4.9,10], whereby every point constitutes a scattering center; the resulting scattering surface is called the corrugation function $\zeta(\boldsymbol{R})$. Atoms with higher energies penetrate more deeply into the sea of surface electrons and the corrugation function corresponds to a contour of larger electron density. *Esbjerg* and *Nørskov* [4.9] established a practically linear relation between the energy of the atoms $E(r)$ (in eV) and the charge density $\rho(r)$ [in atomic units (a.u.), i.e. electrons per cubic Bohr radius a_0]

$$E(r) = \beta\rho(r) \quad . \tag{4.1}$$

Manninen et al. [4.11] quoted $\beta(\text{He}) = 160\ \text{eVa}_0^3$, which means that with He energies between 15 and 300 meV, densities between about 1×10^{-4} and 2×10^{-3} a.u. can be covered. Equation (4.1) implies that the classical turning points are about 2–3 Å away from the surface atom cores.

If the surface under investigation is single crystalline and well ordered, the main effect for light particles like He is elastic scattering into diffraction channels that are determined by the periodicity of the corrugation function and the particle wavelength. The allowed discrete energy levels in the attractive part give rise to the phenomenon of "selective adsorption" via resonant scattering [4.4,5] which allows determination of the bound states. Due to the thermal vibration of the surface atoms and the consequent modulation in time of the surface corrugation, the particles can exchange energy with surface vibrational modes. For light atoms inelastic processes are dominated by one-phonon events, so that measurements of surface phonon dispersion relations are possible [4.12,13]. For the heavier Ne, diffraction is still pronounced [4.14], but due to the deeper potential well (which scales roughly with the particle polarizability [4.5]) selective adsorption becomes more important and multiphonon processes are more probable in inelastic scattering [4.15]. For still heavier particles, like Ar, where the momenta involved are much larger and the particle masses match those of the surface atoms more closely, diffraction no longer plays an important role and multiphonon inelastic processes dominate [4.16,17]; due to the deeper potential well a gradual transition occurs from direct inelastic scattering to trapping–desorption inelastic scattering [4.16]. Finally, for molecules, internal degrees of freedom and even dissociation will play a role, leading to the possibility of investigating such interesting surface phenomena as energy exchange with rotations and vibrations [4.16], trapping [4.18] and substrate-mediated chemical reactions [4.19]. For the lightest molecules, H_2, HD and D_2, diffraction effects are dominant but rotational diffraction plays an important role especially for the asymmetric HD [4.20]. Diffraction of atomic hydrogen is naturally restricted to the most inert systems such as alkali halides [4.21] or rare gas overlayers [4.22].

4.2 Diffraction of Light Particles

4.2.1 Diffraction Kinematics for 3D and 2D Systems. A Simple Derivation of a Useful Intensity Formula

For radiation penetrating sufficiently deep into a crystal (neutrons, X-rays and high energy electrons), diffraction arises from the three-dimensional periodic array of atoms [4.8]. The Laue condition relates the incoming wavevector $\boldsymbol{k}_i$ to the outgoing wavevector $\boldsymbol{k}_g$ via

$$\boldsymbol{k}_g - \boldsymbol{k}_i = \boldsymbol{g} \quad . \tag{4.2}$$

The reciprocal lattice vectors $\boldsymbol{g} = h\boldsymbol{b}_1 + k\boldsymbol{b}_2 + \ell\boldsymbol{b}_3$ are specified by the relation $\boldsymbol{a}_i \cdot \boldsymbol{b}_j = 2\pi\delta_{ij}$ between the direct ($\boldsymbol{a}_i$, $i = 1$–3) and the reciprocal ($\boldsymbol{b}_j$, $j = 1$–3) bulk unit-cell vectors. In the case of particle beams the de Broglie wavelength is given by $\lambda_i = 2\pi/|\boldsymbol{k}_i|$ and the beam energy by $E_i = \hbar^2 k_i^2/2m$, m being the particle mass. The energy of the particles remains unchanged during diffraction, so that $k_i^2 = k_g^2$. The Laue condition can be represented graphically by the Ewald construction as shown schematically in Fig.4.1. A cut through a high index plane of a monatomic (orthorhombic) crystal is shown in Fig.4.1a with the individual atoms plotted as circles and the bulk unit cell indicated (the unit cell vector $\boldsymbol{a}_2$ is perpendicular to the drawing). The corresponding cut through the bulk reciprocal lattice is indicated in Fig.4.1b by the points $(hk\ell)$. A beam impinging in the direction shown in Fig.4.1a with the wavelength λ_i drawn gives rise to the $\boldsymbol{k}_i$-vector of Fig.4.1b. Diffraction occurs if the Ewald sphere with radius $\boldsymbol{k}_i$ coincides with a reciprocal lattice point. For the situation of Fig.4.1 in an X-ray or neutron experiment a detector rotating in the plane of the drawing would pick up just the $(\bar{1}03)$-diffraction beam. Depending on the magnitude of $\boldsymbol{a}_2$, out-of-plane diffraction may also occur.

The situation is quite different for beams sensitive only to the topmost layer (like He or Ne atoms) or penetrating just a few layers (low energy electrons): As the periodicity normal to the surface (assumed to be (001) in Fig.4.1) is broken, the reciprocal lattice points along z become infinitely close and thus degenerate to the lattice rods (hk). In Fig.4.1a a (2×1)-reconstruction of the

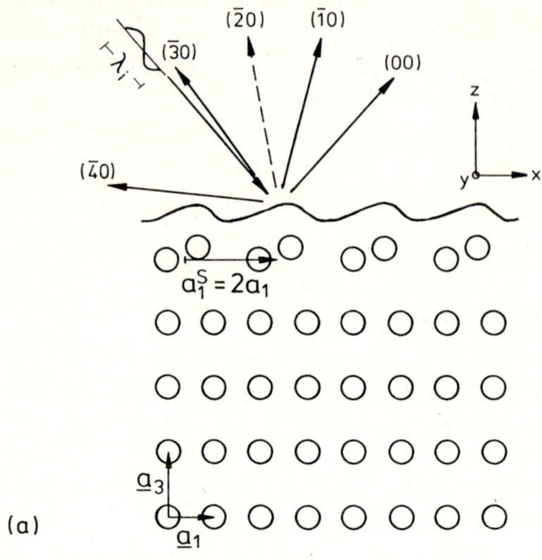

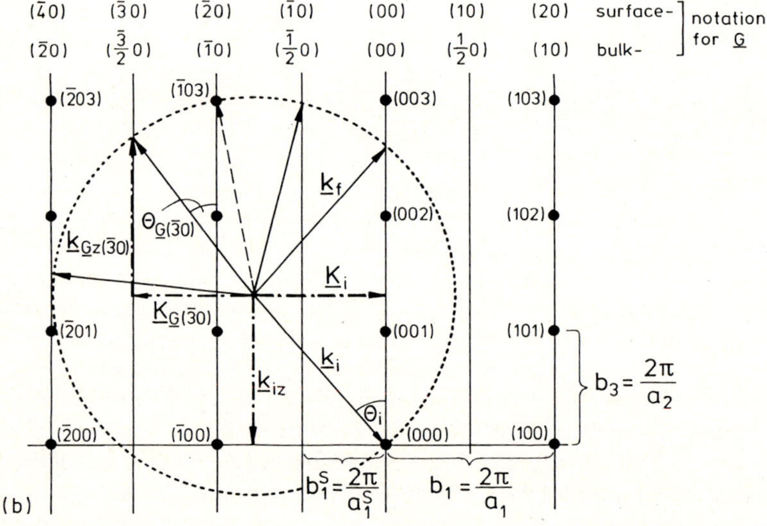

Fig.4.1. (a) Cut through an orthorhombic crystal exhibiting a reconstruction on its (001)-surface which doubles the bulk periodicity along a_1. A contour of constant electron density is shown and the directions of the incoming and (surface) diffracted beams are indicated. (b) The Ewald construction for diffraction from surfaces. For the wavelength indicated in (a) bulk diffraction occurs only in the direction of the dashed arrow, where the Ewald sphere intersects through the reciprocal lattice point $(\bar{1}03)$. Surface diffraction occurs in all directions in which the Ewald sphere cuts the 2D-lattice rods G (the bulk and surface notations for G are both given)

44

(001)-surface is assumed leading to a doubling of the surface periodicity along a_1; consequently in reciprocal space the number of lattice rods doubles and rods with half-order h-indices occur between those corresponding to the truncated bulk plane (reference to the actual reconstructed surface unit cell leads to the integer number surface-notation also given in Fig.4.1b). Surface (2D) diffraction occurs in all directions in which the Ewald sphere cuts the lattice rods and the Laue condition for surface diffraction relaxes to

$$\boldsymbol{K}_G - \boldsymbol{K}_i = \boldsymbol{G} \quad , \tag{4.3}$$

whereby $\boldsymbol{k}_i = (\boldsymbol{K}_i, \; k_{iz}) = (k_i \sin\theta_i, \; -k_i \cos\theta_i)$ and $\boldsymbol{k}_G = (\boldsymbol{K}_G, \; k_{Gz}) = (k_G \sin\theta_G, \; k_G \cos\theta_G)$. θ_i denotes the angle of incidence of the incoming beam and θ_G the angle of emergence of the diffracted beam $\boldsymbol{G}$, both measured from the surface normal z. Notice that slight changes in θ_i cause the 3D-diffraction beam to vanish, whereas the 2D-beams undergo only slight angular shifts.

Generally, the surface reciprocal lattice vectors $\boldsymbol{G} = h\boldsymbol{b}_{1s} + k\boldsymbol{b}_{2s}$ are specified by the relation $\boldsymbol{a}_{is} \cdot \boldsymbol{b}_{js} = 2\pi\delta_{ij}$ between the direct ($\boldsymbol{a}_{is}$, $i = 1,2$) and the reciprocal ($\boldsymbol{b}_{js}$, $j = 1,2$) surface unit-cell vectors. From Fig.4.1b the Bragg relation for in-plane diffraction connecting θ_i and θ_G [with $\boldsymbol{G} = (h\ 0)$, h integer] is easily verified:

$$\sin\theta_h = \sin\theta_i + h\frac{\lambda_i}{a_{1s}} \quad . \tag{4.4}$$

Generalization for out-of-plane diffraction requires specification of an out-of-plane angle and is straightforward but tedious [4.4]. Measurements of the angular locations of the Bragg peaks thus allow the determination of the dimensions of the surface unit cell and its orientation relative to the incoming beam.

The intensities of the Bragg reflections are determined by the distribution of the scattering centers within the unit cell as can be seen by analogy with the structure factor for three-dimensional diffraction [4.8]

$$F(\boldsymbol{g}) = \sum_j f_j \exp[\mathrm{i}(\boldsymbol{k}_g - \boldsymbol{k}_i) \cdot \boldsymbol{r}_j] = \sum_j f_j \exp(\mathrm{i}\boldsymbol{g} \cdot \boldsymbol{r}_j) \quad , \tag{4.5}$$

where the sum is over all atoms j located at $\boldsymbol{r}_j$ within the unit cell and having scattering lengths f_j. The Laue condition for 3D-diffraction (4.2) is incorporated in (4.5). In the case of surface diffraction the Laue condition is relaxed according (4.3) so that (4.5) has to be modified to

$$F(\boldsymbol{G}, q_{Gz}) = \sum_j f_j \exp\{\mathrm{i}\,[\boldsymbol{G} \cdot \boldsymbol{R}_j + (k_{Gz} - k_{iz})z_j]\} \quad , \tag{4.6}$$

with $\boldsymbol{r}_j = (\boldsymbol{R}_j, z_j)$ and $q_{Gz} = k_{Gz} - k_{iz}$ being the component of the scattering vector $\boldsymbol{q}_G = \boldsymbol{k}_G - \boldsymbol{k}_i$ perpendicular to the surface. This form of the structure factor is important for grazing incidence X-ray diffraction [4.23] as well as for kinematical LEED calculations [4.24]. In the case of atom diffraction there is a continuous distribution of equivalent (same f_j) scattering centers along the corrugation function $\zeta(\boldsymbol{R})$, so that the sum in (4.6) transforms into an integral over the surface unit cell (uc) with area $S = \boldsymbol{a}_{is} \cdot \boldsymbol{a}_{2s}$. The proper normalization factor S^{-1} is obtained by realizing that for a completely flat corrugation all the intensity scattered back from the surface goes into the specular beam (00):

$$F(\boldsymbol{G}) = \frac{1}{S} \int_{\text{uc}} \exp\left\{ i \left[\boldsymbol{G} \cdot \boldsymbol{R} + (k_{Gz} - k_{iz})\zeta(\boldsymbol{R}) \right] \right\} d\boldsymbol{R} \quad . \tag{4.7}$$

This is the so-called eikonal formula, which constitutes a convenient method to calculate diffraction intensities $P_G = (k_{Gz}/k_{iz})|F(\boldsymbol{G})|^2$ (the prefactor k_{Gz}/k_{iz}, which corresponds to $\cos\theta_G/\cos\theta_i$ for in-plane scattering, normalizes beam cross sections for different angles of emergence). The applicability range of the eikonal approximation can be checked via the unitarity condition $\sum_G P_G = 1$ requiring that the intensities of all diffracted beams add up to the intensity of the incoming beam. Being derived from kinematic theory, the eikonal formula neglects multiple scattering and is consequently reliable only for shallow corrugations and small angles of incidence; it yields the same intensities for $\zeta(\boldsymbol{R})$ and $-\zeta(-\boldsymbol{R})$, so that for surfaces with 2D-inversion symmetry, $\zeta(\boldsymbol{R}) = \zeta(-\boldsymbol{R})$, $\zeta(\boldsymbol{R})$ cannot be distiguished from $-\zeta(\boldsymbol{R})$, see [Ref.4.4, Fig.4]. The eikonal formula can also be derived within the hard corrugated wall (HCW) model [4.25] as will be shown in the next section.

4.2.2 Diffraction Intensity Calculations Based on the Hard Corrugated Wall Model

A number of simplifying assumptions about the particle–surface interaction potential are made in the HCW-model:

(a) The attractive part of the potential is neglected. This is reasonable as long as the particle energy is much higher than the potential depth D; for He, D values are usually between 5 and 10 meV, so that with a 63 meV room temperature beam this condition is well fulfilled.

(b) The steeply rising repulsive part of the potential is assumed to be infinitely steep.

(c) The surface atoms are taken to be at rest.

Consequently, the potential is zero outside the solid and infinite inside and the surface is described by the time-independent corrugation $\zeta(\boldsymbol{R})$.

With the so-called Rayleigh assumption, which treats the incoming and outgoing beams as plane waves up to the surface, the total particle wavefunction is

$$\psi(\boldsymbol{r}) = \exp(i\boldsymbol{k}_i \cdot \boldsymbol{r}) + \sum_G A_G \exp(i\boldsymbol{k}_G \cdot \boldsymbol{r}) \quad , \tag{4.8}$$

where $\boldsymbol{r} = (\boldsymbol{R}, z)$ denotes the space coordinates. In principle, the sum extends over all $\boldsymbol{G}$'s, i.e. over propagating ($k_{Fz} > 0$) as well as over evanescent ($k_{Ez} < 0$) waves. The intensities of the former, $P_F = (k_{Fz}/k_{iz})|A_F|^2$, must obey the unitarity condition. With the boundary condition $\psi\{\boldsymbol{R}, \zeta(\boldsymbol{R})\} = 0$, stating

46

simply that the particles cannot penetrate the hard wall, one arrives at the equation

$$\sum_{G} A_G \exp\{\mathrm{i}\boldsymbol{G} \cdot \boldsymbol{R} + \mathrm{i}k_{Gz}\zeta(\boldsymbol{R})\} = -\exp\{\mathrm{i}k_{iz}\zeta(\boldsymbol{R})\} \quad , \tag{4.9}$$

which must be fulfilled for any $\boldsymbol{R}$ within the unit cell. Using *Garcia*'s GR-method [4.26], this equation can be solved numerically by relating a finite number N of $\boldsymbol{R}$-vectors uniformly distributed within the surface unit cell to the same number of $\boldsymbol{G}$-vectors, and solving the corresponding matrix equation. Depending on particle wavelength, unit-cell dimensions and corrugation amplitude, the dimensions of the matrices required to obtain convergent solutions may be quite large ($N \approx 250$).

Multiplying (4.9) by $\exp\{-\mathrm{i}[\boldsymbol{G}' \cdot \boldsymbol{R} + k_{G'z}\zeta(\boldsymbol{R})]\}$ and integrating over the surface unit cell yields the matrix equation

$$\sum_{G} M_{GG'} A_G^0 = A_{G'} \tag{4.10}$$

with

$$M_{GG'} = \frac{1}{S} \int_{\mathrm{uc}} \exp\{\mathrm{i}(\boldsymbol{G} - \boldsymbol{G}') \cdot \boldsymbol{R} + \mathrm{i}(k_{Gz} - k_{G'z})\zeta(\boldsymbol{R})\}d\boldsymbol{R} \tag{4.11}$$

and

$$A_G^0 = \frac{1}{S} \int_{\mathrm{uc}} \exp\{\mathrm{i}\boldsymbol{G} \cdot \boldsymbol{R} + \mathrm{i}(k_{Gz} - k_{iz})\zeta(\boldsymbol{R})\}d\boldsymbol{R} \quad . \tag{4.12}$$

Equation (4.10) can be solved either by matrix inversion [4.27] or by iteration [4.28] for the A_G's. Within the Rayleigh ansatz, convergence is limited by the smallest radius of curvature of the corrugation [4.4]: the theoretical maximum for the top to bottom amplitude is 14.3 % of the lattice constant for a one-dimensional sinusoidal corrugation and 18.8 % for a two-dimensional quadratic corrugation described by a sum of cosines in both x- and y-directions. The numerical procedures described above actually yield reliable intensities also for amplitudes somewhat above the Rayleigh limit; this can be judged by comparisons with exact HCW-calculations applicable for large corrugation amplitudes (RR'-method [4.29]).

Neglect of the off-diagonal terms in (4.8), which implies neglect of evanescent waves, yields $A_G = A_G^0$, the eikonal formula (4.7) (Sect. 4.2.1). Comparisons of intensities calculated with the GR-method and the eikonal approximation for several model corrugations presented in [Ref.4.4, Figs.5–8] of confirm the reliability of the eikonal approximation for small angles of incidence and shallow corrugations.

4.2.3 Influences of Realistic Potentials: Limitations of Approximations

For corrugation amplitudes exceeding $\sim 10\,\%$ of the lattice constant in the pertinent direction, the influences of the attractive potential and the finite steepness of the repulsive part are generally no longer negligible and intensity calculations with realistic periodic potentials

$$V(\boldsymbol{r}) = \sum_{G} V_G(z)\exp(\mathrm{i}\boldsymbol{G}\cdot\boldsymbol{R}); \quad V_G(z) = \int_{\mathrm{uc}} V(\boldsymbol{r})\exp(-\mathrm{i}\boldsymbol{G}\cdot\boldsymbol{R})d\boldsymbol{R} \qquad (4.13)$$

are required. The coupled channels method (CCM) [4.30] starts from the Schrödinger equation for particles with mass m and wavevector $\boldsymbol{k}_i$

$$\left(-\frac{\hbar^2}{2m}\nabla^2 + V(\boldsymbol{r}) - \frac{\hbar^2 k_i^2}{2m}\right)\psi(\boldsymbol{r}) = 0 \qquad (4.14)$$

and uses Bloch's theorem in 2D to write for the wavefunction

$$\psi(\boldsymbol{r}) = \sum_{G}\psi_G(z)\exp\left[\mathrm{i}(\boldsymbol{K}_G + \boldsymbol{G})\cdot\boldsymbol{R}\right] \quad, \qquad (4.15)$$

so that upon substitution of (4.15) into (4.14) a set of nonlinear coupled differential equations of second order is obtained

$$\left(\frac{d^2}{dz^2} + k_{Gz}^2\right)\psi_G(z) - \frac{2m}{\hbar^2}\sum_{G'} V_{G-G'}\psi_{G'}(z) = 0 \qquad (4.16)$$

which has to be solved for the ψ_G, whereby the asymptotic form $(z \to \infty)$ for the open (Bragg) channels $(k_{\mathrm{F}z}^2 = k_i^2 - (\boldsymbol{K}_i + \boldsymbol{F})^2 \geq 0)$ is

$$\psi_{\mathrm{F}}(z) = \exp(-\mathrm{i}\boldsymbol{k}_{iz}z) + A_{\mathrm{F}}\exp(\mathrm{i}\boldsymbol{k}_{\mathrm{F}z}z) \qquad (4.17)$$

and for the evanescent waves $(\boldsymbol{k}_{\mathrm{E}z}^2 < 0)$

$$\psi_{\mathrm{E}}(z) = A_{\mathrm{E}}\exp(-\kappa_{\mathrm{E}}z) \quad \text{with} \quad \kappa_{\mathrm{E}} = (-k_{\mathrm{E}z}^2)^{1/2} \quad . \qquad (4.18)$$

The intensities P_{F} have to add up to unity. A fast and stable numerical method has been developed by *Blake* [4.31]. A modified version of his computer program is used for the following model calculations [4.32], which serve to illustrate the range of applicability of the methods based on the HCW. Since atom–surface interactions have been shown for numerous metal surfaces [4.33] to be well described by the Morse potential, the form

$$V(x,z) = D\left\{\exp\left[-2\alpha(z - z_0 - \phi)\right] - 2\exp\left[-\alpha(z - z_0 - \phi)\right]\right\} \qquad (4.19a)$$

(with D denoting the potential well depth, α the reciprocal range parameter and z_0 the position of the laterally averaged potential minimum) is assumed in connection with a (1D) sinusoidal corrugation ϕ exponentially decreasing along z with decay constant $\beta(10)$

$$\phi(x, z) = \frac{d(10)}{2} \cos \frac{2\pi a}{x} \exp\left[-\beta(10)(z - z_0)\right] \quad . \tag{4.19b}$$

Typical values $D = 8$ meV, $\alpha = 1$ Å^{-1}, $\beta(10) = 0.4$ Å^{-1} and $a(\text{Ni}) = 3.52$ Å are assumed. The dependence of beam intensities as a function of corrugation amplitude $\zeta(10)$ is shown in Fig.4.2 for $\theta_i = 0°$ and $\lambda_i = 0.57$ Å; these data were calculated with the CC-method using the potential (4.19) and with the GR- and RR$'$-methods in the HCW approximation (the value of $\zeta(10)$ in Fig.4.2 refers to the potential contour corresponding to the incoming energy $E_i = 63$ meV; according to the value of $\beta(10)$ the corrugation amplitude at the potential minimum d(10) is ~ 60 % of $\zeta(10)$). Notice that the HCW curves are systematically shifted with respect to those calculated with the full interaction potential. Fitting the CC-results with the HCW thus systematically overestimates the corrugation amplitudes, the deviation becoming larger with increasing corrugation amplitude. Figure 4.3 shows the angular dependence of the diffraction intensities for $\zeta(10) = 0.32$ Å; the systematic deviations between the curves calculated with the full potential and the HCW become more pronounced at larger θ_i, so that application of the HCW is reasonable only for small θ_i. A detailed discussion is presented in [4.32].

4.2.4 The Problem of Data Inversion

The derivation of the corrugation function from a given set of measured intensities is seriously hampered by the fact that the measured intensities P_G do not contain information on the phases ϕ_G of the respective scattering amplitudes $A_G = |A_G| e^{i\phi_G}$. Therefore, up to now trial-and-error procedures have usually been used within the HCW. Assuming for example for the corrugation the Fourier representation

$$\zeta(\boldsymbol{R}) = \sum_G \zeta(\boldsymbol{G}) \exp(i\boldsymbol{G} \cdot \boldsymbol{R}) \quad , \tag{4.20}$$

with the number of possible parameters $\zeta(\boldsymbol{G})$ reduced according to the symmetry properties of the unit cell, the best-fit coefficients can be determined by varying their amplitudes until optimum agreement with the measured intensities is obtained; it is important to note that in this way data analyses can be performed entirely free of any model assumptions on the surface structure. It is, however, also legitimate to start from a model structure (as usual in LEED [4.7]) and to describe the topmost adatoms and substrate atoms by — for example — Gaussian hills whereby heights, widths and relative locations are varied [4.34]. As a measure for the agreement between measured P_G^{exp} and calculated intensities P_G^{calc}, "reliability factors" such as

$$R = \frac{1}{N} \left\{ \sum_G \left(P_G^{\text{calc}} - P_G^{\text{exp}}\right)^2 \right\}^{1/2} \tag{4.21}$$

can be used (N denotes the number of beams $\boldsymbol{G}$ measured).

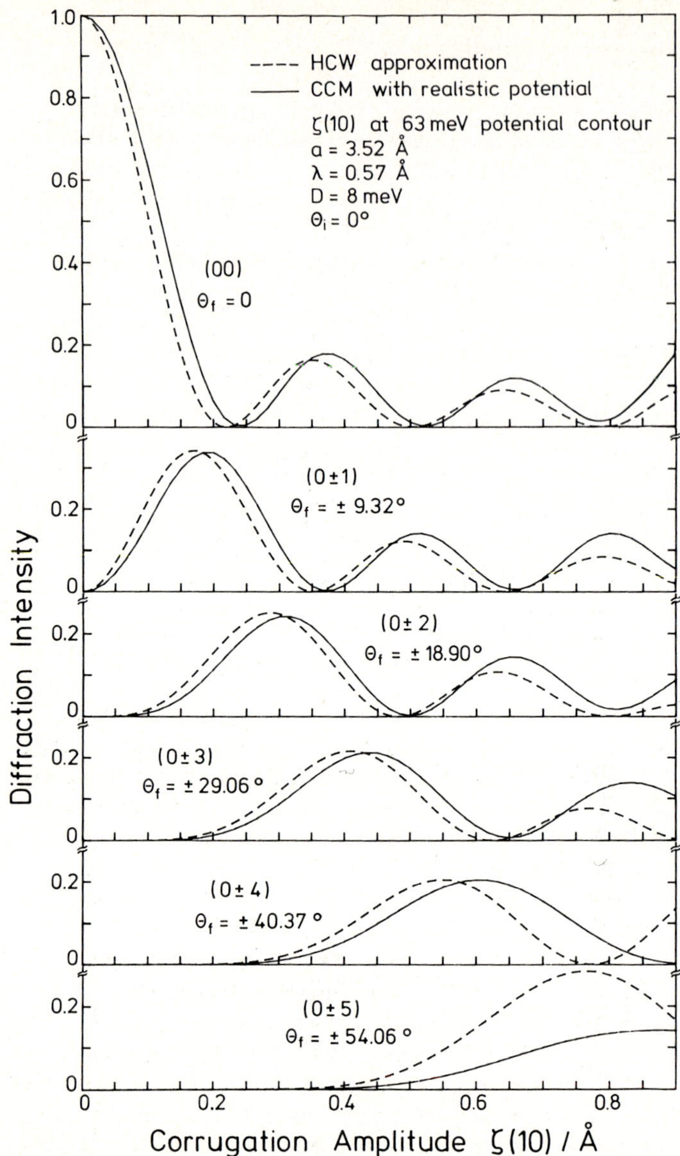

Fig.4.2. Intensity variation of diffraction beams for a sinusoidal corrugation as a function of corrugation amplitude calculated with the HCW-GR method and the CC method with a realistic Morse potential (D = 8 meV). After [4.32]

Effective inversion procedures on the basis of the HCW using both eikonal [4.35] and GR-methods [4.36] have been proposed, in which trial phases are refined in iterative steps until agreement with experimental intensities is achieved. Since these procedures seem to require starting conditions not too far from the

50

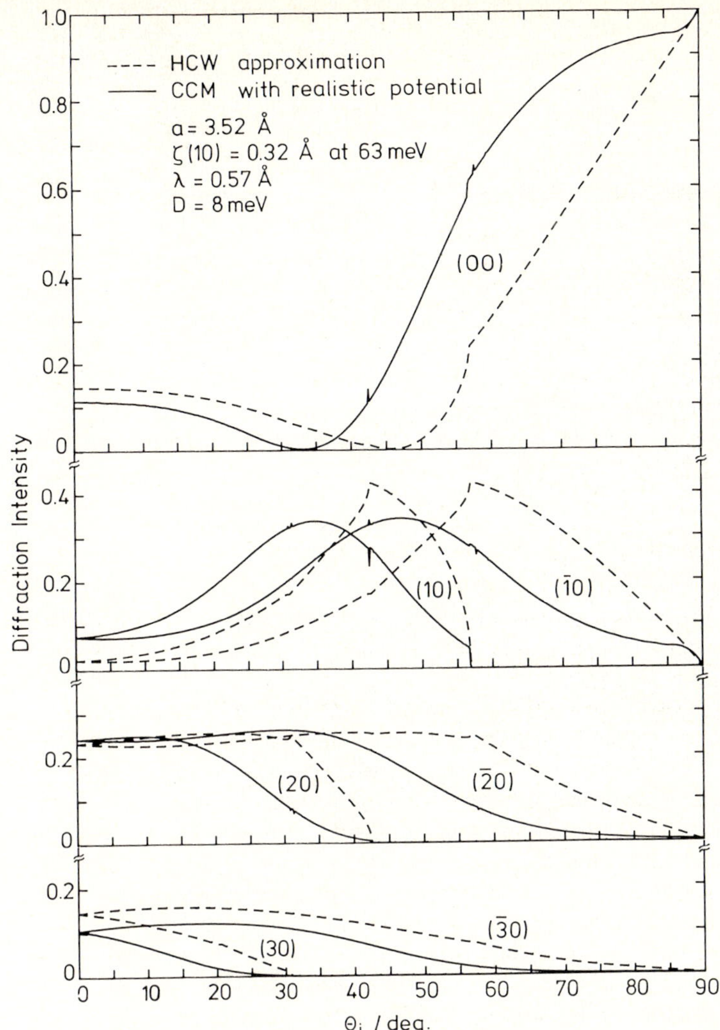

Fig.4.3. Intensity variation of diffraction beams for a sinusoidal corrugation with corrugation amplitude 0.32 Å at the 63 meV potential contour as a function of the angle of incidence calculated with the HCW-GR method and the CC method with a realistic Morse potential ($D = 8$ meV). After [4.32]

desired answer [4.37], they can at least be used to refine the final corrugations point-by-point without relying on analytical forms [4.37].

Having realized that the eikonal formula is identical to the kinematic structure factor, it is tempting to apply the inversion procedures used in X-ray and neutron crystallography. One of the most widely used methods is the Patterson synthesis, from which the distances of all atoms to all others in the unit cell can be deduced. *Cantini* et al. [4.38] have derived the following relation specifically for the case of atomic beam diffraction

$$\sum_{G} |A_G|^2 \exp(-i\boldsymbol{G}\cdot\boldsymbol{R}) = \int_{uc} \exp\{i\overline{q}_z\,[\zeta(\boldsymbol{r}) - \zeta(\boldsymbol{r}-\boldsymbol{R})]\}\,d\boldsymbol{r} \quad . \tag{4.22}$$

In principle, the intensities of all beams used on the left side of (4.20) have to be measured for the same $\overline{q}_z$. However, since the eikonal formula is only valid for small corrugations, all beams with appreciable intensities are close to the specular so that for small θ_i the mean value $q_z(00)$ for $\overline{q}_z$ can be used in connection with data sets for fixed θ_i and λ_i [4.37]. Although not much experience is available up to now, (4.22) has at least been used to compare the Patterson function derived directly from experimental intensities [left-hand side of (4.22)] with that calculated with the best-fit corrugation [right-hand side of (4.22)] in order to check the reliability of the latter; an example is given in Sect.4.4.2.

4.3 Experimental

In their historic experiments in 1929 to prove the validity of the de Broglie relation for particles other than and heavier than electrons, *Stern* and *Estermann* [4.39] used low-pressure (10^{-3} Torr) effusion (Knudsen) sources delivering He beams with low intensity and broad Maxwellian velocity distribution, so that only first-order diffraction beams could be resolved upon scattering from LiF(100). Several decades of research in gas dynamics resulted in the development of sources yielding thermal beams with high intensity and narrow velocity distribution [4.2]. In modern high-pressure supersonic nozzle sources, an interplay of thermodynamic cooling and hydrodynamic flow leads to appreciable narrowing of the velocity distribution and to an enhancement of the mean energy of the He atoms by a factor of 5/3 relative to the most probable energy of the atoms in the reservoir. Such sources are now commonly in use in gas–surface scattering experiments. Fig.4.4 shows a schematic sketch of the system used in the author's laboratory for diffractive studies.

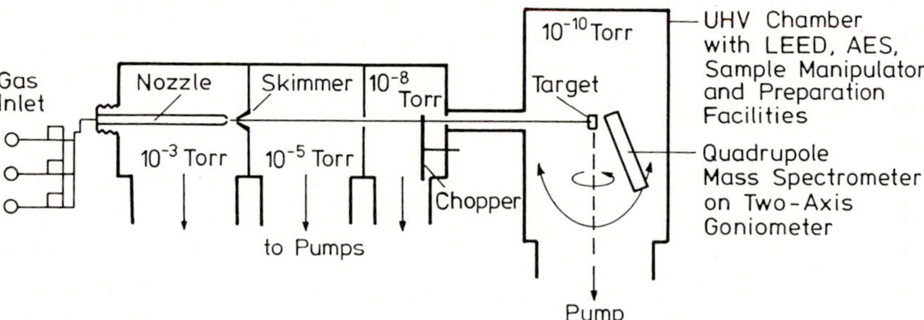

Fig.4.4. Schematic sketch of the atomic beam scattering apparatus used in the author's laboratory; the system is designed primarily for diffraction studies

The gas load due to the high pressure behind the 10 μm nozzle is handled by several differential pumping stages using high-speed pumps. With such arrangements, pressures of $\lesssim 5 \cdot 10^{-10}$ Torr can be maintained in the scattering chamber with the beam on (base pressure in the sample chamber $\sim 2 \cdot 10^{-11}$ Torr). A well-adjusted nozzle-skimmer geometry is essential for good performance of the beam system. Collimation of the beam is obtained by the skimmer and several apertures placed between the differentially pumped stages. Typical He beam parameters for a nozzle at room temperature are: $E_i = 63$ meV or $\lambda_i = 0.57$ Å; with a pressure behind the nozzle of 3 atm, a velocity spread of about 10 % (FWHM) of the mean velocity is readily achieved; increase in pressure decreases the velocity spread and in our system we reach about 1.5 % at 80 atm. The mean energy (and wavelength) of the He beam can be varied easily by heating and cooling the nozzle and He energies between 5 and 270 meV [4.12,13] have been reported. The velocity spread becomes narrower at lower temperatures with the same pressure behind the nozzle [4.2] and with the nozzle at liquid nitrogen temperature spreads below 0.5 % were obtained [4.12]. For diffractive work, the particles scattered from the target are best analyzed with a quadrupole mass spectrometer mounted on a two-axis goniometer which allows in-plane as well as out-of-plane detection [4.4]. The possibility to cover a large part of a sphere with proper goniometer movements is important for determination of the symmetry elements of the surface under investigation as well as for an as complete as possible measurement of diffraction intensities for a given scattering geometry (λ_i, θ_i) to guarantee a reliable determination of the surface morphology. The pressure increase in the sample chamber due to the particle beam of about one order of magnitude gives rise to a continuous background in the scattering chamber which limits the signal-to-noise ratio; the latter can be improved by modulation of the beam with a mechanical chopper in the pumping stage before the scattering chamber combined with phase sensitive detection. With the system used in the author's laboratory diffraction beams of the order of 2×10^{-4} of the intensity of the incoming beam can be measured. Recently quadrupoles in differentially pumped housings have also been constructed without appreciable reduction of goniometer movements [4.40], so that even beams with an order of magnitude smaller intensity are accessible. Rotation of the sample around several axes as well as linear motions can usually be performed with modern commercial high-precision manipulators. The possibility of heating and cooling the sample is desirable for all cleaning and experimental procedures. Surface preparation also requires ion bombardment and reactions with gases; careful surface characterization is necessary in every experiment and may be performed with combinations of standard methods like low energy electron diffraction (LEED), Auger electron spectroscopy (AES), photoelectron emission (XPS and UPS) or secondary ion mass spectroscopy (SIMS) [4.24].

Figures 4.5–7 show examples of He diffraction spectra for clean and hydrogen-covered (110)-surfaces of Rh and Ni. Exact determination of the diffraction intensities requires that both the increase in beam width relative to the specular due to the finite velocity spread of the incoming beam, and Debye–Waller

(DW) attenuation due to the thermal motion of the surface atoms are taken into account. For the latter an effective surface Debye temperature can be determined experimentally by measurements at different sample temperatures and extrapolation to 0 K [4.4,8]; however, since for θ_i between 20 and 40° beams $(h\ k)$ and $(\overline{h}\ k)$ — lying symmetrically around the specular — are influenced oppositely by the DW–factor, the error introduced by neglecting the DW–correction usually averages out in the best-fit corrugations provided as many symmetry-related beams as possible are included in the intensity analyses.

4.4 Examples of Surface Structure Determinations

4.4.1 Unreconstructed Surfaces

Figure 4.5 shows the He diffraction spectra of clean Rh(110) [4.41] and Ni(110) [4.42] measured with the room temperature He beam. The more complex He diffraction patterns of hydrogen overlayers on Rh(110) [4.43] and Ni(110) [4.34] both corresponding to a coverage of 1/2 monolayer (ML) are shown in Figs.4.6 and 4.7. In all cases the beam impinges perpendicular to the close packed metal rows, i.e. along the [001]-direction (Fig.4.10a). The best-fit peak intensities are indicated by crosses in Figs.4.5 and 4.6 and a full best-fit spectrum is shown in Fig.4.7; the corresponding best-fit corrugations for the clean Rh(110) surface and the two H adsorption phases are shown in Fig.4.8. Hard sphere models of the clean surface and the respective rather complicated H adatom arrangements are shown in Figs.4.9a, 4.9c and 4.10b.

Figure 4.5 shows that the first order diffraction intensities (0 ± 1) relative to the specular intensity are much stronger for Rh(110) than for Ni(110) indicating that the corrugation parameter $\zeta(01)$ is appreciably larger for the system with the larger lattice constant. This systematic trend for $\zeta(01)$ is nicely visible in the compilation of all hitherto investigated fcc(110)-surfaces presented in Table 4.1. Notice that the corrugation amplitudes along the close packed rows $\zeta(10)$ do not follow this trend at all; whether this is connected with different strengths of anticorrugating effects [4.44] is not clear at present. It seems possible, however, that due to peculiarities in the electronic structures of transition metal surfaces even such gentle probes as He atoms can give rise to electron redistributions during impact so that the maxima in the corrugation along the close packed directions do not correspond to underlying atoms but rather to the short-bridge sites [4.44]. It is also important to notice that with Ne diffraction systematically much larger corrugation amplitudes than with He are obtained [4.14] as can also be seen from Table 4.1. Anticorrugating effects have been stated to be smaller for Ne than for He [4.45], although experimentally there is no such evidence up to now; we will return to the question of anticorrugating effects in our discussion of H chemisorption structures (Sect.4.4.2.). Table 4.1 also comprises the experimentally determined He–surface potential well depths. The corrugation amplitudes of the closest-packed fcc-metal surfaces (111) and (100) are extremely small: for example for Ni(100) and Ni(111) values of 0.015 Å and 0.01 Å , respectively, were measured [4.46]. More open metal surfaces like Ni(113) exhibit larger corrugation amplitudes between the close-packed

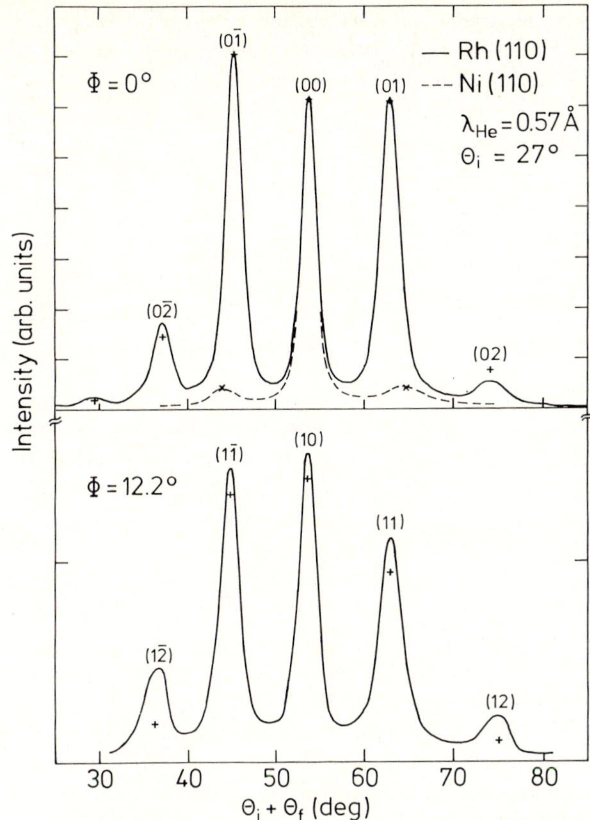

Fig.4.5. Diffraction scans for the clean (110)-surfaces of Rh [4.41] and Ni [4.42]; the intensities of the specular beams are normalized to the same height. The best-fit peak intensities are indicated as crosses

rows due to the larger distance between them; for this particular surface an interesting corrugation asymmetry was observed via a one-wing out-of-plane rainbow pointing to an appreciable charge transfer from the (111)- to the more open (100)-microfacets [4.46].

In concluding this section we mention that the clean (100)-surfaces of the hitherto investigated alkali halides exhibit peak-to-peak corrugation amplitudes which are in good agreement with the differences in the ionic radii of the respective constituents, but that the corrugation values of the oxides MgO and NiO are appreciably smaller; since ion rumpling can be excluded [4.47], these observations point to heavy charge redistributions and different bonding character at (and near) the oxide surfaces [4.48].

Table 4.1. Dependence of corrugation amplitudes on lattice spacing for different fcc(110) surfaces. Notice the strong increase in corrugation amplitudes perpendicular to the close-packed rows, $z(01)$, with increasing distance observed with both He and Ne diffraction. The corrugation amplitudes parallel to the close-packed rows, $z(10)$, measured up to now with He diffraction exhibit rather irregular behaviour. Values of the potential well depths as derived from selective adsorption measurements are also included

Surface	Lattice constant $d(\text{Å})$	Corrugation amplitudes [Å] He $z(10)$	$z(01)$	Ne $z(01)$	Potential depth for the $D[\text{meV}]$
Ni(110)	3.52	0.025	0.075	0.17	—
Cu(110)	3.61	≈ 0.0	0.13	0.21	6.35*
Rh(110)	3.80	0.06	0.15	0.29	8.2
Pd(110)	3.89	0.02	0.21	0.42	8.05
Ag(110)	4.09	≈ 0.0	0.29	—	6.1

* extrapolated from Cu(11n) surfaces (n=3–7)

4.4.2 Adsorption Systems. Quantitative Aspects: Does Helium Diffraction Allow Determination of Bond Lengths?

Figure 4.6 shows diffraction from Rh(110)+(1×2)H [4.43]. In the corresponding corrugation, Fig.4.8b, the H adatoms give rise to the larger hills (amplitude 0.3 Å) and the basic corrugation hills of the substrate atoms with smaller amplitude are still discernible; the linear arrangement of the adatoms in nearly threefold-coordinated sites alongside every second close-packed Rh row is directly visible. Two different domains with adatom chains to the left or to the right of the close-packed metal rows are possible and are taken into account in the best-fit intensity calculations. A sphere model of the left domain of this phase is shown in Fig.4.9c. Notice that according to this adatom arrangement, which has been quantitatively determined by dynamical LEED calculations [4.49], one would expect the substrate atom hills in the corrugation of Fig.4.8b to be shifted by $a_1/2$ relative to the adatom hills. A similar observation of the 'wrong phase' of substrate atoms was also made with the corrugation of the $\theta = 1/3$ ML phase of H on Ni(110) (see Fig.4.10a for a sphere model); it thus appears possible that anticorrugating effects as discussed in Sect.4.4.1 may indeed give rise to this phenomenon in the case of fcc(110) transition metal surfaces [4.43,44]. Further investigations are necessary to clarify this important point.

Figure 4.7 shows a typical diffraction pattern [4.34] for the H phase with coverage $\theta = 1/2$ ML on Ni(110); the corrugation of this c(2×4) phase, Fig.4.8c, shows clearly that — in contrast to the case of Rh(110) — on Ni(110) the H adatoms (corrugation height 0.25 Å) form alternating zigzag and zagzig chains along every second close-packed metal row, whereby the H adatoms again occupy nearly threefold-coordinated sites; notice that the systematic absence of (n,e/4)-beams and (n/2,o/4)-beams (n integer, e even, o odd) indicates that this structure is centered [4.34].

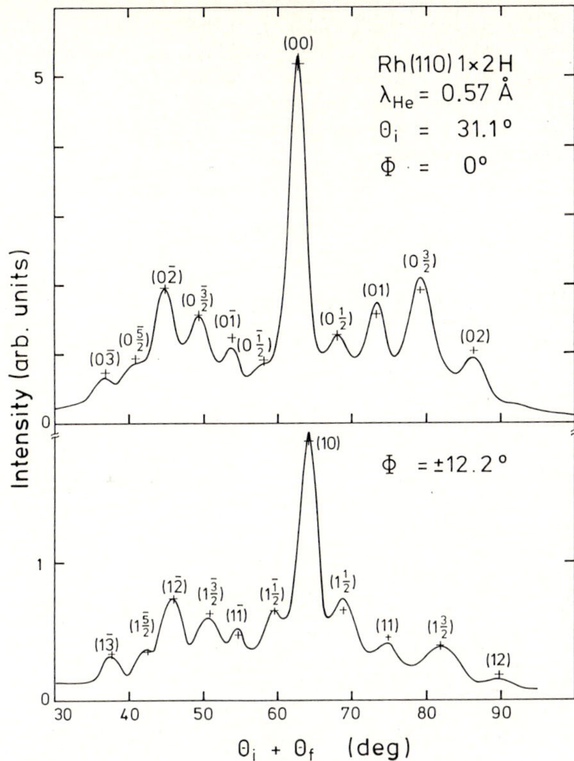

Fig.4.6. In-plane and out-of-plane He diffraction traces for the (1×2) phase of hydrogen on Rh(110) corresponding to a coverage of 1/2 ML. The full line corresponds to the experimental result, the crosses indicate best-fit peak intensities [4.43]. The corresponding corrugation function is shown in Fig.4.9c

Figures 4.9 and 4.10 show via hard-sphere models the fascinating sequences of the adatom configurations of all H-phases observed with increasing coverage on Rh(110) and Ni(110) with the sample at 100 K. On Ni(110) there is a quite regular sequence of phases corresponding to coverages of 1/3, 1/2, 2/3, 5/6, 1.0 and 1.5 ML. Notice that the zigzag and zagzig configurations are maintained up to $\theta = 2/3$ ML, but that all H chains are parallel for $\theta = 1.0$ ML. The phase with $\theta = 5/6$ ML occurs only up to temperatures of ~ 140 K and seems to ease the transition of all zagzig chains to zigzag ones at low temperatures: the elongated hexagonal pattern typical for the (2×1)-phase is preformed in the $\theta = 5/6$ ML phase by adsorption of 1/6 ML hydrogen in the energetically unfavourable short-bridge sites (shaded circles in Fig.4.10d), so that upon further H uptake the (2×1)-structure can form by local compression of the hexagons along the [001]-direction. At temperatures exceeding 140 K the mobility of the zagzig chains is high enough to allow a direct transition from the $\theta = 2/3$ to the $\theta = 1$ ML phase. Notice that the saturation phase with $\theta = 1.5$ ML shows a H-induced pairing-row reconstruction of the metal substrate;

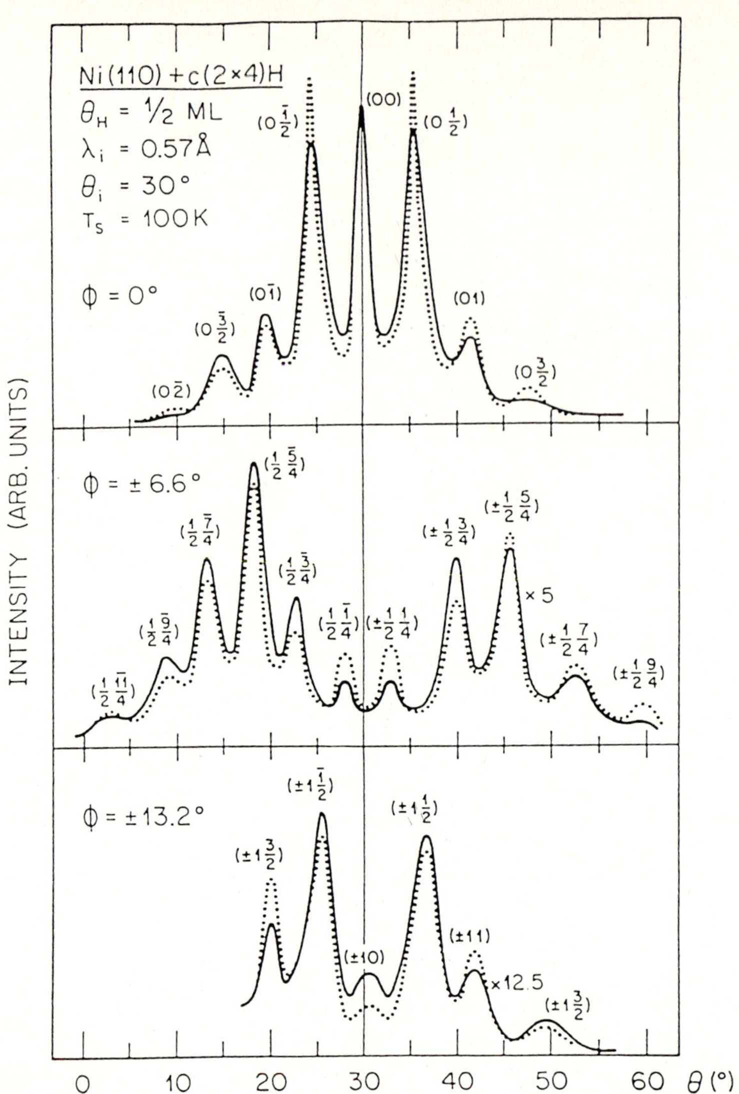

Fig.4.7. In-plane and out-of-plane He diffraction traces for the c(2×4) phase of hydrogen on Ni(110) corresponding to a coverage of 1/2 monolayer. The full line corresponds to the experimental result, and the broken line to the best-fit calculation [4.34]. The corresponding corrugation function is shown in Fig.4.10b

the model shown in Fig.4.10f was preferred on the basis of the corrugation determined and was recently confirmed by dynamical LEED calculations [4.50] and high resolution electron energy loss spectroscopy (HREELS) investigations [4.51]; the latter require the H atoms within the paired rows (shaded circles in Fig.4.10f) to be shifted slightly away from the high symmetry locations.

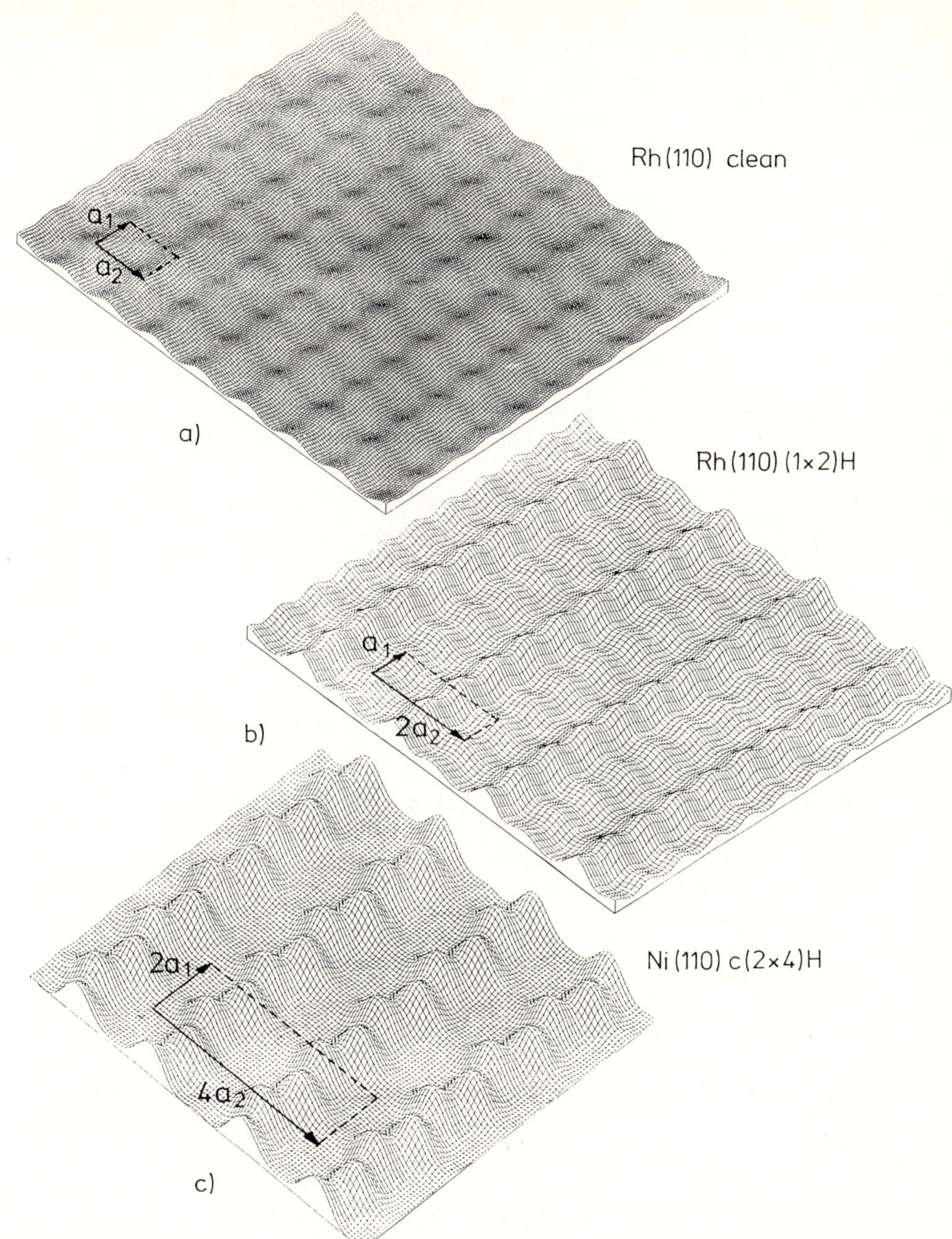

Fig.4.8. Corrugation functions of (a) the clean Rh(110) surface [4.43], (b) the (1×2)H phase on Rh(110) [4.43] and (c) the c(2×4)H phase on Ni(110) [4.34]. Both H phases correspond to a coverage of 1/2 ML. Note that the corrugations (b) and (c) yield a direct picture of the adsorbate configurations: every pronounced hill corresponds to a H atom. The formation of linear H chains on Rh(110) and of zigzag H chains on Ni(110) is clearly visible. The corrugations are expanded by about a factor of five in the vertical direction

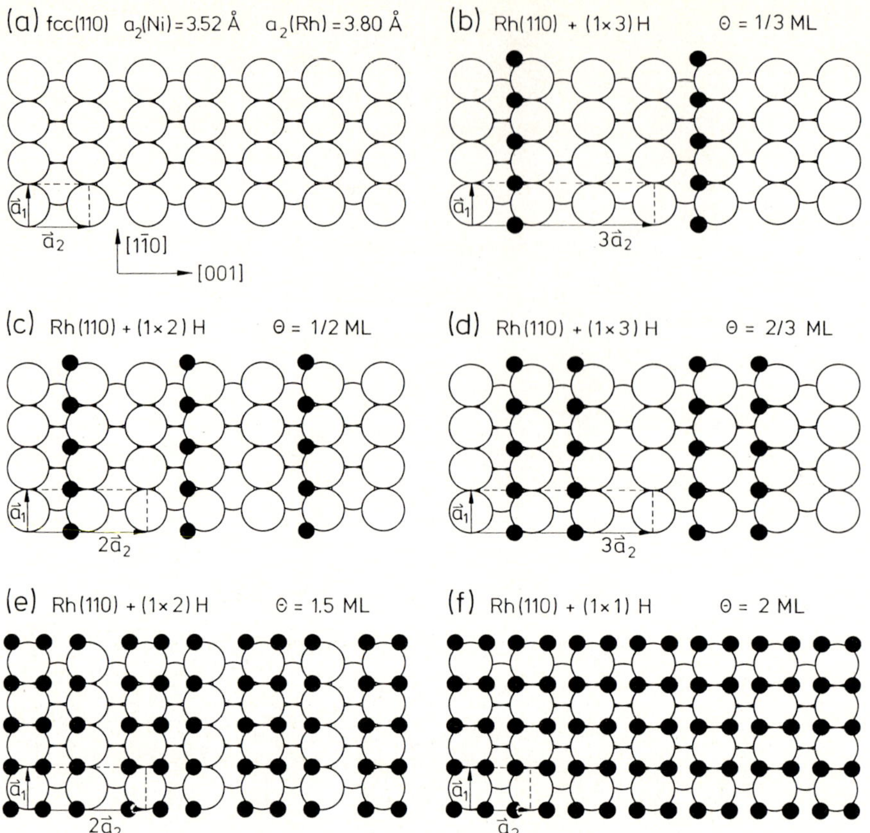

Fig.4.9. Sphere models of (a) the clean fcc(110)-surface and (b) - (f) of the sequence of ordered H phases on Rh(110). The small dark spheres denote H atoms. After [4.54]

In the sequence of ordered H phases on Rh(110) with $\theta = 1/3,\ 1/2,\ 2/3$, 1.5 and 2 ML the absence of a phase with 1 ML coverage is remarkable and presently not understood. The saturation phase has the highest adatom density known up to now for a fcc surface. The formation of linear adatom chains on Rh(110) and zigzag chains on Ni(110) indicates subtle differences in the adsorbate–substrate and adsorbate–adsorbate interactions on the two metals.

The intensity analyses of all H phases on Ni(110) were performed free of structural model assumptions within the HCW by Fourier ansatz for the corrugations and systematic searches for the best-fit coefficients; the adatom configurations shown schematically in Fig.4.10 thus became directly visible via the corrugations [4.34]. Recently, dynamical LEED calculations for the (2×1)-phase confirmed the adatom configuration in threefold sites and gave detailed information about the hydrogen bond lengths to the substrate and adsorbate-induced changes in substrate relaxations [4.52]. In order to support the previous He diffraction intensity analyses for the low coverage phases, calculations of the

Patterson functions were recently performed based on the experimental data on the one hand [left-hand side of (4.18)] and on the best-fit corrugations on the other [right-hand side of (4.18)]. For the c(2×4)-phase both results are shown in Fig.4.11. The agreement between the two functions convincingly supports the correctness of the best-fit corrugation; notice especially that the plaiting pattern around the maxima in the Patterson function indicates the alternation of zigzag and zagzig chains in the corrugation, Fig.4.8c [4.37].

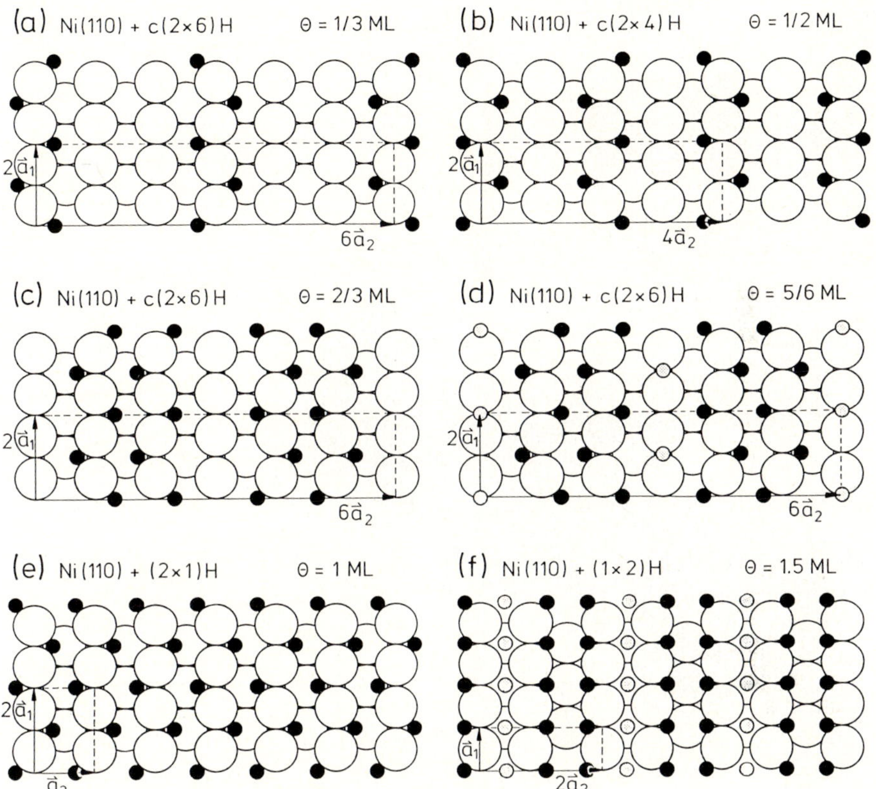

Fig.4.10a–f. Sphere models of the sequence of ordered H phases on Ni(110). The small dark and dotted spheres denote H atoms. After [4.34]

The adatom configurations for four of the five ordered H structures on Rh(110) were analyzed with dynamical LEED by the group of *Mueller* and *Heinz* [4.49,53] and the He diffraction corrugations [4.54] in the main support these results. It should be noted that the development of video-LEED techniques [4.55] has made it possible to measure LEED intensities with great accuracy and — due to the rapid data acquisition — also in a practically non-destructive manner, so that adsorbates even as light as hydrogen adsorbed on

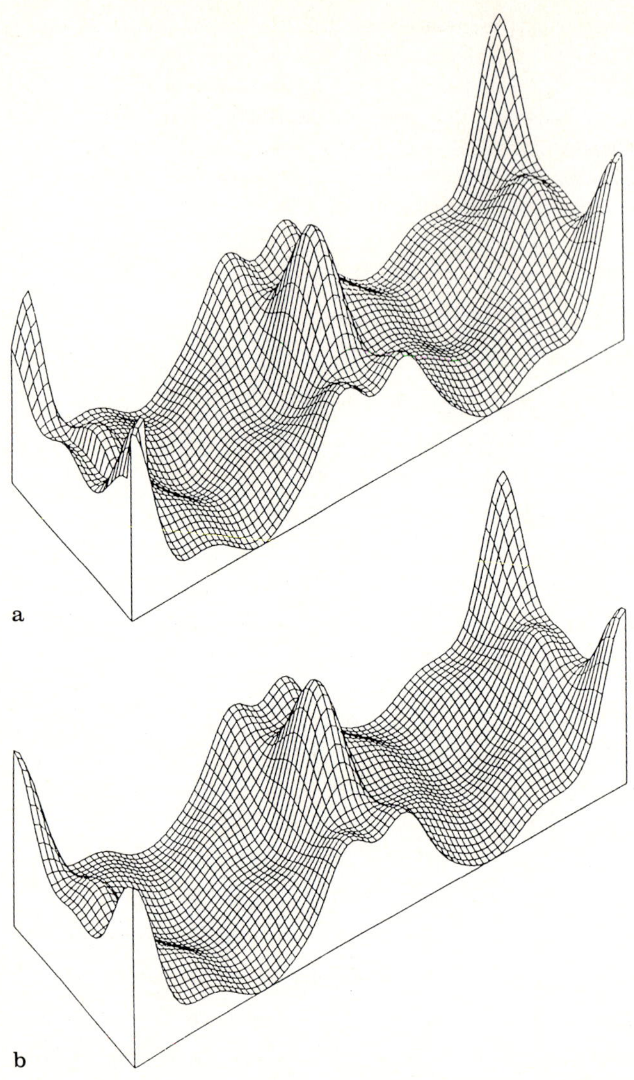

Fig.4.11. Patterson functions (real parts) of the c(2×4)H structure on Ni(110).The area shown corresponds to the nonprimitive centered unit cell of Figs.4.8c and 4.10b.(a) Patterson function derived from the left-hand side of (4.22) using the experimental data shown in Fig.4.7. (b) Patterson function derived from the right-hand side of (4.22) using the best-fit corrugation function [4.34] displayed in Fig.4.8c. After [4.37]

heavy substrates can be analyzed. As there are now several H adsorption systems for which both He diffraction and LEED or ion scattering analyses have been performed, it becomes possible to aim for quantitative information on adsorbate bond lengths from He diffraction data by an empirical procedure which

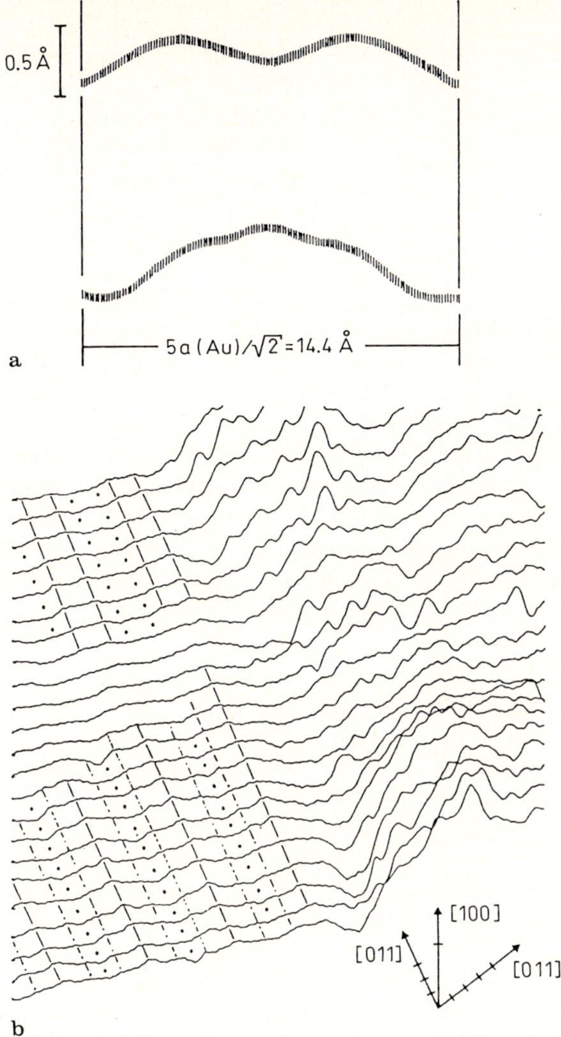

Fig.4.12. (a) Dominant (1×5)-corrugations on reconstructed Au(100) derived from He diffraction intensity analyses [4.65]. (b) High resolution STM scans on Au(100) near a carbon island; the broken lines indicate main maxima and the dots minima of (1×5)-ribbons [4.67]. Notice the similarity of the two dominant corrugation types with those of (a)

is illustrated for the case of (1×2)H on Rh(110): Surface charge densities calculated simply by overlapping atomic charge densities [4.56] are compared with respect to heights and locations of the maxima with the corresponding features in the best-fit corrugation; as reference material data on the H positions for the systems (2×1)H on Ni(110) [4.52] and Pd(110) [4.57] obtained with LEED and H(1×1) on Ni(100) [4.58] obtained with ion transmission channeling can be used in connection with the respective corrugations obtained from He diffraction [4.34,59,60]. Fixing the positions of the H nuclei according to the LEED and ion scattering results, the reproduction of the increase of the corrugation amplitudes from the clean to the H covered surfaces requires a negative excess charge of $0.4e$ on the H adatoms. Assuming the same charge transfer for the H(1×2)Rh(110) phase, one arrives at an adsorption height of 0.82 ± 0.1 Å and a lateral distance to the close-packed Rh rows of 0.97 ± 0.1 Å. These values are in very good agreement with the respective values of 0.84 and 1.08 Å deduced from the dynamical LEED analyses [4.49].

It should be noted that the excess charge on the H adatoms is necessary to reach the observed corrugation amplitudes but — compared to the experimental widths — gives lateral charge distributions that are too broad. A similar effect was also observed for oxygen in the p(2×2) and c(2×2) phases on Ni(100), where an excess charge of $1e$ was necessary to arrive at the measured corrugation amplitudes [4.61]. It is gratifying, however, that with O^- in the system (2×1)O on Ni(110), not only the adsorbate-induced missing row reconstruction perpendicular to the close-packed metal rows, but also the correct height of the oxygen adatoms above the Ni atoms in the long-bridge sites was deduced [4.62] in good agreement with a very recent LEED analysis [4.63] (the small lateral shift of the oxygens 0.1 Å away from the long bridges in the [1$\bar{1}$0] direction could not be derived from the He data).

4.4.3 Reconstructions with Small Corrugation Amplitudes

The tendency of the close-packed surfaces of Au, Ir and Pt to form surface reconstructions has been known since the earliest LEED observations [4.64]. The first attempt to analyze a reconstructed surface quantitatively on the basis of He diffraction data concerned Au(100) [4.65], whose reconstruction was claimed to be c(26×68) [4.66]; the analogous surfaces of Pt and Ir were found to have (5×20) and (1×5)-periodicities, respectively [4.64]. Due to the limited instrumental resolution the He diffraction data indicated a dominant (1×5)-periodicity for Au(100), so that the large unit cell observed with LEED could be thought of a long range modulation of one or more basic (1×5) motifs. Intensity analyses for many angles of incidence and wavelengths yielded two different types of corrugations as shown in Fig.4.12a. The success in STM in resolving atomic structures revealed the full periodicity of the complicated surface structure as shown in Fig.4.12b [4.67]. Notice that the two basic corrugations deduced from the He diffraction data both appear in the STM picture and that the long-range modulation is also visible; *Binnig* et al. [4.67] concluded from their observations that the overlayer should also exhibit a small rotation relative to the substrate, resulting in four different domains.

The full solution of the Au(100) was, in principle, only hampered by the limited instrumental resolution of the He diffraction system used in this early study [4.65]. That complete analyses of surface structures with large unit cells are possible with He diffraction was shown by *Harten* et al. [4.68] for the reconstruction of the Au(111) surface. These authors used the high resolution time-of-flight apparatus designed to study surface phonons [4.69] and measured diffraction

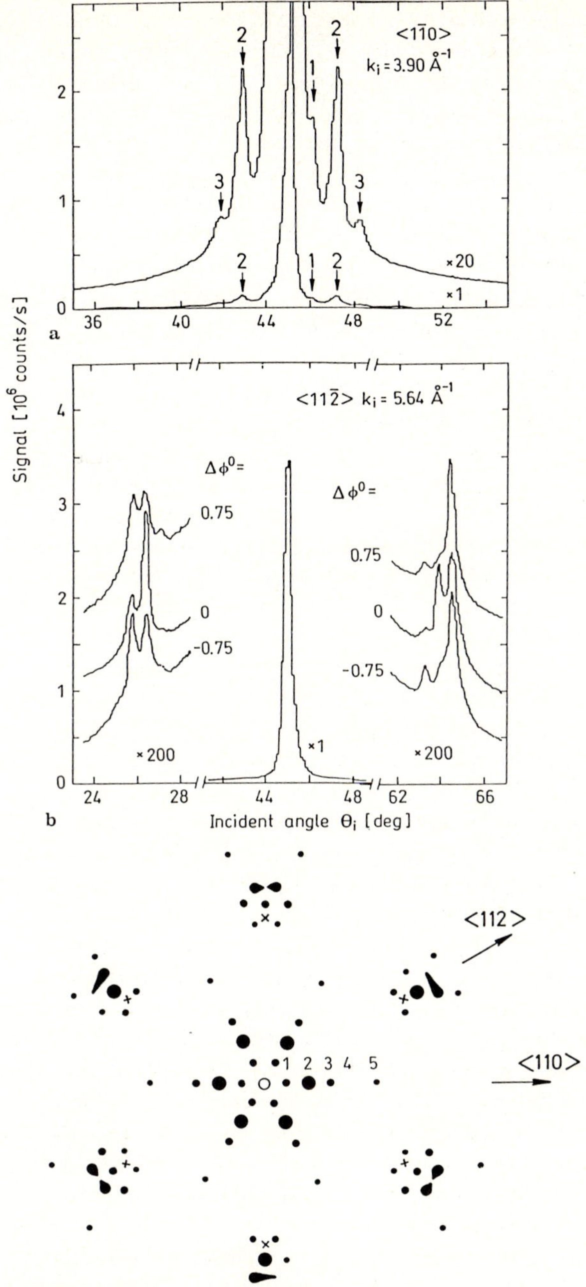

Fig.4.13. (a,b) High resolution He diffraction scans along two different crystal azimuths of the reconstructed Au(111) surface. (c) Schematic plot of full diffraction pattern obtained for Au(111); crosses indicate diffraction peaks expected for the unreconstructed surface. After [4.12a]

intensities along several different azimuths of the Au(111) surface, Figs.4.13a,b. The corresponding schematic diffraction pattern is shown in Fig.4.13c. Intensity analyses performed with the eikonal approximation led to the corrugation displayed in Fig.4.14a [4.68], which shows directly that 24 surface atoms are accommodated above 23 atoms of the (undisturbed) second layer along $\langle 110 \rangle$; in the resulting compressed structure the stacking changes from ABC (fcc) to ABA (hcp) as can be seen from Fig.4.14b. A molecular dynamics study by *El-Batanouny* et al. [4.70] based on a double-Sine-Gordon model using a potential felt by the surface atoms as shown in Fig.4.14c led to the refined surface structure shown in the model of Fig.4.14b. Recent STM measurements of *Wöll* et al. [4.71] made both the soliton walls and the slight shift of the surface atoms along $\langle 112 \rangle$ directly visible, Fig.4.14d . These results constitute the most direct evidence on soliton wall formation on crystal surfaces. For solitary lattice distortions in rare gas overlayers on Pt(111), see [4.13].

4.4.4 Systems with Large Corrugations

The (110) surfaces of Au, Ir and Pt exhibit (1×2) reconstructions of the missing row type as is now well established [4.72,73]. The first strong evidence for the missing row structure came from a He diffraction investigation on Au(110) [4.74]. In this study a pronounced rainbow structure at a large angular distance from the specular allowed an estimation of the corrugation amplitude of ~ 1.5 Å, which excluded the other proposed reconstruction types, namely the pairing row and shifted row models. In a very recent extensive study of (1×2)Pt(110) an attempt was made to fit a large number of diffraction spectra measured with various θ_i and λ_i [4.75]. It was found that, due to the large corrugation of the system, consistent fits of all data were only possible if all parts of the He surface interaction potential were taken into account: The corrugation and steepness of the repulsive part as well as the depth, width and shape of the attractive part; the corrugations of both the bottom of the potential well and the attractive part had to be determined accurately to arrive at this goal.

Figure 4.15 compares experimental and best-fit spectra for three different θ_i; notice that not only the stable rainbow structure, which is determined mainly by the corrugation of the repulsive part, but also the rapidly varying intensities of the lower index beams, which are very sensitive to all parts of the potential, are reproduced very well by the calculations. The full He/Pt(110) interaction potential is shown in Fig.4.16. This potential also reproduces very well the resonant scattering structures observed as a function of θ_i in the specular intensity for various λ_i [4.5b]. Along z the potential shape is close to the classical Morse form; as this form is also obtained for several other metal surfaces from analyses of experimentally determined bound state energies [4.33,76], it certainly provides a good starting point for quantitative intensity analyses. The corrugation contour corresponding to the incident He energy of 63 meV has an amplitude as large as 1.5 Å (~ 20 % of the lattice constant). Comparison of the potential contours with charge density calculations allowed the He physisorption distance to be estimated as ~ 3.4 Å from the topmost metal ion cores and also supports

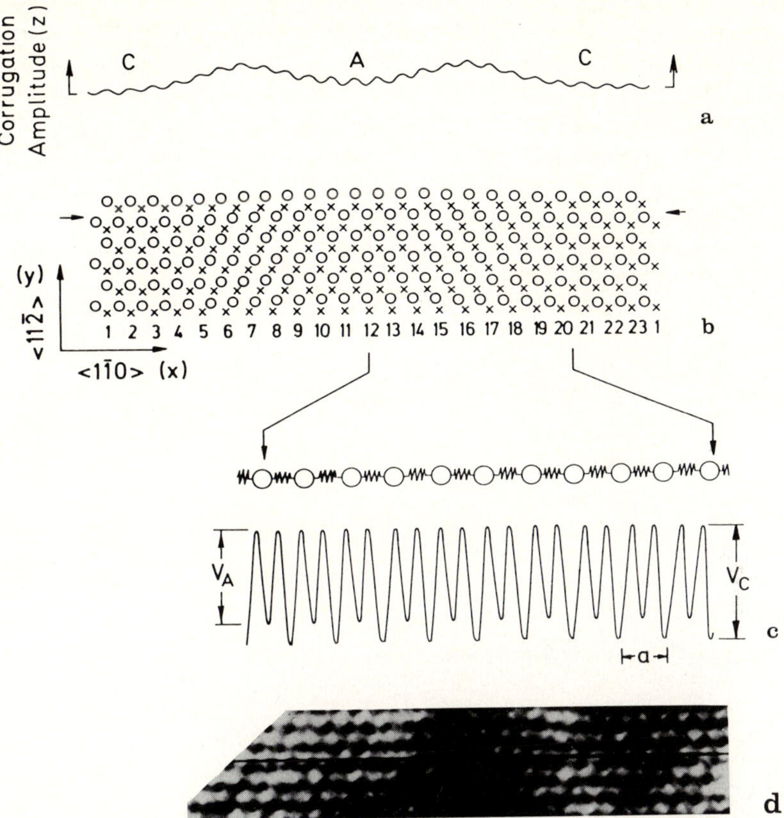

Fig.4.14. (a) Best-fit corrugation for Au(111) along x for fixed y derived from HCW intensity analyses. (b) Top-view of locations of surface atoms (circles) on second layer atoms (crosses). (c) Postulated double Sine-Gordon potential with two different potential depths for fcc and hcp top sites used in [4.70]. (d) High-resolution STM picture of soliton domain wall showing the deviation from a straight line along $\langle 110 \rangle$. (a-c) After [4.12a], (d) after [4.71]

an inward relaxation of the topmost atoms of ~ 15 % as derived earlier on the basis of LEED [4.72] and first principles calculations [4.73].

Upon H chemisorption on Pt(110)1×2 the repulsive corrugation amplitude increases to 2.0 Å (25 % of the relevant lattice constant!) as the more strongly bound β_2-state is filled. Taking into account the shape of the corrugation as well as the unchanged potential depth, it could be concluded that β_2-H occupies the first available octahedral subsurface sites whereby the substrate inward relaxation changes into an outward relaxation of about the same magnitude [4.77]; the H-Pt bond length so found is practically identical with that in Pt-hydride [4.78]. Upon saturation of the surface with H (filling of the more weakly bound β_1-state), the corrugation decrease signals occupation of chemisorption sites in the missing row troughs [4.77].

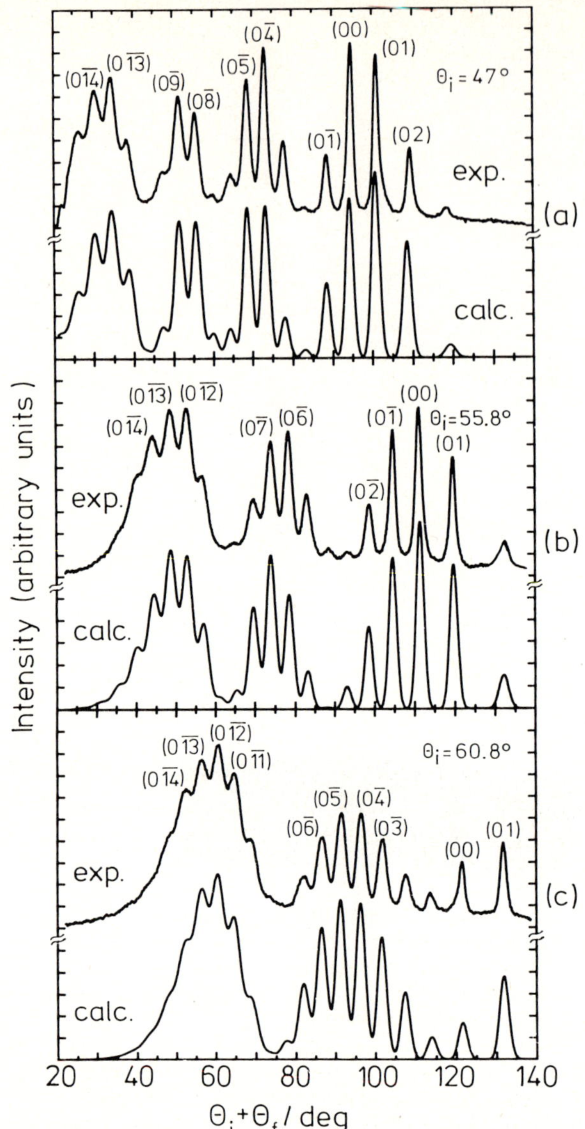

Fig.4.15. Comparison of experimental and best-fit He diffraction patterns from clean Pt(110)1×2 for $\lambda_i = 0.57$ Å and three θ_i. Notice the stable rainbow structure at the beams (0 $\overline{11}$) to (0 $\overline{15}$) and the rapidly varying intensities of the lower order beams [4.75]

For He diffraction from reconstructed semiconductor surfaces, which usually exhibit medium and large corrugations, see [4.79].

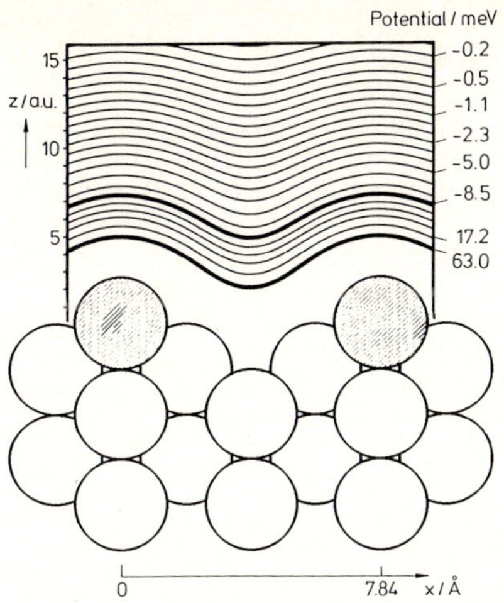

Fig.4.16. Lower part: side view of hard sphere model of the Pt(110)1×2 missing row structure. Upper part: the He–Pt(110)1×2 interaction potential contours derived from intensity analyses. Contours corresponding to the energy of the particle beam (63 meV) and the potential minimum are emphasized [4.75]

Acknowledgments

The author wishes to express his thanks to Dipl. Phys. E. Kirsten for providing Figs.4.2 and 4.3 as well as for carefully reading the manuscript. G. Gross and Dipl. Phys. G. Parschau contributed Figs.4.8–11. Dr. Ch. Wöll is acknowledged for making Fig. 4.14d available. Thanks are also due to Mrs. Angelika Koffinke for her expert preparation of the manuscript.

References

4.1 J.P. Toennies: Appl. Phys. **3**, 91 (1974)

4.2 J.P. Toennies, K. Winkelmann: J. Chem. Phys. **66**, 3965 (1977); G. Brusdeylins, H.D. Meyer, J.P. Toennies, K. Winkelmann: Prog. Astronautics Aeronautics **51**, 1047 (1977)

4.3 J.F. O'Hanlon: *A Users Guide to Vacuum Technology*, 2nd Ed. (Wiley, New York 1989)

4.4 T. Engel, K.H. Rieder: In *Structural Studies of Surfaces*, Springer Tracts in Modern Physics, Vol **91** (Springer, Berlin, Heidelberg 1982)

4.5 (a) H. Hoinkes: Rev. Mod. Phys. **52**, 933 (1980);
     (b) H. Wilsch, H. Hoinkes, this volume, Chap. 7.

4.6 G. Binnig, H. Rohrer: IBM J. Res. Develop. **30**, 355 (1986)

4.7 M.A. Van Hove, W.H. Weinberg, C.-M. Chan: *Low-Energy Electron Diffraction*, Springer Ser. Surf. Sci. Vol. **6** (Springer, Berlin, Heidelberg 1986).

4.8 Ch. Kittel: *Introduction to Solid State Physics*, 6th ed. (1987)

4.9 N. Esbjerg, J. Norskov: Phys. Rev. Lett. **45**, 807 (1980).

4.10 J. Harris, A. Liebsch: J. Phys. **C15**, 2275 (1982); Phys. Rev. Lett. **49**, 341 (1982)

4.11 M. Manninen, J.K. Norskov, M.J. Puska, C. Umrigar: Phys. Rev. B **29**, 2314 (1984)

4.12   (a) J.P. Toennies: J. Vac. Sci. Technol. **A5**, 440 (1987);
       (b) Physica Scripta **T19**, 39 (1987)

4.13   K. Kern, G. Comsa: In *Molecule Surface Interactions*, ed. by K. Lawley, Advances in Chemical Physics (Wiley, New York 1989)

4.14   K.H. Rieder, W. Stocker: Phys. Rev. Lett. **52**, 352 (1984); K.H. Rieder, M. Baumberger, W. Stocker: ibid. **55**, 390 (1985)

4.15   E. Semerad: Surf. Sci. **211/212**, 343 (1989); E. Semerad, E.M. Hörl: Surf. Sci. **115**, 346 (1982)

4.16   J.A. Barker, D.J. Auerbach: Surf. Sci. Rep. **4**, 1 (1984)

4.17   E.K. Schweitzer, C.T. Rettner: Phys. Rev. Lett. **62**, 3085 (1989)

4.18   A. Moedl, T. Gritsch, F. Budde, T.J. Chuang, G. Ertl: Phys. Rev. Lett. **57**, 384 (1986)

4.19   M.P. D'Evelyn, R.J. Madix: Surf. Sci. Rep. **3**, 314 (1983)

4.20   G. Boato, P. Cantini, L. Mattera: Jpn. J. Appl. Phys. Suppl. **2**, Part 2, p.553 (1974); J. Lapujoulade, J. Perreau: Phys. Scripta **T4**, 138 (1986); K.B. Whaley, C.F. Yu, C.S. Hogg, J.C. Light, S.J. Sibener: J. Chem. Phys. **83**, 4235 (1985); R. Berndt, J.P. Toennies, Ch. Woell: J. Chem. Phys. **92**, 1468 (1990)

4.21   S. Ianotta, G. Scoles, U. Valbusa: Surf. Sci. **161**, 411 and 429 (1985)

4.22   T.H. Ellis, G. Scoles, U. Valbusa, H. Jonsson, J.H. Weare: Surf. Sci. **155**, 499 (1985); H. Jonsson, J.H. Weare, T.H. Ellis, G. Scoles: Surf. Sci. **180**, 353 (1987).

4.23   K.H. Rieder: In *Structure and Dynamics of Surfaces I*, ed. by W. Schommers, P. von Blanckenhagen: Topics Curr. Phys. **41** (Springer, Berlin, Heidelberg 1986) p. 17

4.24   G. Ertl, J. Kueppers: *Low Energy Electrons and Surface Chemistry* (Verlag Chemie, Weinheim 1974)

4.25   U. Garibaldi, A.C. Levi, R. Spadacini, G.E. Tommei: Surf. Sci. **48**, 649 (1975)

4.26   N. Garcia: J. Chem. Phys. **67**, 897 (1977)

4.27   H. Chow, E.D. Thompson: Surf. Sci. **82**, 1 (1979)

4.28   R.H. Swendsen, K.H. Rieder: Surf. Sci. **114**, 405 (1982)

4.29   N. Garcia, N. Cabrera: Phys. Rev. B **18**, 576 (1978)

4.30   A. Tsuchida: Surf. Sci. **14**, 375 (1969); **46**, 611 (1974); G. Wolken Jr.: J. Chem. Phys. **58**, 3047 (1973)

4.31   R.J. Blake: Computer Phys. Commun. **33**, 425 (1982)

4.32   E. Kirsten: Doctoral Thesis, Freie Universitaet Berlin (1991)

4.33   E. Kirsten, G. Parschau, K.H. Rieder: Surf. Sci. Lett. **236**, L365 (1990)

4.34   K.H. Rieder, W. Stocker: Surf. Sci. **164**, 55 (1985)

4.35   K.H. Rieder, N. Garcia, V. Celli: Surf. Sci. **108**, 169 (1981)

4.36   D.S. Kaufmann, L.R. Allen, E.H. Conrad, R.M. Aten, T. Engel: Surf. Sci. **173**, 517 (1986)

4.37   G. Gross: Diploma Thesis, Freie Universitaet Berlin (1990)

4.38   P. Cantini, R. Tatarek, G.P. Felcher: Phys. Rev. B **19**, 1161 (1979)

4.39   L. Estermann, O. Stern: Z. Physik **61**, 95 (1930); see also N.F. Ramsey: *Molecular Beams* (Clarendon, Oxford 1985)

4.40   Vacuum Science Workshop, Warwick Road South, Old Trafford, Manchester M16 OJT, data sheet on request

4.41   G. Parschau, E. Kirsten, A. Bischof, K.H. Rieder: Phys. Rev. B **40**, 6012 (1989)

4.42   T. Engel, K.H. Rieder: Surf. Sci. **109**, 140 (1981)

4.43   G. Parschau, E. Kirsten, K.H. Rieder: Surf. Sci. **225**, 367 (1990)

4.44   J.F. Annett, R. Haydock: Phys. Rev. Lett. **53**, 838 (1984); Phys. Rev. B **34**, 6860 (1986)

4.45   J.F. Annett: Daresbury Laboratory Information Quarterly for Surface Sciences **14**, 9 (1984)

4.46   K.H. Rieder, M. Baumberger, W. Stocker: Phys. Rev. Lett. **55**, 390 (1985); K.H. Rieder: Phys. Rev. B **39**, 10708 (1989)

4.47   M. Prutton, J.A. Walker, M.R. Welton-Cook, R.C. Felton, J.A. Ramsey: Surf. Sci. **89**, 95 (1979)

4.48   K.H. Rieder: Surf. Sci. **118**, 57 (1982)

4.49   W. Puchta, W. Nichtl, W. Oed, N. Bickel, K. Heinz, K. Mueller: Phys. Rev. B **39**, 1020 (1988)

4.50   G. Kleinle, V. Penka, R.J. Behm, G. Ertl, W. Moritz: Phys. Rev. Lett. **58**, 148 (1987)

4.51   B. Voigtlaender, S. Lehwald, H. Ibach: Surf. Sci. **208**, 113 (1989)

4.52   W. Reimer, V. Penka, M. Skottke, R.J. Behm, G. Ertl, W. Moritz: Surf. Sci. **186**, 45 (1987)

4.53   M. Michl, W. Nichtl-Pecher, W. Oed, H. Landskron, K. Heinz, K. Mueller: Surf. Sci. **220**, 59 (1989)

4.54   G. Parschau, E. Kirsten, K.H. Rieder: Phys. Rev. B, in press

4.55   K. Heinz, K. Mueller: In *Structural Studies of Surfaces*, Springer Tracts in Modern Physics, Vol. **91** (Springer, Berlin, Heidelberg 1982)

4.56   M. Manninen, J.K. Norskov, C. Umrigar: Surf. Sci. **119**, L393 (1982); D. Haneman, R. Haydock: J. Vac. Sci. Technol. **21**, 330 (1982)

4.57   M. Skottke, R.J. Behm, G. Ertl, V. Penka, W. Moritz: J. Chem. Phys. **87**, 6191 (1987)

4.58   I. Steensgaard, F. Jacobsen: Phys. Rev. Lett. **54**, 711 (1985)

4.59   K.H. Rieder, M. Baumberger, W. Stocker: Phys. Rev. Lett. **51**, 1799 (1983)

4.60   K.H. Rieder, H. Wilsch: Surf. Sci. **131**, 245 (1983)

4.61   K.H. Rieder: Phys. Rev. B **27** (RC), 6978 (1983)

4.62   T. Engel, K.H. Rieder, I.P. Batra: Surf. Sci. **148**, 321 (1984)

4.63   G. Kleinle, J. Wintterlin, G. Ertl, R.J. Behm, F. Jona, W. Moritz: Surf. Sci. **225**, 171 (1990)

4.64   P.J. Estrup: In *Chemistry and Physics of Surfaces V*, Springer Ser. Chem. Phys. Vol. **35** (Springer, Berlin, Heidelberg 1984) p. 205

4.65   K.H. Rieder, T. Engel, R.H. Swendsen, M. Manninen: Surf. Sci. **127**, 223 (1983)

4.66   J.F. Wendelken, D.M. Zehner: Surf. Sci. **71**, 178 (1978); D. Wolf, H. Jagodzinski, W. Moritz: Surf. Sci. **77**, 283 (1978)

4.67   G. Binnig, H. Rohrer, Ch. Gerber, E. Stoll: Surf. Sci. **144**, 321 (1984)

4.68   U. Harten, A.M. Lahee, J.P. Toennies, Ch. Wöll: Phys. Rev. Lett. **54**, 2629 (1985)

4.69   G. Brusdeylins, R.B. Doak, J.P. Toennies: Phys. Rev. Lett. **46**, 437 (1981), Phys. Rev. B **27**, 3662 (1983); J.P. Toennies: In Springer Ser. Chem. Phys., Vol. **21** (Springer, Berlin, Heidelberg 1982) p. 208

4.70   M. El-Batanouny, S. Burdick, K.M. Martini, P. Stancioff: Phys. Rev. Lett. **58**, 2762 (1987)

4.71   Ch. Wöll, S. Chiang, R.J. Wilson, P.H. Lippel: Phys. Rev. B **39**, 7988 (1989)

4.72   W. Moritz, D. Wolf: Surf. Sci. **163**, L641 (1985)

4.73   K.M. Ho, K.P. Bohnen: Phys. Rev. Lett. **59**, 1833 (1987)

4.74   K.H. Rieder, T. Engel, N. Garcia, in: Proc. 4th Int. Conf. on Solid Surf., 3rd Europ. Conf. on Surf. Sci., Cannes(1980); Suppl. à la Revue "Le Vide, les Couches Minces", Nr. 201, p. 861 (1980)

4.75   E. Kirsten, K.H. Rieder: Surf. Sci. **222**, L837 (1989)

4.76   E. Kirsten, G. Parschau, K.H. Rieder: Phys. Rev. B **41**, 5392 (1990)

4.77   E. Kirsten, G. Parschau, W. Stocker, K.H. Rieder: Surf. Sci. Lett. **231**, L183 (1990)

4.78   *Hydrogen in Metals I and II*, Ed. by G. Alefeld, J. Voelkl; Topics Appl. Phys. **28**, **29** (Springer, Berlin, Heidelberg 1982)

4.79   M. Cardillo: In *Chemistry and Physics of Surfaces IV*, Springer Ser. Chem. Phys. Vol. **20** (Springer, Berlin, Heidelberg 1982) p. 149

5. Investigation of Surface Imperfections by Diffuse Scattering of He Atoms

Ch. Wöll and A.M. Lahee

Helium atom scattering (HAS) is a well established technique for studying the physical properties of atoms and molecules [5.1] in the gas phase. The decrease in intensity of a He atom beam caused by the target molecules allows the determination of the integral cross-section of a molecule, which is related to its total size. Since in a typical molecular beam experiment the de Broglie wavelength of the He atoms is of the order of the diameter of the target molecule (several Å), interference effects will generally cause a variation of the integral cross-section with the energy of the He atoms. This dependence can be used to draw conclusions about the shape of the intermolecular potential. With a more sophisticated molecular beam machine it is additionally possible to measure the angular distribution of the scattered He atoms. The differential cross-section, which can be evaluated from these data, contains a rich structure, again originating from interference between different scattering paths. A careful analysis of these structures allows one to draw much more detailed conclusions on the form of the potential, e.g. one can determine the contour of zero potential [5.2]. In some special cases high precision data for the differential cross-section has even allowed a potential inversion [5.1].

If the scattering of helium atoms takes place from a perfect surface of a solid single crystal the specular peak is the analogue of the undeflected beam in a gas-phase experiment. But because the scattering is now dominated by the interference between the highly ordered and correlated scattering centres of the uppermost surface plane, the angular distributions of He atoms will exhibit sharp additional features, diffraction peaks. Only surface defects, such as steps, vacancies and adatoms, will give rise to smoothly varying structures in the angular distributions of elastically scattered He atoms. In special cases inelastic scattering may be intense enough to cause distinct features in the angular distributions, but this is rare and will not be dealt with here.

One of the simplest systems to compare concepts in gas-phase scattering and surface scattering is a single CO molecule adsorbed on an otherwise perfect Pt(111) surface. The (111) surface of the fcc metal platinum consists of a densely packed hexagonal array of atoms, and the corrugation sensed by a thermal energy He atom is extremely small, resulting in diffraction intensities of less than 10^{-4} of that of the specular peak. If CO molecules are now deposited

Springer Series in Surface Sciences, Vol. 27   **Helium Atom Scattering from Surfaces**
Editor: E. Hulpke   © Springer-Verlag Berlin, Heidelberg 1992

on the surface in a random, uncorrelated way, the intensity of the specular peak will decrease and the determination of the integral cross section of a CO molecule on a surface is possible. Indeed, for adsorbed molecules similar values are obtained as for the gas phase [5.3]. A detailed investigation of this integral cross-section can be used to obtain more information about special properties of this and similar systems; the reader is referred to Ref. [5.4] for a detailed discussion on this (restricted) aspect of diffuse He atom scattering.

As long as a point defect on a surface is being considered, the scattering mechanism remains comparable to that for the gas phase. The situation becomes more interesting if one takes into account scattering from surface defects that have a higher dimensionality than zero (zero dimensional defects correspond to point defects). Step edges, i.e. line defects, have a dimensionality of one, whereas small islands represent two-dimensional scattering centres. In principle there is also the possibility of a non-integral dimension between 1 and 2, corresponding to an island with 'fractal' boundaries, although the latter has not yet been identified experimentally with He atom scattering.

5.1 Scattering of He Atoms from an Ideal Surface

For the perfect surface of a single crystal, be it a metal, semiconductor or insulator, the elastic scattering event for a He atom is governed by conservation of energy and of the momentum component parallel to the surface plane:

$$E_f = E_i \tag{5.1}$$

and

$$\boldsymbol{K}_f = \boldsymbol{K}_i + \boldsymbol{G} \ , \tag{5.2}$$

where E_f and E_i ($\boldsymbol{K}_f$ and $\boldsymbol{K}_i$) are the final and initial energies (parallel wave vectors), respectively. $\boldsymbol{G}$ is a vector from the surface reciprocal lattice, which is defined by the periodicity of the surface. Fig.5.1 indicates the distribution of elastically scattered He atoms expected for a very weakly corrugated (111) surface of a fcc metal like Pt(111). The diffraction pattern is very similar to that obtained in a LEED experiment, with the position of the diffraction peaks depending only on the incident energy and the periodicity of the surface. The relative intensities of the various peaks, however, are very different for He atoms and electrons due to the completely different interactions involved: He atoms do not penetrate the surface, i.e. they only interact with the very first layer; thus multiple scattering events are generally not important. Electrons, on the other hand, depending on their incident energy, penetrate at least 2–3 layers. Thus multiple scattering events are very important and there is no direct interpretation of the intensity of diffraction peaks.

Studying Fig.5.1 in more detail, however, reveals a major drawback of HAS as compared to LEED. Whereas for electron detection a fluorescent screen

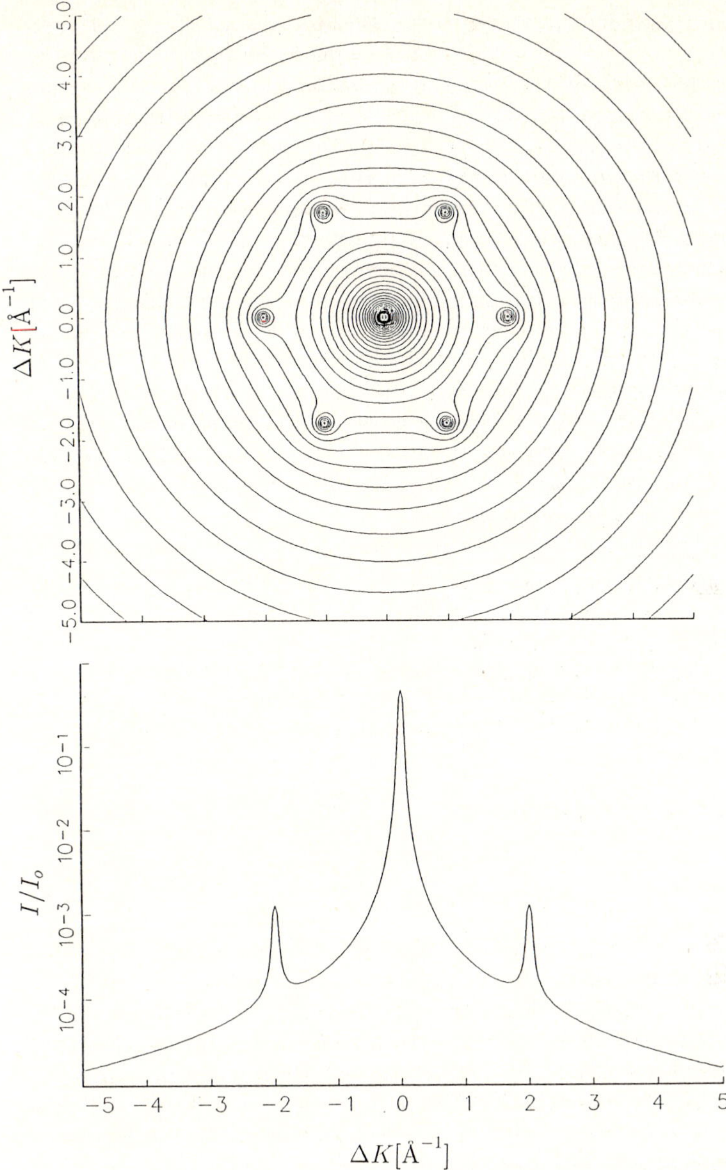

Fig.5.1. Upper panel: Logarithmic contour plot from a model calculation for the angular distribution of He atoms scattered from a highly ordererd fcc(111) metal surface. Lower panel: Angular scan along the dotted line in the contour plot shown above

makes it possible to obtain the intensities for all diffraction peaks simultaneously (at a given energy), for HAS an image such as the upper part of Fig.5.1 has to be constructed from many scans as shown in the lower part of Fig.5.1.

This is due to the fact that the detection of He atoms is much less efficient than that of electrons and a mass-spectrometer has to be used. Therefore at any one time the scattered intensity can be determined only for a single point in $\boldsymbol{K}$-space. Usually one maps out the intensity in the scattering plane either by varying the position of the detector or by rotating the crystal. A single angular scan, as shown in the lower part of Fig.5.1, takes about 10 min. Depending on the desired resolution many scans are needed to construct a complete two-dimensional diffraction pattern as shown in the upper part of Fig.5.1.

For an ideal surface of a single crystal the peaks at the positions given by the above equation should be delta-function like, with a vanishing halfwidth. The finite halfwidth of the experimentally observed peaks would then correspond to the angular resolution of the detector, which can amount to 0.1° or better. This high angular resolution, by the way, is a strong advantage of HAS over LEED. For electron scattering a considerable effort is needed to obtain high angular resolution, involving electrostatic lenses and the use of a Faraday cup; but the latter introduces the drawback mentioned earlier in connection with HAS, namely of detecting at only one point in $\boldsymbol{K}$-space at a given time.

Between the diffraction peaks no elastic intensity is expected, either for HAS or for LEED. Any elastic intensity scattered in these regions of $\boldsymbol{K}$-space is due to a lack of perfection at the surface, finite coherence length, and the presence of vacancies, adatoms and steps, i.e. lattice defects. The specific features in the distribution in $\boldsymbol{K}$-space due to these various types of defects will be discussed in the next section.

5.2 Scattering from Surfaces with Defects

5.2.1 Finite Coherence Length and Two-Dimensional Defects

A finite coherence length can be visualized as the presence of domains on the surface, which, although perfectly ordered, have only a finite diameter. For a clean surface these perfect domains may be separated by small angle grain boundaries and steps. For the case of an adsorbate, islands can occur in early stages of the adsorption process. To discuss the basic features for He diffraction patterns from surfaces with a small coherence length we will first use a very simple model for an island with a fcc(111) surface, namely a circular disc with diameter R_0 with a corrugation function

$$\xi(\boldsymbol{R}) = \sum_{\boldsymbol{G}_{mn}}^{|G_{mn}|=2\pi/a} \xi \cos(\boldsymbol{G}_{mn} \cdot \boldsymbol{R}) \quad , \tag{5.3}$$

with

$$\boldsymbol{G}_0 = \frac{2\pi}{a}(1,0) \, , \ \boldsymbol{G}_1 = \frac{2\pi}{a}(\frac{1}{2}, \frac{\sqrt{3}}{2}) \ \text{and} \ \boldsymbol{G}_{mn} = m\boldsymbol{G}_0 + n\boldsymbol{G}_1 \quad . \tag{5.4}$$

76

If the disc diameter R_0 is much larger than the lattice constant a and the corrugation $\xi(\boldsymbol{R}) \ll a/(2\pi)$ we can approximate the potential between He atom and surface by a hard wall potential (Sect.5.2.4), and a simple expression for the angular distribution can be given:

$$I(\varDelta\boldsymbol{K}) \propto \left| \sum_{\boldsymbol{G}_{mn}} a_{mn} \frac{J_1(\boldsymbol{G} \cdot \varDelta\boldsymbol{K})}{\varDelta\boldsymbol{K}} \right|^2 \quad , \tag{5.5}$$

with:

$$a_{mn} = \begin{cases} \xi & \text{if } |\boldsymbol{G}_{mn}| = \dfrac{2\pi}{a} \\ 0 & \text{otherwise} \end{cases} \quad . \tag{5.6}$$

Here a denotes the lattice constant and J_1 the Bessel function of first order.

The lower part of Fig.5.2 shows the distribution of elastic intensity in $\boldsymbol{K}$-space expected for these islands with finite size. The halfwidth of the specular beam and the diffraction peaks increase due to the finite coherence length, the halfwidth being proportional to $1/R_0$. For a single island there will be no strong dependence of peak widths on incident energy.

The behaviour for an island on an otherwise perfect surface will be different from the situation considered above. If, for example, the island consists of substrate atoms (e.g. a terrace), all atoms are at a lattice position of the perfect 3d crystal lattice and there is no reduction in the coherence length. Trajectories from the upper and the lower terraces will interfere, causing changes in the diffraction pattern which will depend strongly on the incident energy. If the interference is constructive, the specular and diffraction peaks will be very sharp, as sharp in fact as for a "perfect" surface, and thus limited by the resolution of the apparatus only. For the case of destructive interference, however, all peaks will be broadened with a maximum halfwidth again proportional to $\bar{R}_0$, now the average terrace size. Note, that for a crystal surface with regularly spaced steps the behaviour is different: in that case a splitting of all peaks is expected, with peak intensities depending strongly on incident energy, see [5.5].

If the islands are formed by an adsorbate there will also be a variation of the halfwidth with incident energy. In this situation even for the case of constructive interference the halfwidth of the diffraction peaks corresponding to diffraction from the adsorbate lattice will not be as low as for the clean surface, as the upper part of the terrace (the island) has only a finite width. When varying the incident energy, interference effects will again broaden all peaks. Note that because the height of the islands will in general be different from the height of a substrate monatomic step, the detailed dependency of halfwidths on incident energy can be used to distinguish between adsorbate islands and terraces (or islands formed by substrate atoms).

Generally, 2d defects will give rise to two different features in the distribution of diffusely scattered intensity: The finite width of the islands or domains will give rise to a broadening of diffraction peaks. This has been discussed above and will be referred to as small angle scattering. But of course the boundaries

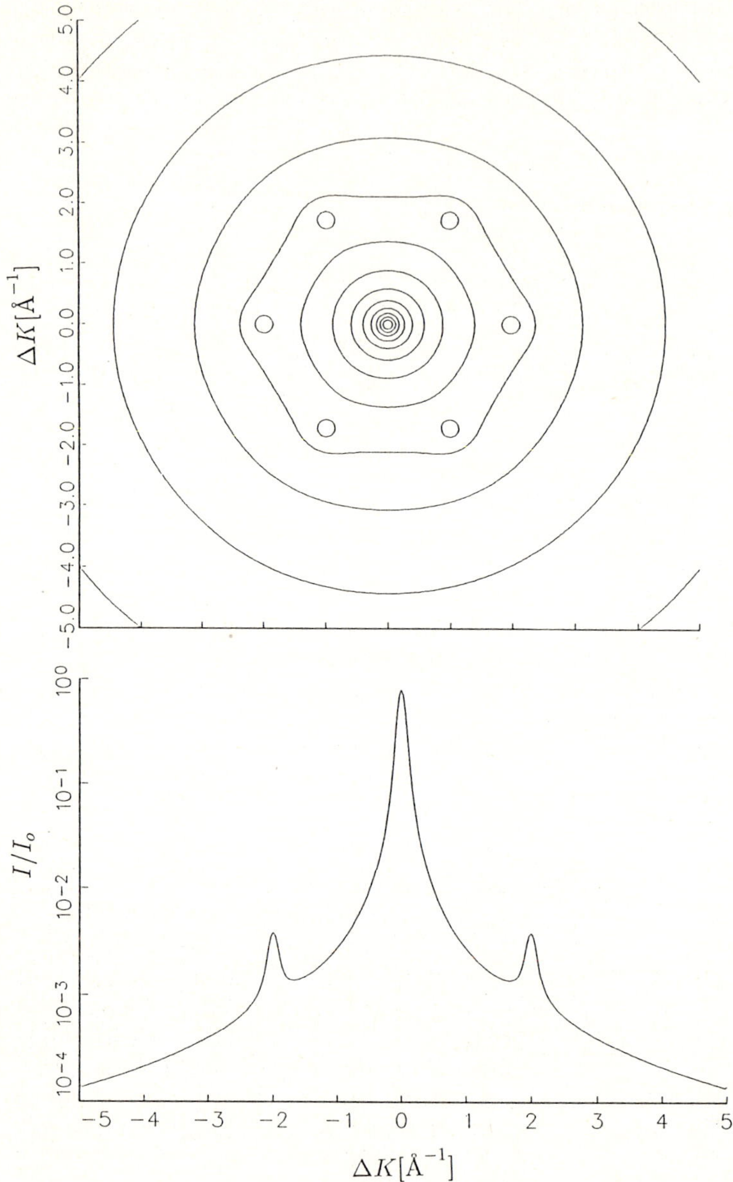

Fig.5.2. Upper panel: Logarithmic contour plot from a model calculation for the angular distribution of He atoms scattered from a stepped surface with an average terrace size of 100 Å. Lower panel: Angular scan along the dotted line in the countour plot shown above. Compare to Fig.5.1 and note that the scattering for small values of ΔK is affected

themselves, steps, dislocations, or, in the case of the islands, the edge atoms will scatter diffusely. This latter mechanism will produce features at positions not necessarily close to diffraction peaks. This type of scattering will be referred to as large angle scattering and discussed in the next section, since the edges of 2d defects can be considered as one dimensional. In the simulations used to produce Fig.5.2 this type of scattering has been neglected by ignoring the scattering from the boundaries between different domains, and at the edges of the islands.

5.2.2 Steps or One-Dimensional Defects

As mentioned above, a single step on a otherwise perfect surface will affect the scattering of He atoms in two ways: first it will give rise to interference between different paths from the upper and lower terrace and, second, the step-edge itself will give rise to diffuse scattering intensity, as illustrated in Fig.5.3.

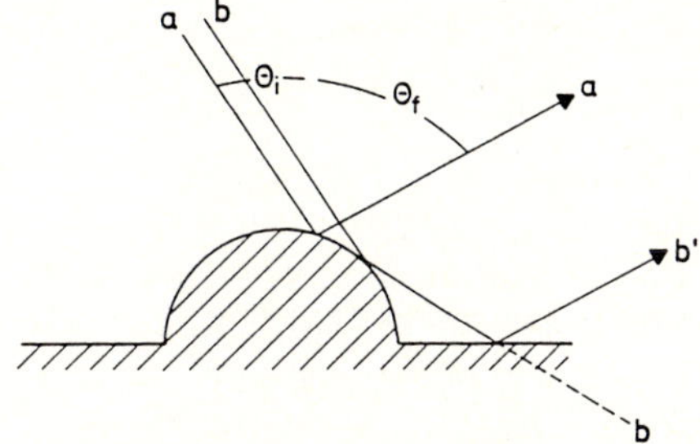

Fig.5.3. Schematic geometry for scattering of He atoms from a defect on a surface. The different trajectories shown will interfere and cause different angular distributions, which depend on the dimensionality of the defect

We have discussed the interference phemomena in the last section and will now illustrate the effects of the presence of steps on the angular distributions by investigating the scattering from a half-cylinder on a otherwise perfect surface. Because the parts of the surface to the left and right of the half-cylinder have the same height, interference effects between trajectories deflected from these areas are absent. This half-cylinder model (Fig.5.4) is of course an oversimplification. It nonetheless serves for illustrative purposes, and has the advantage that, together with the hard-wall approximation for the He-surface potential (Sect.5.2.4) analytical solutions can be obtained for the intensity distribution in $\boldsymbol{K}$-space. One finds [5.6]:

$$I(\theta_i) = |f_s(|\theta_i - \theta_f - \pi/2|) - f_s(\pi - |\theta_i + \theta_f|)|^2 \quad , \tag{5.7}$$

with

$$f_s(\theta_i) = -\left[\frac{R_o \sin(\theta_i/2)}{2}\right]^{\frac{1}{2}} \exp(-2ikR_o \sin(\theta_i/2)) - $$

$$\frac{\exp(-i\pi/4)}{(2k\pi)^{\frac{1}{2}}} \frac{\cos(\theta_i)+1}{\sin(\theta_i)} \sin(kR_o \sin(\theta_i)) \tag{5.8}$$

where R_0 is the radius of the half-cylinder. Note that the step height does not enter in this simple expression, the spacing of the interference maxima is given by the actual shape of the step. Fig.5.4 shows a logarithmic contour-plot of the intensity expected for this type of defect, together with a cross-section in a plane perpendicular to this surface. Note that due to the one-dimensional character of the imperfection on the surface, the diffuse intensity in K-space is focussed perpendicular to the step edge. On a microscopic scale step edges tend to show preferential orientations in real space, as step edges consisting of densely packed atoms are thermodynamically favoured. Depending on the crystal symmetry there will be in general different possible orientations (e.g. rotated by multiples of 60° for a fcc(111) surface) of a step edge, leading to an incoherent superposition of the corresponding intensity distribution of a single step, as displayed in Fig.5.4.

5.2.3 Zero-Dimensional Defects—Adatoms and Vacancies

There are various possible reasons for the existence of point-defects on a surface. They can be due to the presence of adatoms, either from the substrate or as a contamination, the adsorption of molecules, and vacancies. We will demonstrate the effect of scattering from a zero-dimensional defect (or point-defect) by using the same idea outlined in the last section: we will model the adatom by a hard hemisphere on an otherwise perfect surface. In contrast to the case of steps, it turns out that this is quite a realistic model for the potential due to an adsorbate molecule. Note that the geometry is the same as depicted in Fig.5.4 for the case of a cut through the step edge, only the dimensionality is different. If we again assume the hard-wall potential approximation (Sect.5.2.4), we can obtain analytical expressions for the intensity distribution in K-space [5.7]:

$$I(\theta_i) = |f_s(|\theta_i - \theta_f|) - f_s(\pi - |\theta_i + \theta_f|)|^2 \quad , \tag{5.9}$$

with

$$f_s(\theta_i) = \left(\frac{i\exp(-2ikR_0 \sin(\theta_i/2))}{2R_0} - \frac{(1+\cos\theta_i)J_1(kR_0 \sin\theta_i)}{2R_0 \sin\theta_i}\right) ,$$

$$\tag{5.10}$$

where R_0 denotes the radius of the sphere and J_1 the first order Bessel function. The diffraction pattern will have the same symmetry as the scattering object, therefore the total diffraction pattern will have rotational symmetry with respect to the surface normal, see Fig.5.5.

Note that these symmetry features of the intensity distribution in K-space can be exploited to distinguish between diffuse scattering from line defects such as step edges and point defects such as adatoms.

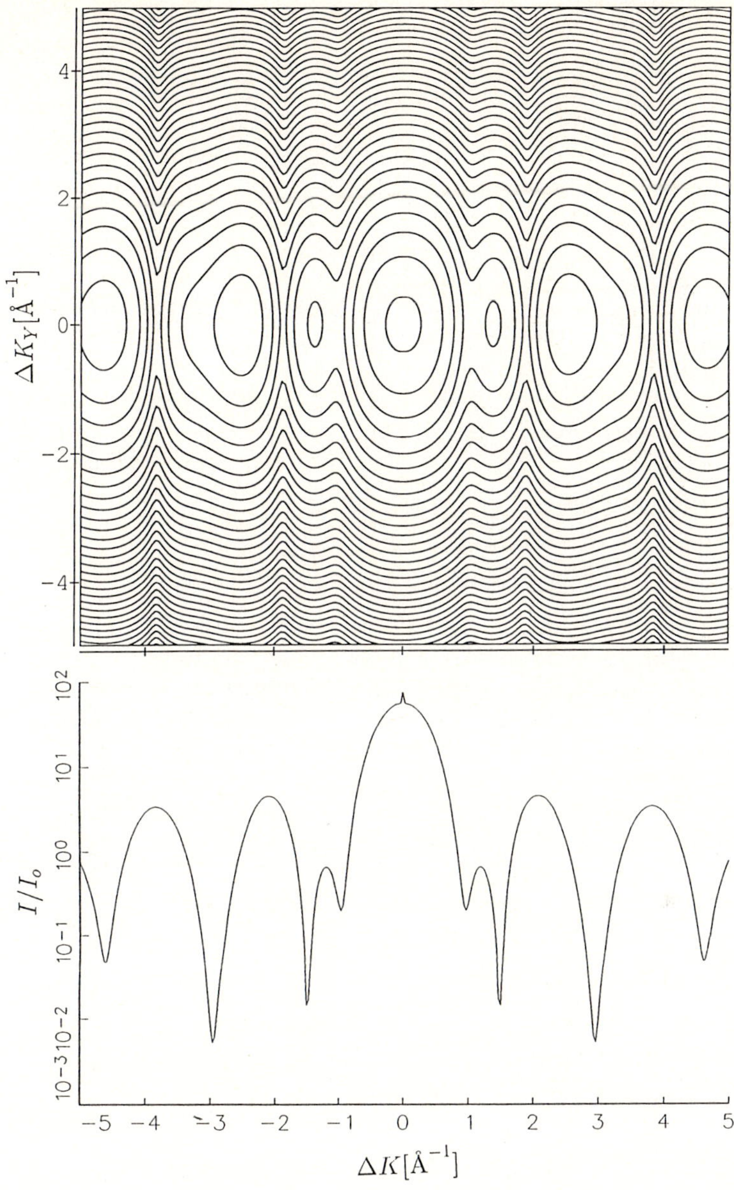

Fig.5.4. Logarithmic contour plot from a model calculation for the angular distribution of He atoms scattered from a single step edge on an otherwise flat surface ($k_i = 9.7\text{Å}^{-1}$, $R_o = 2.2\text{Å}.$). Note the two-fold symmetry of the diffraction pattern. In order to account for experimental resolution and a finite step-length the dependence on ΔK_y (which is that of a delta-function, see text) has been smeared out by a Gaussian in ΔK_y.

Lower panel: Angular scan along the dotted line in the contour plot shown above. Compare to Fig.5.3 and note the undulations for large values of ΔK

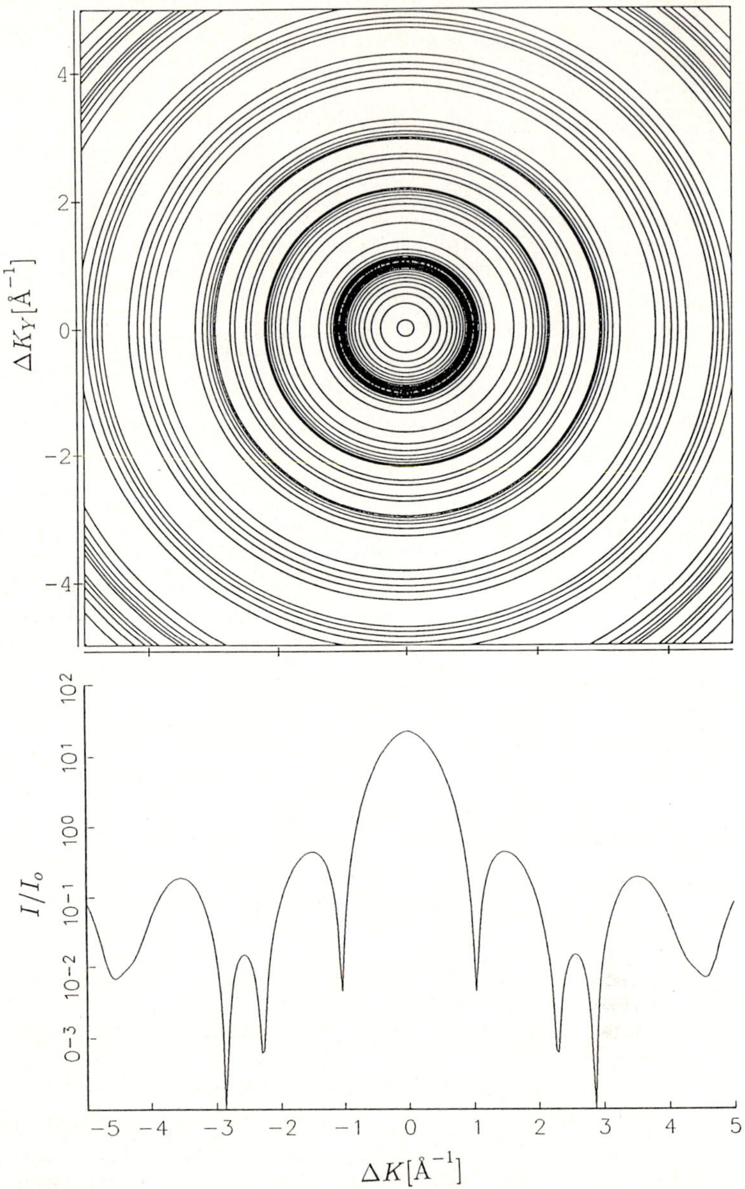

Fig.5.5. Upper panel: Logarithmic contour plot from a model calculation for the angular distribution of He atoms scattered from a hard hemisphere ($R_o = 2.6\,\text{Å}$, $k_i = 9.7\,\text{Å}^{-1}$) on an otherwise flat surface. Note the rotational symmetry of the diffraction pattern. Lower panel: Angular scan along the dotted line in the countour plot shown above. Compare to Fig.5.4 and note the undulations for large values of ΔK.

5.2.4 Potentials

For the calculations used to illustrate the scattering from the different types of defects in the sections above a hard-wall potential has been used. For a more detailed analysis of experimental data more realistic potentials are needed, in particluar attractive contributions to the interaction have to be accounted for. For clean surfaces recent years have seen a significant advance, experimentally as well as theoretically, in the understanding of the interaction between He atoms and surfaces, especially of metals and semiconductors. Note that in the case of molecular adsorbates on surfaces one can often make use of He–molecule interaction potentials known from gas-phase experiments as a first approximation (for the case of CO see Ref. [5.2]). The reader is also referred to the discussion by Celli in Chap. 3.

5.2.5 Diffusion of Defects

All the discussion presented in the previous sections explicitly referred to diffuse *elastic* scattering. Strictly speaking the diffuse scattering is elastic only if the defects are stationary. In the case of a high mobility of the defects, for example for a 2d gas on a surface, the presence of a thin liquid layer or even the surface of a liquid, the diffuse peak will show a broadening in energy, which in a favourable case has been demonstrated experimentally. Naively, one can picture this effect as being due to a Doppler shift. The energetic broadening of the diffuse elastic peaks is related to the diffusion coefficient D_s by [5.8]:

$$\Delta E = 2\hbar D_s K^2 \quad . \tag{5.11}$$

For clean surfaces the diffusion coefficient is very small, but when the melting point is approached, the energetic halfwidth of the diffuse elastic peak increases. This effect has been successfully demonstrated for the case of Pb(110) [5.9]. In practice, for a clean surface at a temperature far away from its melting point, this effect can be neglected. When adsorbates are present on the surface, however, diffusion of the adsorbate particles has to be expected, which, depending on the diffusion coefficient, should also give rise to a broadening. This effect is discussed in detail in Chap. 12.

5.2.6 Inelastic Scattering of He Atoms

A time-of-flight spectrum of He atoms scattered from a surface will also show inelastic features, in addition to the elastic signal. This reflects the fact that He atoms can exchange energy with the surface, by excitation or annihilation of phonons. This is readily visible in the time-of-flight spectrum shown in Fig.5.6, which has been taken for a small concentration of CO on Pt(111) [5.10]. The diffuse elastic peak (at 0 meV energy transfer) is the dominant feature. The rather sharp peaks at +6 meV and −6 meV correspond to annihilation and creation of low energy normal modes of the adsorbed CO molecule [5.10]. Note the presence of the second overtones. For the similar case of CO on Ni(100), see

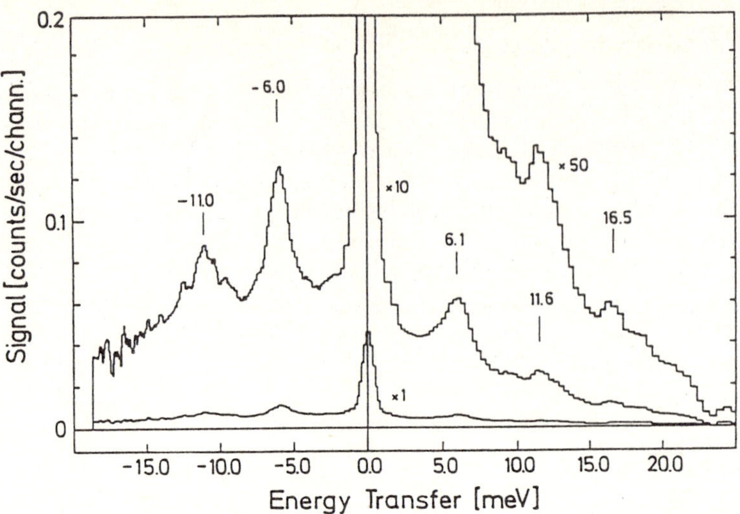

Fig.5.6. He atom time-of-flight spectra revealing the excitation and annihilation of low-energy vibrations of CO molecules adsorbed on a Pt(111) surface, from Ref.[5.10]. Note that in this case the diffuse elastic scattering (peak at zero energy transfer) dominates the diffuse scattering

[5.11]. In principle, a time-of-flight spectrum as shown in Fig.5.6 should also show losses related to vibrations of the clean Pt(111) surface (see Ref. [5.12] for a complete discussion); but in this case phonon creation and annihilation are not detected due to the choice of kinematical conditions.

Although the total intensity of inelastic effects in this spectrum is small compared to the elastic intensity, for a detailed shape analysis of the diffuse elastic component for small defect concentrations a separation of elastic and inelastic intensity is important (Sect.5.3.).

5.3. Experimental Results and Analysis

In the preceding sections a theoretical framework has been presented and illustrated by the results of a simple theory. We have demonstrated the features which are to be expected when He atoms are scattered from a surface containing various types of defects. So far, experimental considerations have only been mentioned in connection with the finite angular halfwidth of the diffraction peaks of a clean surface. For the experimental determination of integral cross-sections the experimental requirements are not so severe, and a number of groups have used this technique for a long time. However, for an experimental investigation of the large-angle scattering, or determination of the differential cross-section, a number of different problems have to be addressed. As a more detailed discussion of experimental problems arising with a UHV molecular

84

beam apparatus is beyond the scope of this article, we would simply like to stress the two major points:

1) All results presented in the previous sections refer to elastic scattering only. But, of course, not all scattering channels are elastic. Especially at higher temperatures the scattering can be dominated by inelastic effects, which can also be described by phonon creation and annihilation. Whereas the theoretical understanding of phonons on clean metal surfaces and the corresponding excitation and annihilation probabilities by HAS has made significant progress in recent years, defects may give rise to completely unexpected inelastic features. Therefore for a detailed analysis of diffuse scattering a separation of the elastic part from the inelastic part is absolutely necessary, even if only qualitative features in the angular distributions are analyzed.

2) For a measurement of low defect concentrations the diffuse intensity may be several orders of magnitude lower than that of the diffraction peaks. This is particularly true in cases where the diffuse intensity is not focused in certain directions of K-space (point-defects in contrast to line defects). A dynamical range of the apparatus of 5–6 orders of magnitude is therefore necessary. This, together with the need to discriminate between elastic and inelastic scattering even for these low intensities, makes a rather sophisticated design of the apparatus necessary. Whereas the ability to measure inelastic events upon the scattering from He atoms from a solid surface has been demonstrated by several groups, data which fulfill the above criteria for the analysis of diffusely scattered intensities have so far only been reported from the group of Toennies in Göttingen. Detailed descriptions of a typical apparatus used for these experiments can be found in Refs. [5.12,13].

In the following sections we will discuss experimental results for the different types of imperfections.

5.3.1 Two-Dimensional Defects

A typical example for a two-dimensional defect are terraces on a monocrystalline surface. As discussed above, the presence of a finite coherence length on the surface will cause a broadening of the specular and the diffraction peaks, with the amount of the broadening depending on the incident energy. This behaviour is exemplified in Fig.5.7 for an Al(110) surface with a high step density. The upper angular distribution in Fig.5.7 has been recorded for an incident energy at which, for the specular beam, there is a constructive interference between He trajectories reflected from two adjacent terraces. For the first order diffraction peaks, however, the interference is destructive, causing the corresponding peaks at about 1.8 Å^{-1} to exhibit a rather large halfwidth. In the lower part of Fig.5.7 the situation is reversed, in this case the interference is destructive for the specular and almost constructive for the first-order diffraction peaks. The dependence of the halfwidth of the various peaks can actually be used to determine the average terrace width and thus the step density [5.14].

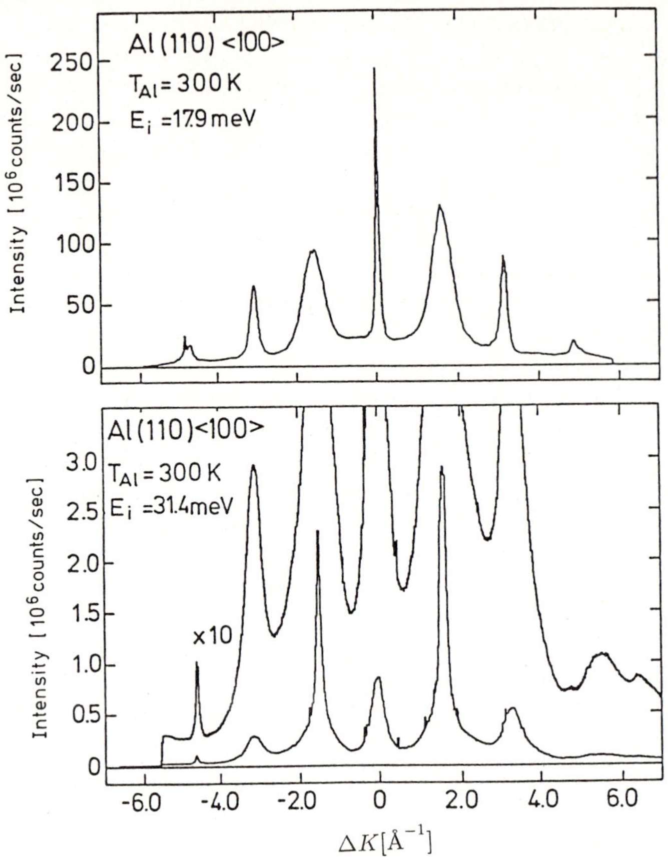

Fig.5.7. Two experimental angular distributions of He atoms scattered from a stepped Al(110) surface, from Ref.[5.29]. Note the change of the halfwidth of the various diffraction peaks with incident energy

5.3.2 Line-Defects, One-Dimensional Defects

The ideal realisation of a line-defect on a surface is a step edge. Other possibilities are (sharp) grain boundaries or in general boundaries between adsorbate islands and the bare substrate. Even a well-aligned single crystal surface will have a certain density of steps, as a perfect alignment of the polishing plane with respect to the desired crystallographic orientation of the surface is not possible. Steps will affect the scattering of He atoms in two ways. First, there will be a phase shift between trajectories scattered from the upper terrace and those scattered from the lower terrace. The amount of this phase shift depends on the step height, the angle of incidence and the He atom wavelength. Changing the incident energy will therefore give rise to strong intensity variations of the specular beam, with maximum intensity for constructive interference and

86

minimum intensity for destructive interference. The usefulness of this method
has been demonstrated by *Poelsma* et al. [5.15].

The position of the constructive and destructive extrema depend on the step
height, which in general is known from the crystallographic data. The intensity
of the modulation allows the determination of step densities. In contrast the
angular distribution of the diffusely scattered He atoms will be sensitive to the
shape of the step, as outlined above. Although the experimental requirements
for measuring the undulations of the diffusely scattered He atoms are much
more stringent the feasibility of this method has been demonstrated. Fig.5.8
shows the experimental data obtained by *Lahee* et al. [5.6] for steps on a Pt(111)
surface. The broad undulations predicted by theory are clearly visible in the
experimental data. Note the log intensity scale. The simple theory outlined in
Sect.5.2.2 contains only one free parameter, R_0, the radius of the half-cylinder
mimicking the step-edge, and the solid line shows the results for a best-fit
value of 2.3 Å. The sharper oscillations occur at the positions expected for the
diffraction peaks. More elaborate calculations [5.16,17] go beyond the simple
half-cylinder model considered so far and, by taking the atomic corrugation
into account explicitly, are also able to reproduce these sharp oscillations.

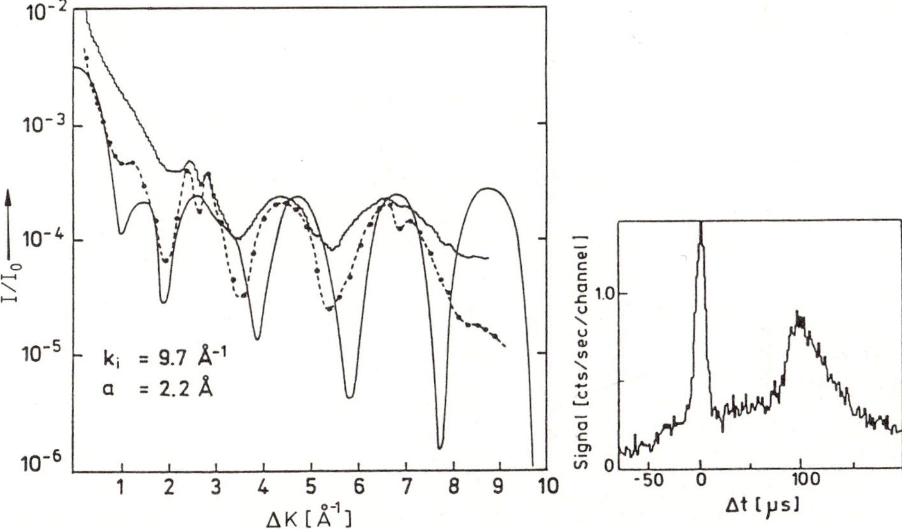

Fig.5.8. Experimental angular distributions of He atoms scattered from a stepped Pt(111)
surface, from [5.6]. In the inset a typical time-of-flight spectrum is shown. The sharp peak at
zero time difference corresponds to the diffuse elastic contribution, whereas the rather broad
peak on the right is related to inelastic processes, in this case excitation of a surface phonon

The diffuse scattering signal from a single step is expected to be strongly
focused in one azimuthal direction (Fig.5.5). Therefore an analysis of the experimental behaviour will give direct information on the orientation of steps.
The experimental finding is that the intensity of the undulations is strongest

in the $[11\bar{2}]$, $[12\bar{1}]$ and $[2\bar{1}1]$ direction, with a strong decrease in intensity for a variation of azimuthal angle as small as 5°. This is in agreement with the expectation that the steps occur mainly in those crystallographic directions where the step edges consist of densely packed rows of atoms. In addition it should also be noted that the precise dependence of the undulation intensity of the azimuthal angle allows a determination of the average distance between kinks [5.18].

Because the diffuse scattering of He atoms is sensitive to the shape of step edges, changes of the shape can be monitored directly. These changes can be caused, for example, by a thermally activated diffusion of admolecules to step edges. A more detailed discussion of these phenomena is given in Ref. [5.18].

5.3.3 Point Defects, Zero-Dimensional Defects

Point defects on a surface have the lowest dimensionality, zero. They can be produced by vacancies or adatoms on a clean surface, or atomic or molecular adsorbates. Note that any condensation of the point defects will give rise to small islands, corresponding to 2d defects rather than a point defect. Therefore, to realize the case of adsorbates, a repulsive interaction and a very small coverage is needed. The adsorption of carbon-monoxide on Pt(111) fulfills these requirements. CO adsorbs non-dissociatively on Pt(111) at room temperature and below, with the molecular axis oriented perpendicular to the surface plane and the carbon atom closer to the surface. The interaction between CO molecules is repulsive, and at low coverages the molecules are thus isolated. Fig.5.9 shows the experimental HAS data obtained by *Lahee* et al. [5.7] for this system. The experimental data clearly show the undulations as expected from the simple calculations assuming a hard wall potential as outlined in Sect.5.2.3. The analytical expression given there for the intensity distribution contains only one free parameter, R_0. By a fit procedure a value of $R_0 = 2.4$ Å is obtained, which agrees very well with available gas-phase data and theoretical calculations (see Ref.[5.7] for a more detailed discussion).

There is another interesting aspect related to this system. Since the CO molecules stand upright rather than lying flat on the surface, the angular distributions as displayed in Fig.5.9 should not show any variation with azimuthal angle. This rotational symmetry has been verified experimentally [5.7]. For CO molecules lying flat on the surface the variation of the azimuthal angle should produce a variation of the distance between the undulations. There is some indication that CO molecules do lie flat on Al(111). An application of HAS to study this system should therefore be interesting.

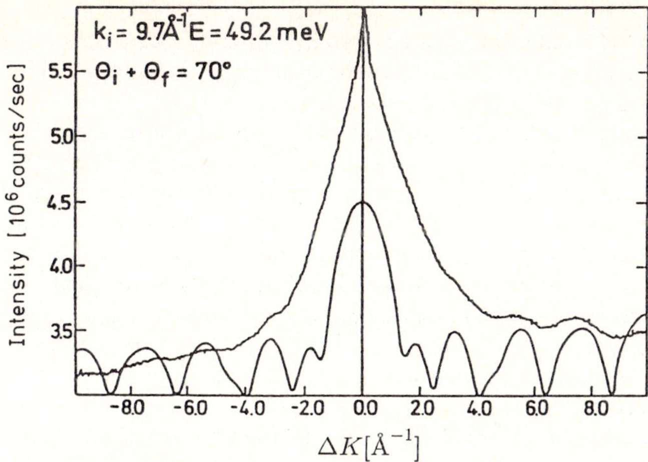

Fig.5.9. Experimental angular He atom distributions for scattering from single CO atoms adsorbed on Pt(111), from Ref.[5.7]

5.4 Comparison with Other Techniques

In order to emphasize the strengths and weaknesses of He atom scattering as a probe of surface imperfections, it will be useful to compare this technique with other methods that are currently used to study surface defects. To best illustrate the wide range of theoretical and experimental problems that enter such investigations, we choose for this comparison one diffraction technique, diffuse LEED (or DLEED), and one real-space probe, scanning tunneling microscopy (STM). Together with He scattering, these are the most powerful methods for studying surface defects, but other methods can also be applied on occasion, for example (neutrons, X-rays, NEXAFS, photo-electron diffraction). None of these, however, have the same sensitivity to low densities of defects.

5.4.1 Diffuse Low Energy Electron Scattering

The technique of low-energy electron diffraction (LEED) has become one of the standard tools of modern surface science. Its diffuse counterpart (DLEED), however, has only been exploited in the last few years. The basic principles of LEED and DLEED are exactly analogous to those of He scattering: When a monoenergetic beam of electrons (E_i=10–1000 eV gives a de Broglie wavelength of the same order as crystal lattice constants) strikes the surface of perfectly periodic solid, diffracted beams are produced at angles corresponding to constructive interference between waves scattered from neighbouring ion cores. The scattered electrons are imaged on a fluorescent screen and, for a perfect surface, one expects to observe a set of extremely sharp spots whose width simply reflects the finite coherence length of the electron beam, and whose positions tell us the surface symmetry and lattice constants.

Even for a perfectly ordered surface, the intensity between diffraction spots does not fall to zero. This is the result of inelastic scattering processes in which electrons exchange energy with surface phonons.

There are two ways in which LEED can be exploited for studying surface imperfections. The first involves simply measuring the broadening and/or splitting of the diffraction spots that is observed for stepped surfaces and surfaces with islands or antiphase domains [5.19]. This will not be discussed further since it does not provide information about simple defects such as adatoms. The second source of information about single defects is the diffuse elastic intensity between the diffraction spots, in exact analogy with the diffuse intensity measured in He scattering.

The fact that this intensity should show a structure (K-dependence) was predicted by calculations of *Pendry* et al. in 1984 [5.20,21]. It should be noted immediately that considerable theoretical effort is required to predict this structure—one may use cluster calculations [5.20] or modified LEED calculations [5.20]. Furthermore, there is no intuitive way of understanding the structure in the diffuse electron scattering in terms of a simple scattering process (Fig.5.2). This is because of the essential role played by multiple scattering in the LEED process.

Nonetheless it is possible to calculate the DLEED intensities using a realistic model, and since 1985 experimental measurements have become available for comparison. As in the case of He scattering a DLEED experiment puts stringent requirements on the experimental sensitivity, but using sensitive video-LEED equipment a number of groups have succeeded in measuring reliable DLEED intensity distributions [5.22–27].

A comparison with experiment is performed using the time-honoured R-factor, used also in LEED for an analysis of $I(V)$ curves. One does not compare intensities directly but uses a so-called Y-factor which is a logarithmic derivative of the intensity that includes a cutoff function. Fig.5.10 shows experimental and theoretical Y-functions for DLEED measured from W(100) with about 0.25 ML oxygen. The theory attempts to reproduce the observed Y-function by varying two parameters — the adsorbate height and the induced reconstruction of the substrate.

Both DLEED and diffuse He scattering are in their infancy. Experimentally DLEED is more straightforward — simply because large quantities of data can be gathered in a short time. From the theoretical point of view, however, DLEED is the more cumbersome to analyse and an intuitive relation between diffuse intensities and the nature of surface imperfections is lacking. DLEED has been shown to be sensitive to adsorbate geometry as well as other features such as the height of an adsorbate and an associated surface reconstruction. Diffuse He scattering has yet to be tested in this respect, but has shown itself sensitive to charge density profiles around adsorbed atoms and at surface steps. Thus, to some extent, the two techniques are complementary. In both cases, the full potential of the method has yet to be exploited, especially in the case of diffuse He scattering.

90

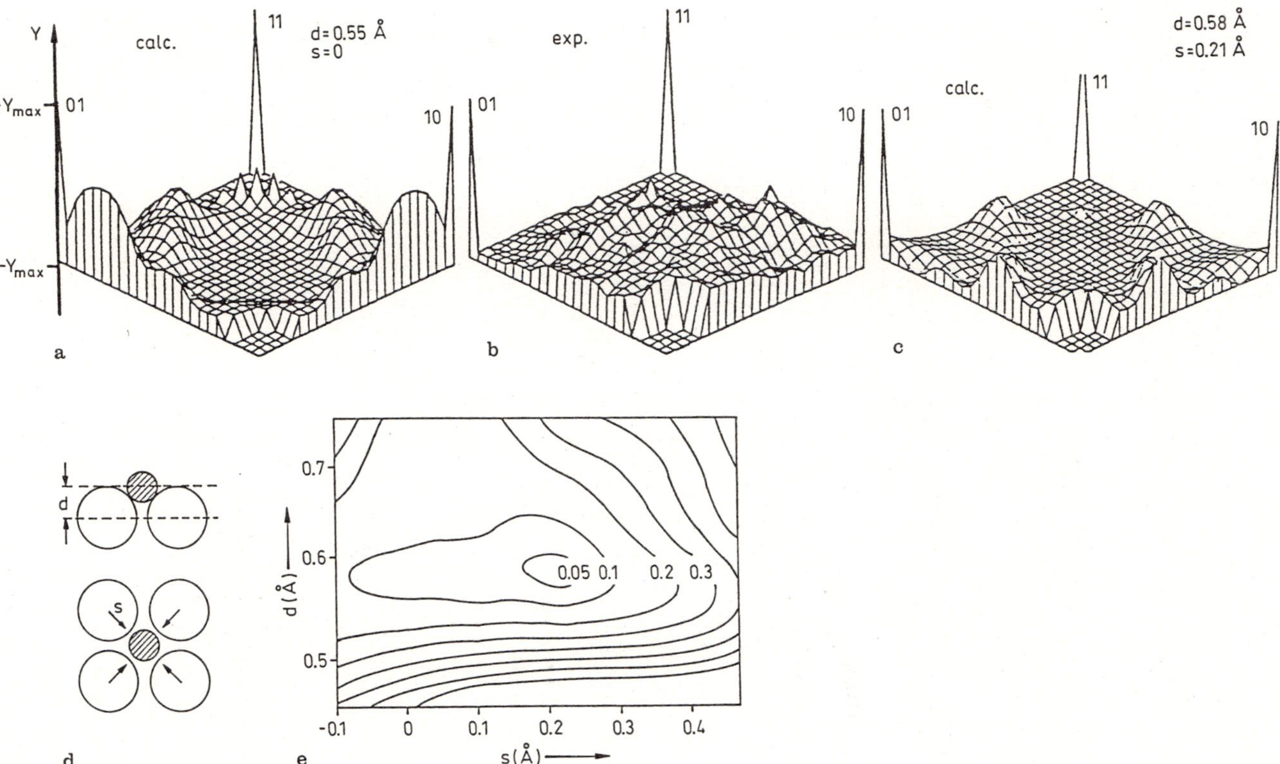

Fig.5.10. Experimental LEED Y-function map (b) for the disordered adsorption system O/W(100) compared to the best fit maps for the unreconstructed (a) and reconstructed (c) substrate according to the model (d). In (e) the R-factor contour map is displayed as function of the adsorption height d and the reconstruction amplitude s. From [5.29]

5.4.2 Scanning Tunneling Microscopy

Scanning tunneling microscopy (STM) is a powerful probe of the local, real-space structure of a conducting surface and on favourable substrates, such as Si, can be used to image areas of 1000 Å × 1000 Å with atomic resolution. Many beautiful STM images have been produced of surfaces with steps. Indeed such images are frequently taken as an assurance that the microscope is working reliably. In a typical constant-current scan taken by an STM a surface step clearly shows up as a vertical movement of the tip by the height of one (or more) lattice planes. Thus steps are easy to identify and one can also determine their orientation and investigate adsorption at step edges. As a probe of step densities and step distributions the STM is limited, simply by the finite size of the area that it can scan with sufficient resolution.

The case of point defects (adsorbates and vacancies) is a typical situation in which the STM may give ambiguous results. If operating with atomic resolution, it will certainly detect the fact that a surface atom is missing, but may not, for example, distinguish a vacancy from a substitutional impurity.

It has been shown that the STM is sensitive to large adsorbed molecules (benzene, phthalocyanin), but a clear identification of a smaller molecule has so far not been reported. A strong limitation arises from the fact that only conducting surfaces can be investigated, and even on conducting surfaces the presence of larger clusters of nonconducting atoms or molecules may make the imaging of the surface impossible.

A considerable drawback of STM is its pronounced sensitivity to diffusion. Adsorbate atoms or molecules moving along the surface will not hamper HAS or DLEED, as the interaction time of a He atom or electron is much shorter than the time in which the adsorbate moves, e.g., a lattice constant. As the imaging in the case of STM is much slower, diffusion effects strongly limit the acquisition of data for single molecules. Only if the interaction to the substrate is strong enough to lock the molecule at a certain position can the STM be expected to deliver reliable results. For the case of CO on a room-temperature Pt(111) surface the STM is therefore not expected to give reliable results.

5.4.3 Summary

Each of the techniques described in this section has great potential for the study of metal and semiconductor surfaces with defects; only He scattering can be applied without any changes to insulators. Since they have different strengths and weaknesses, experimentally, theoretically, and in terms of their sensitivity to various defects, it will be well worthwhile to invest further effort in developing each of these techniques to its full potential.

References

5.1   J.P. Toennies: In *Physical Chemistry, An Advanced Treatise*, Vol. VI A, Kinetics of gas-reactions, ed. by W. Jost (Academic, New York 1974) pp.228-332

5.2   M. Faubel, K.H. Kohl, J.P. Toennies: J. Chem. Phys. **73**, 2506 (1980)

5.3   B. Poelsema, S.T. de Zwart, G. Comsa: Phys. Rev. Lett. **49**, 578 (1982) see also erratum: B. Poelsema, S.T. de Zwart , G. Comsa: Phys. Rev. Lett. **51**, 522 (1983)

5.4   B. Poelsema, G. Comsa: *Scattering of Thermal Energy Atoms from Disordered Surfaces* (Springer, Berlin, Heidelberg 1989)

5.5   B.J. Hinch, A. Lock, H.H. Madden, J.P. Toennies, G. Witte: J. Electron Spectroscopy, in press

5.6   A.M. Lahee, J.R. Manson, J.P. Toennies, Ch. Wöll: Phys. Rev. Lett. **57**, 471 (1986)

5.7   A.M. Lahee, J.R. Manson, J.P. Toennies, Ch. Wöll: J. Chem. Phys. **86**, 1727 (1988)

5.8   A.C. Levi, R. Spadacini, G.E. Tommei: Surf. Sci. **121**, 504 (1982)

5.9   J.M.W. Frenken, J.P. Toennies, Ch. Wöll: Phys. Rev. Lett. **60**, 1727 (1988)

5.10   A.M. Lahee, J.P. Toennies, Ch. Wöll: Surf. Sci. **177**, 371 (1986)

5.11   R. Berndt, J.P. Toennies, Ch. Wöll: J. Electr. Spectr. on Relat. Phen. **44**, 183 (1987)

5.12   V. Bortolani, A. Franchini, G. Santoro, J.P. Toennies, Ch. Wöll, G. Zhang: Phys. Rev. B **40**, 3524 (1989)

5.13   G. Lilienkamp, J.P. Toennies: J. Chem. Phys. **66**, 3965 (1977)

5.14   T.M. Lu, M.G. Lagally: Surf. Sci. **120**, 47 (1982)

5.15   B. Poelsema, R.L. Palmer, G. Mechtersheimer, G. Comsa: Surf. Sci. **117**, 50 (1982)

5.16   B.J. Hinch: Phys. Rev. B **38**, 5260 (1988)

5.17   C.W. Skorupka, J.R. Manson: Phys. Rev. B **41**, 9783 (1990)

5.18   R. Berndt, J. Hinch, J.P. Toennies, Ch. Wöll: J. Chem. Phys. **92**, 1435 (1989)

5.19   M. Henzler: Electron diffraction and surface defect structure, In *Electron Spectroscopy for Surface Analysis*, ed. by H. Ilbach, Topics Curr. Phys. Vol. 4 (Springer, Berlin, Heidelberg 1977)

5.20   J.B. Pendry, D.K. Saldin: Surf. Sci. **145** 33 (1984)

5.21   J.B. Pendry: In *The Structure of Surfaces* Springer Ser. Surf. Sci., Vol. 2, ed. by M.A. Van Hove, S.Y. Tong (Springer, Berlin, Heidelberg 1985) p.124

5.22   K. Heinz, D.K. Saldin, J.B. Pendry: Phys. Rev. Lett. **55**, 2312 (1985)

5.23   K .Heinz, K. Müller, W. Popp, H. Lindner: Surf. Sci. **173**, 366 (1986)

5.24   G. Illing, D. Heskett, E.W. Plummer, H.-J.-Freund, J. Somers, Th. Lindner, A.M. Bradshaw, U. Buskotte, M. Neumann, U. Starke, K. Heinz, P.L. de Andres, D.K. Saldin, J.B. Pendry: Surf. Sci. **206**, 1 (1988)

5.25   P. Heimann, G. Michalk, P. Piersy, D. Menzel: To be published

5.26   G.S. Blackman, M.-L. Xu, D.F. Ogletree, M.A. Van Hove, G.A. Somorjai: Phys. Rev. Lett. **61**, 2352 (1988)

5.27   U. Starke, P.L. de Andres, D.K. Saldin, K. Heinz, J.B. Pendry: Phys. Rev. B **38**, 12277 (1988)

5.28   K. Heinz: Vacuum, in press

6. Scattering from Stepped Surfaces and Roughening

J. Lapujoulade

One of the most interesting aspects of helium diffraction is its outstanding sensitivity to steps on surfaces. Indeed step analysis has became one of the major applications of helium diffraction in surface structure analysis. Steps are almost always present on surfaces as a consequence of either an imperfect preparation or thermal equilibrium. Thus the characterization of residual steps on a sample is a necessary operation prior to any serious surface study.

Steps divide the surface into two regions: the terrace region where the atomic structure is the same as in the perfect surface and the step edge region where this structure is drastically modified. For a randomly stepped surface the interference effect between the various terraces was experimentally evidenced in 1977 by *Lapujoulade* et al. [6.1, 2] and a hard wall model was proposed in 1981 by the same author [6.3] in order to make a quantitative description. It was later generalized by *Spadacini* et al. [6.4], *Levi* [6.5] and *Blatter* et al. [6.6] to a larger class of step statistics. The hard wall restriction has been recently removed by *Armand* et al. [6.7].

The integral scattering by step edges can be characterized by a cross section which was experimentally measured by *Verheij* et al. in 1981 [6.8]. The diffuse differential scattering cross section from these edges was first calculated by *Blatter* [6.9] in 1985 and experimentally observed by *Lahee* et al. in 1986 [6.10]. A more realistic calculation was proposed later [6.11].

The diffraction of helium from a periodic array of steps (vicinal surface) was first observed by *Lapujoulade* et al. in 1977 [6.12, 13] and the interaction potential shape extensively studied by *Gorse* et al. in 1984 [6.14]. The observation of an anomalous behaviour of the Debye-Waller factor on such surfaces was interpreted in 1983 by *Lapujoulade* et al. as evidence for the occurrence of a roughening transition [6.15]. A model was developed by *Villain* et al. in 1985 [6.16] which confirms this interpretation and subsequent experiments soon definitely established and characterized this transition [6.17, 18].

In the following we first explain the basic physics of the scattering of helium by stepped surfaces (Sect. 6.1). Then we consider the case of randomly distributed steps and we distinguish between the scattering by step edges alone (Sect. 6.2) and by terraces alone (Sect. 6.3). We discuss in Sect. 6.4 how information on real surfaces can be obtained from these models and we give a

Springer Series in Surface Sciences, Vol. 27   **Helium Atom Scattering from Surfaces**
Editor: E. Hulpke   © Springer-Verlag Berlin, Heidelberg 1992

typical example. Section 6.5 is devoted to the case of vicinal surfaces and their roughening transition.

6.1 The Physics of Helium Scattering by Stepped Surfaces

For the sake of clarity we shall first use a hard wall approximation for the helium–surface potential. This hard wall may be thought to represent some appropriate equipotential curve of a more realistic soft potential. It is schematically represented in Fig.6.1. Far from the step edges the wall corrugation function has the same stucture as the step free surface but translated by an integer multiple of unit vector t (tangential component T, normal component t_z). There is around every step edge a region of width l where the corrugation function steadily varies. For a monatomic step, l is expected to be of the order of a few lattice constants.

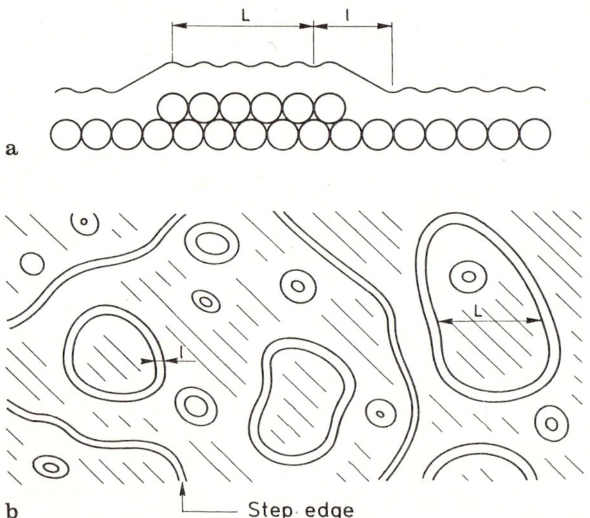

Fig.6.1. Schematic view of a surface with random steps. (a) Section normal to the surface; (b) top view

Assuming that the terrace size L is much larger than l one can neglect multiple scattering from terrace to terrace and from step edge to step edge. Then one can separate edge scattering from terrace scattering. Terrace to step edge multiple scattering which is of course not negligible can be included into the edge scattering by replacing the edge width l by an effective width l_{eff} such that $l_{\text{eff}} > l$. If the steps are randomly distributed the edge scattering due to the steep variation of the corrugation is expected to be very diffuse. On the

96

contrary terrace scattering due to the periodicity of the corrugation function is expected to be concentrated in the vicinity of Bragg directions related to the perfect surface.

The intensity in the Bragg peaks is dominated by the interference between the waves scattered from the different terraces. For instance in the specular direction the phase shift between two adjacent terraces is

$$\varphi = 2t_z k_i \cos\theta_i$$

where t_z denotes the step height as defined above, k_i the incident wave vector, and θ_i the angle of incidence. When $\varphi = 2n\pi$ (in-phase condition) beams specularly reflected from the various terraces interfere constructively. Then the specular peak is not affected by the steps and its shape remains a δ-function. Its intensity is simply decreased since a fraction of the total flux is now diffusely scattered by the step edges. On the other hand, when $\varphi = (2n+1)\pi$ (anti-phase condition) the interferences are destructive. The intensity of the peak is then strongly reduced and its shape is now dependent upon the statistics of step positions as will be discussed in Sect. 6.3.

An example is given in Fig.6.2 which is the first observation of this effect. It shows helium diffraction from a chemically etched Cu(100) surface. The intensity of the specular peak intensity is plotted versus incident angle. The intensity oscillations due to the alternate in-phase and anti-phase conditions are clearly seen.

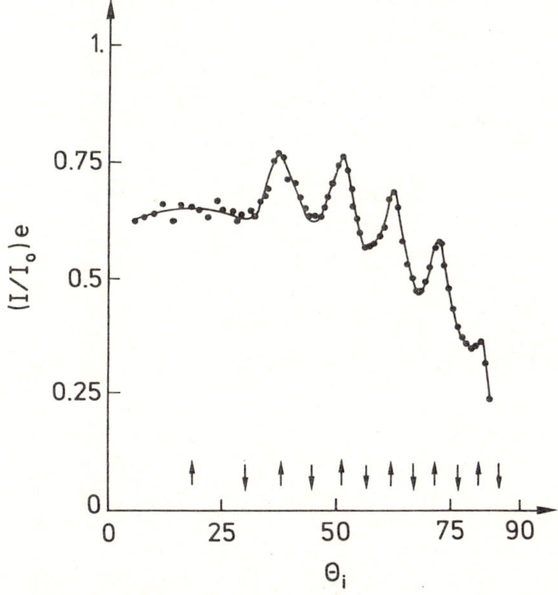

Fig.6.2. Intensity of the specular peak versus incidence angle for an etched Cu(100) surface. Arrows indicate the positions of in- and anti-phase conditions

If the terrace size L is not much larger than the step width l, the separation between terrace and step scattering can no longer be achieved. Provided that steps occur at crystal lattice positions one can still expect to observe maxima in the scattering around Bragg positions. The shape in the vicinity of these positions is certainly more closely related to the long range correlation function of step positions than to the step edge shape but no exact prediction could be made to date. The case of vicinal surfaces is different since the multiple scattering can be calculated exactly by numerical techniques as shown in Sect. 6.5.

6.2 The Scattering by Step Edges

For simplicity we suppose that in-phase conditions are achieved so that terrace scattering is concentrated into Bragg δ-function peaks. Then the problem of edge scattering is basically identical to the scattering by a surface covered with randomly distributed impurities which was discussed by *Levi* et al. [6.19] within the hard wall approximation and by *Armand* et al. [6.20] for unrestricted potentials. Here the impurity has a small extension l in one direction and a very large one in the the other.

This edge scattering has been experimentally observed in the Toennies laboratory by *Lahee* et al. [6.10] on a Pt(111) surface where random steps were created by a soft Ne ion bombardment at 600K. A typical result is shown in Fig.6.3. The main feature of this pattern is its strong oscillatory behaviour with some additional superstructure in the vicinity of Bragg positions.

These oscillations were first explained in the original paper by the assumption that the step edge potential is a half-cylinder hard boss. This potential is not very realistic but has the advantage of yielding exactly solvable scattering equations. Later *Hinch* [6.11] used a probably more realistic shape as indicated in Fig.6.3. The fit with the data is really improved, the use of the Eikonal approximation can, however, be seriously questioned for such a large corrugation (2.27 Å).

6.3 Scattering by Terraces

We shall now neglect edge scattering and consider only the effect of terraces. The physical basis of the scattering by terraces separated by random steps was formulated by *Lapujoulade* [6.3] in the hard wall approximation and for a particular step statistics. It was then extended to a much larger number of statistical cases by different authors [6.4–6]. More recently *Armand* et al. [6.7] have generalized the calculation to an arbitrary soft potential. We shall use their formalism here. Under the assumption of large terraces ($L \gg l$) the differential scattering cross section is given by

98

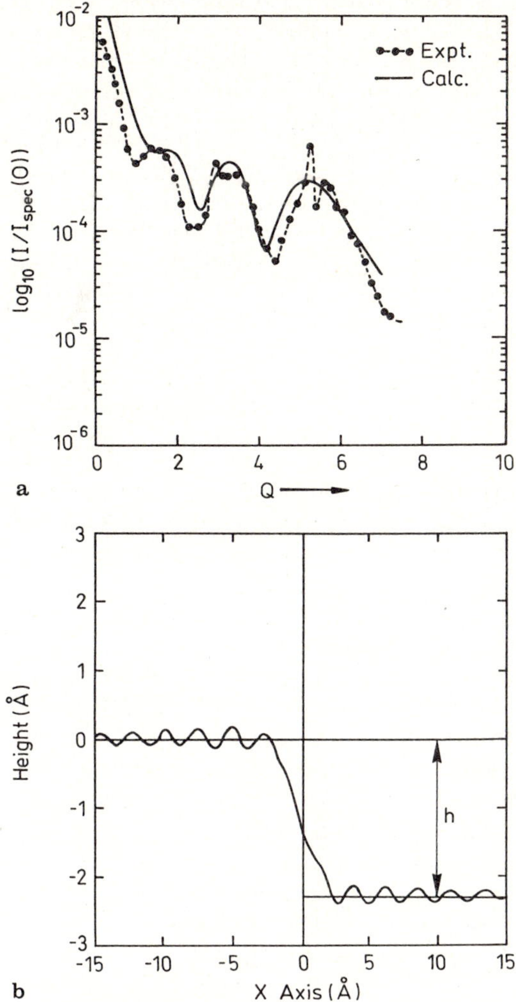

Fig.6.3. (a) Diffuse scattering cross section versus tangential momentum exchange for the scattering of helium on Pt(111). In-phase condition. Dashed curve: experimental data, solid curve: calculated results. (b) Potential used for the calculation. From [6.11]

$$\frac{dR}{d\Omega} = \frac{\pi^2}{A_G \cos\theta_i}\, I_0(\boldsymbol{Q})$$

$$\times \sum_{c,c'} \exp\left[-i\boldsymbol{Q}\cdot(\boldsymbol{R}_c - \boldsymbol{R}_{c'})\right]\, \langle\langle\exp[i\varphi(m - m')]\rangle\rangle \quad , \tag{6.1}$$

where A_G is the area of the reciprocal lattice unit cell, θ_i the angle of incidence,

Q the parallel momentum exchange, $R_c, R_{c'}$ the in-plane vector defining the position of the c, c' unit cell, m, m' integers defining the displacement of the c, c' unit cell due to the steps in terms of the elementary vector displacement $t(T, t_z)$ defined previously; m and m' are functions of R and R' respectively, $I_0(Q)$ is the form factor which only depends upon the corrugation within the unit cell, φ the phase shift between the various terraces

$$\varphi = -[Q \cdot T + q_z t_z] \quad . \tag{6.2}$$

q_z denotes the normal momentum exchange, $\langle\langle\ \rangle\rangle$ the statistical average over step configurations, and $\sum$ the sum over all the surface unit cells. Thus the cross-section is the product of a form factor which contains all the scattering properties of the perfect surface with a structure factor which is the Fourier transform of a phase factor $\Phi = \langle\langle\exp[i\varphi(m - m')]\rangle\rangle$ which only depends upon the statistics of the step positions.

It can be shown that the phase factor Φ is a real and even function of the phase shift φ.

- For $\varphi = 2\pi n$ (n integer) Φ is maximum and equal to 1 (in-phase condition),
- For $\varphi = (2n + 1)\pi$ the phase factor $\Phi = \langle\langle(-1)^{m-m'}\rangle\rangle$ (anti-phase condition).

If the phase factor remains finite for infinite distances $R_c - R_{c'}$ then

$$\langle\langle\exp[i\varphi(m - m')]\rangle\rangle_\infty = \Phi_\infty \tag{6.3}$$

and the above formula can be split into a coherent and an incoherent term. The coherent term is made of δ-functions located at the Bragg positions defined by the terrace lattice:

$$I_{\mathrm{coh}} = \sum_G \delta_{Q+G}\, I_G\, \Phi_\infty \quad , \tag{6.4}$$

where I_G is the intensity of the diffraction peaks given by the perfect infinite terrace lattice. The incoherent part, assuming stationarity in space (i.e. Φ only depends upon the difference $R_c - R_{c'}$), is given by:

$$\left(\frac{dR}{d\Omega}\right)_{\mathrm{inc}} = \frac{\pi^2}{A_G\, \cos\theta_i}\, I_0(Q)$$

$$\times \sum_c \exp[-iQ \cdot (R_c - R_0)]\ \langle\langle\exp[i\varphi(m - m_0)]\rangle\rangle - \Phi_\infty \quad . \tag{6.5}$$

In order to go further we shall now distinguish between systems with a finite small number of terrace levels and systems with an infinite range.

6.3.1 Surfaces with a Small Number of Terrace Levels

Let us assume here that the level shifts become uncorrelated when the separation $\boldsymbol{R}_c - \boldsymbol{R}_0$ becomes infinite. Then the calculation of Φ_∞ is easy. We give the result for a two–level model where m can only take the values 1 with probability α or 0 with probability $(1 - \alpha)$. Then:

$$I_{\text{coh}} = \sum_G \delta_{(\boldsymbol{Q}+\boldsymbol{G})} I_{\boldsymbol{G}} \Phi_\infty (\alpha) \quad , \tag{6.6}$$

with

$$\Phi_\infty(\alpha) = [1 - 4\alpha(1 - \alpha)\sin^2(\varphi/2)] \tag{6.7}$$

and

$$\left(\frac{dR}{d\Omega}\right)_{\text{inc}} = \frac{\pi^2}{A_{\boldsymbol{G}}\cos\theta_i} \, I_0(\boldsymbol{Q})$$

$$\times \, 4\sin^2(\frac{\varphi}{2}) \sum_c \exp[i\boldsymbol{Q} \cdot (\boldsymbol{R}_c - \boldsymbol{R}_0)](\langle\langle mm_0\rangle\rangle - \alpha^2) \quad . \tag{6.8}$$

The intensity of the coherent scattering only depends upon the level occupation number α. The correlation function $\langle\langle mm_0\rangle\rangle$ of terrace shifts is contained in the shape of the incoherent scattering.

We now look at the scattering in a Bragg direction: For the in-phase condition we get simply: $I_{\text{coh}} = I_{\boldsymbol{G}}$ and $(dR/d\Omega)_{\text{inc}} = 0$, and we recover the same result as for the perfect structure. For the anti-phase condition one has $\sin^2(\varphi/2) = 1$. Then the coherent intensity goes through a minimum while the incoherent one goes through a maximum.

In Fig.6.3 the variation of I_{coh} is plotted versus φ. When the number of allowed levels is increased the dependence of Φ_∞ versus φ becomes more and more complicated and the curve presents an increasing number of extrema as shown in Fig.6.4, but the general trend remains the same.

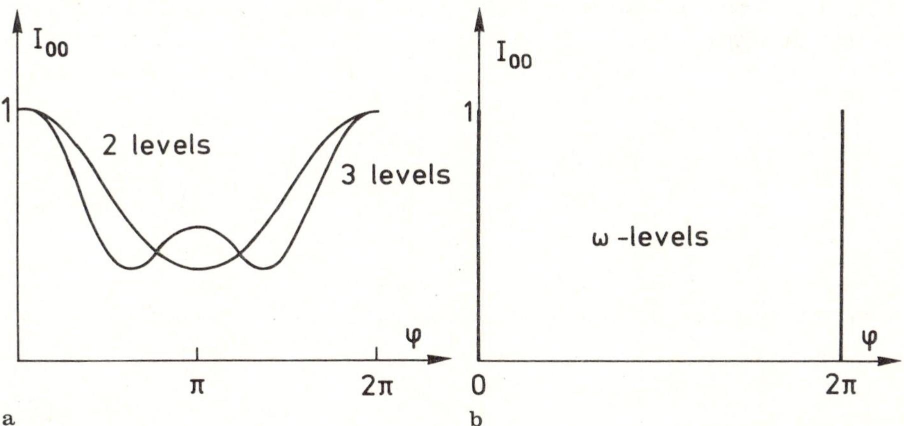

Fig.6.4. Intensity of the specular peak versus phase shift. (a) 2 or 3 levels; (b) infinite number of levels

6.3.2 Surfaces with an Infinite Number of Terrace Levels

If the surface is rough enough, it is reasonable to assume that for any distance $\boldsymbol{R}-\boldsymbol{R}_0$, $d-d_0$ is a nearly Gaussian random variable. In such a case it is possible to further evaluate the statistical average in the structure factor:

$$\Phi = \langle\langle\exp[i\varphi(m-m_0)]\rangle\rangle \approx \exp[W(\varphi)\langle\langle-(m-m_0)^2\rangle\rangle] \quad . \tag{6.9}$$

$W(\varphi)$ is even and periodic with a period 2π and is equal to $\varphi^2/2$ for small argument φ. One then obtains

$$\Phi_\infty = \begin{cases} \exp[\langle\langle-(m-m_0)^2\rangle\rangle W(\varphi)] & \text{for } 0 < \varphi < 2\pi \\ 1 & \text{for } \varphi = 0 \text{ or } 2\pi \end{cases} \quad . \tag{6.10}$$

If $\langle\langle(m-m_0)^2\rangle\rangle_\infty$ is infinite for infinite $\boldsymbol{R}-\boldsymbol{R}_0$ then Φ_∞ is equal to zero everywhere except for $\varphi = 2n\pi$ (Fig.6.4). Thus there is no longer any coherent scattering except for the exact in-phase condition. Otherwise only incoherent scattering takes place and the cross section is given by:

$$\left(\frac{dR}{d\Omega}\right)_{\text{inc}} = \frac{\pi^2}{A_G\cos\theta_i} I_0(\boldsymbol{Q})$$

$$\times \sum_c \exp[-i\boldsymbol{Q}\cdot(\boldsymbol{R}_c - \boldsymbol{R}_0)] \exp[\langle\langle-(m-m_0)^2\rangle\rangle W(\varphi)] \quad . \tag{6.11}$$

The function $\langle\langle(m-m_0)^2\rangle\rangle$ is closely related to the correlation function $\langle\langle mm_0\rangle\rangle$. Assuming that it is a smooth function of $\boldsymbol{R}$, the diffraction spectrum has the shape schematically represented in Fig.6.5. It is composed of enlarged peaks centered around Bragg positions. Their width varies according to the value of the phase shift φ ranging from zero (δ-function) for the in-phase condition to a maximum for anti-phase conditions. It has been shown that a correlation function which depends linearly upon distance (this is sometimes called the geometrical distribution) leads to a Lorentzian peak shape. A logarithmic dependence, as occurs above the roughening transition (see below), brings about an inverse power law shape.

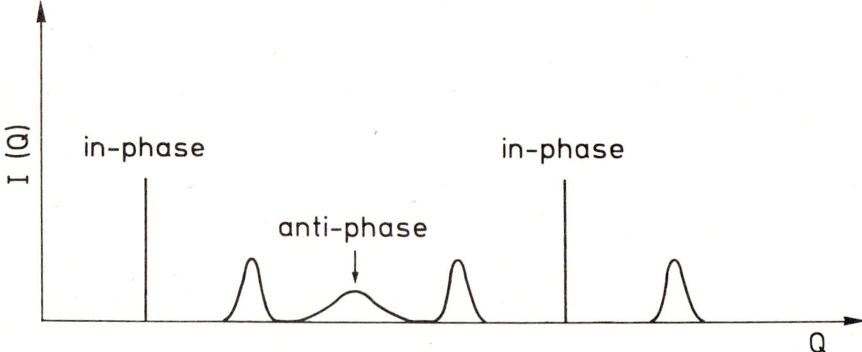

Fig.6.5. Diffraction pattern for a rough surface

6.4 Scattering by Randomly Stepped Surfaces: Some Applications

For a randomly stepped surface with large terraces, the total scattering is the superposition of edge and terrace scattering:

$$\left(\frac{dR}{d\Omega}\right)_{tot} = \left(\frac{dR}{d\Omega}\right)_{edge} + \left(\frac{dR}{d\Omega}\right)_{terr} \tag{6.12}$$

and the unitarity condition must be fulfilled:

$$\int d\Omega \left(\frac{dR}{d\Omega}\right)_{tot} = 1 \quad . \tag{6.13}$$

By analogy with the treatment of impurity scattering (Chap. 5) the total flux scattered by the edges can be expressed as:

$$\int d\Omega \left(\frac{dR}{d\Omega}\right)_{edge} = \Theta al \quad , \tag{6.14}$$

with Θ being the probability that a step edge occurs at a given unit cell, a the unit cell length along the step, and l the effective width of the step.

This expression is consistent with the large terrace assumption since it neglects step overlapping and crossing. Then in order to satisfy the unitarity condition one must write:

$$\left(\frac{dR}{d\Omega}\right)_{terr} = (1 - \Theta al) \left(\frac{dR}{d\Omega}\right)_{terr}^{0} \quad . \tag{6.15}$$

The last term is the terrace scattering with $l = 0$ which obviously fulfills

$$\int d\Omega \left(\frac{dR}{d\Omega}\right)_{terr}^{0} = 1 \quad . \tag{6.16}$$

Thus for in-phase scattering on a non–corrugated surface the specular intensity is simply:

$$I_{00} = 1 - \Theta al \quad . \tag{6.17}$$

We are now in a position to analyse the information that can be obtained from helium scattering from a randomly stepped surface.

1. The measurement of the intensity of coherent peaks for in-phase condition allows one to determine the product Θal.
2. The dependence of the intensity of coherent peaks on phase shift φ gives the number of occupied terrace levels. A cosine dependence indicates two levels while a $\delta_{2n\pi}$-like dependence is evidence for a large number of levels.
3. For a two–level situation the ratio of anti-phase to in-phase coherent peak intensity gives the level occupancy. This ratio is zero for an equal occupancy.

4. From the behaviour of the incoherent scattering under in-phase conditions one can obtain the shape of the surface corrugation in the vicinity of the step. Its integration over the whole space can provide an estimate of the effective step width l.

5. The information about the step statistics can be derived from the behaviour of the incoherent scattering for anti-phase conditions provided that:
   (a) the edge scattering can be subtracted
   (b) the structure factor can be separated from the form factor due to the terrace corrugation

For large terraces condition 5a is easily fulfilled since edge scattering is then negligible with respect to terrace scattering in anti-phase conditions. Condition 5b is easy to achieve when the terrace is flat; it can be more difficult in a corrugated case.

This full analysis has never been completely carried out in practice. Only partial results have been obtained. For instance combining the measurement of anti-phase incoherent scattering with the measurement of in-phase coherent peak intensity *Verheij* et al. [6.8] were able to determine the step density Θ and the step width l. For Pt(111) they found $l = 12$Å. This figure confirms the high sensitivity of helium scattering to steps. Moreover the analysis is often complicated by additional effects which have not been taken into account in the previous discussion:

– Except at very low temperature inelastic scattering is present which decreases the intensities of coherent elastic peaks (Debye-Waller factor) and adds inelastic diffuse scattering to the incoherent elastic scattering. Thus a Debye-Waller correction is needed in order to analyse correctly the intensities of coherent elastic peaks. An energy–resolved experiment (time-of-flight) is highly desirable for the determination of the incoherent elastic scattering when the temperature is not very low.

– The finite angular resolution of the experiment makes it sometimes quite difficult to separate the coherent from the incoherent scattering. Especially when a large number of surface levels are present (rough surfaces), the structure factor is generally strongly peaked at Bragg positions. Then, within a finite angular resolution, it is almost impossible to decide whether a peak is a δ-function-shaped coherent peak or a narrow incoherent peak or a mixture of both. For this reason it is very difficult to characterize a roughening transition (see below) by the vanishing of a coherent peak intensity.

A good example of the application of helium scattering to the characterization of steps on surfaces is the study of layer-by-layer crystal growth. The first expriment was carried out by *Gomez* et al. [6.21] for the nucleation and growth of Cu(100) from its vapour. Figure 6.6 reproduces typical results [6.22]. The specular peak intensity is plotted versus coverage for various surface temperatures and for in-phase and anti-phase conditions. The quasi-cosine oscillations observed in anti-phase are typical of the nucleation of monolayer islands leading to layer-by-layer growth. The authors deduced a step width $l = 13$Å and a step density which decreases with increasing temperature as a result of an increasing diffusion rate which reduces the number of nucleation centers and thus increases the island size.

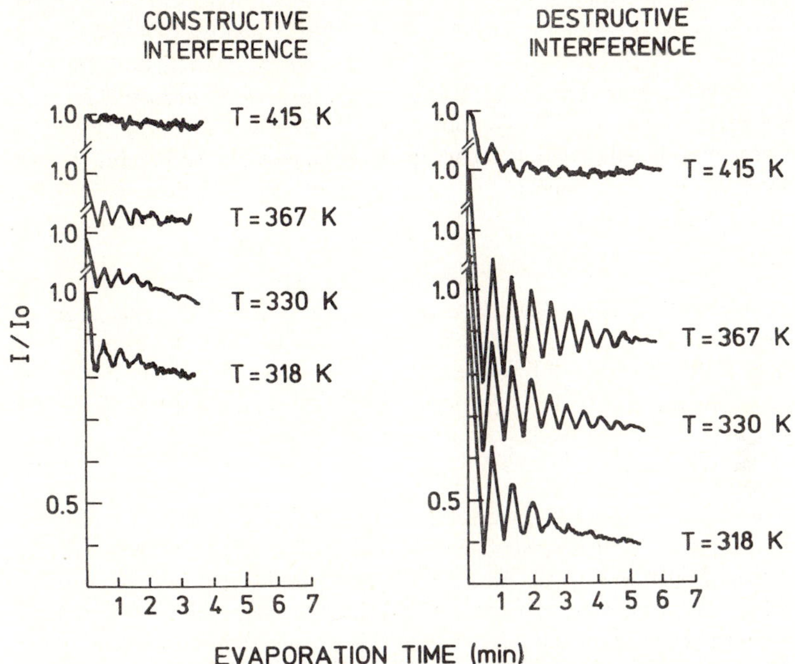

Fig.6.6. Vapour deposition of Cu on Cu(100). Specular intensity versus time of deposition. The maxima correspond to completing of a monolayer. From [6.22]

6.5 Scattering from Vicinal Surfaces

6.5.1 Perfectly Ordered Vicinal Surfaces

On a perfect vicinal surface the steps are parallel and equidistant. The terraces have a constant width L. If again $L \gg l$ the preceding analysis remains valid but now the edge scattering splits into δ-functions located at the Bragg positions corresponding to the reciprocal superlattice of the step array. There is, of course, an infinite number of terrace levels. So the terrace scattering gives, in-phase only, a strong coherent peak whose position corresponds to one of the step superlattice peaks; it is precisely the rainbow angle of this vicinal staircase corrugation. There is no other contribution from the terrace since obviously no incoherent component can appear on this perfectly periodic surface.

However, most vicinal surfaces studied to date are such that $L \approx l$. Then the separation into terrace and edge scattering is no longer justified. The surface is now described by a periodic corrugated potential and exact numerical methods have been developed in order to calculate the diffraction peak intensities. The case of copper vicinal surfaces Cu(113), (115), (117) has been carefully studied by *Gorse* et al. [6.23]. The structure of Cu(115) is shown in Fig.6.7. The diffraction pattern has been recorded for a large variety of incidence energies and incidence angles. The diffraction peak intensities were corrected for the Debye-Waller factor by a zero-T extrapolation. These data are used to fit the parameters of a model potential: the corrugated Morse potential (CMP). It is defined as follows:

$$V(\boldsymbol{R}, z) \;=\; V_0(z) \;+\; \sum_{G} V_G(\boldsymbol{R}, z)\exp(i\boldsymbol{G} \cdot \boldsymbol{R}) \quad , \tag{6.18}$$

with

$$V_0(z) \;=\; D[\exp(-2\chi z) \;-\; 2\exp(-\chi z)] \tag{6.19}$$

and

$$V_G(\boldsymbol{R}, z) \;=\; Dv_G\exp(-2\chi z) \quad . \tag{6.20}$$

$\boldsymbol{G}$ is the vector of the step array reciprocal lattice and z is now the coordinate along the normal to the vicinal surface. The well depth D and the laterally averaged potential stiffness χ are fitted with the bound states derived from selective adsorption data. They do not depend upon surface orientation within experimental accuracy. The amplitude v_G of the Fourier components is fitted to the diffraction peak intensities. It can be concluded from this analysis that the corrugation of the equipotential surface is almost sine-shaped for Cu(113), a small asymmetry arises for Cu(115) but the staircase shape begins only to appear for Cu(117). This empirical potential $V(\boldsymbol{R}, z)$ can be related to the surface electron density $N(\boldsymbol{R}, z)$ through the effective-medium calculation of *Esbjerg* and *Nørskov* [6.23]:

$$V(\boldsymbol{R}, z) \;=\; \beta_{\mathrm{EN}} N(\boldsymbol{R}, z) \quad , \tag{6.21}$$

with $\beta_{\mathrm{EN}} = 176$ eV $\cdot$ Bohr^3.

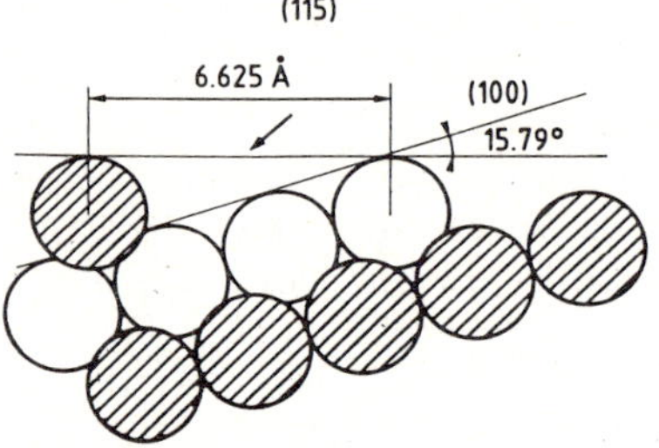

Fig.6.7. Structure of Cu(115)

The electron density $N(\boldsymbol{R}, z)$ has been calculated assuming no relaxation of surface atoms from their crystal positions and using a superposition of atomic densities. A good agreement is found using an effective proportionality constant $\beta_{\text{eff}} = 600$ eV $\cdot$ Bohr^3 which is somewhat larger than the theoretical value. This discrepancy has been removed by *Annett* [6.24] who included into the calulation hybridization effects and then obtained a very good agreement with the experimental data.

6.5.2 Thermally Disordered Vicinal Surfaces

One of the most important successes of the helium diffraction technique is to have clearly demonstrated the occurrence of a roughening transition on vicinal surfaces. Let us recall the physical basis of this transition.

(a) The Statistics of Steps on a Surface We consider first a single isolated straight step on a simple cubic lattice with a lattice constant a (Fig.6.8). When the temperature is raised, thermally excited structural defects can appear. We assume that the energetic cost of a defect is proportional to the number of broken nearest neighbour bonds. A kink on the step costs much less energy (one broken bond) than a defect on the terrace, i.e. an adatom or advacancy (four broken bonds). Thus one can expect a large temperature range where kinks on the step are the only defects created. Let x and y be the surface coordinates parallel and perpendicular to the step respectively. A kink may occur at any lattice position with a probability σ. Positive and negative kinks have the same probability. The step displacement in the y–direction is $u(x)$. The calculation of $u(x)$ is a simple one-dimensional random walk problem from which it is easy to obtain the correlation function of step displacements:

$$\langle\langle [u(x) - u(0)]^2\rangle\rangle = x\sigma \quad . \tag{6.22}$$

Thus the correlation function has a linear divergence for infinite distances. If the kink formation energy is W_0, the probability σ is easily obtained from the canonical distribution:

$$\sigma = [1 + (1/2)\exp(W_0/kT)]^{-1} \quad . \tag{6.23}$$

Thus, if one calls a step "smooth" or "rough" according to the behaviour of its correlation function for large distances (bounded or unbounded), we deduce from the above results that an isolated step is always rough if the temperature is not strictly zero.

When many steps are present their interaction must be taken into account. It is essential to emphasize that the well known stability (or at least metastability) of vicinal surfaces could not be understood in the absence of repulsive interaction between steps. This interaction may have three origins:

— statistical, coming from the topological impossibility of step crossings;
— elastic, coming from the overlapping of the lattice relaxation fields present around every step;

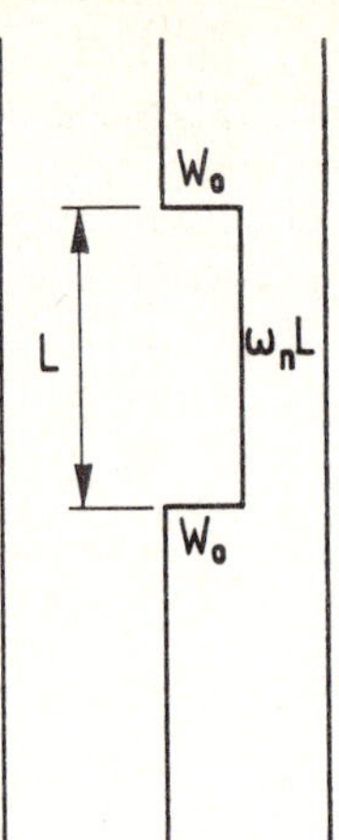

Fig.6.8. Elementary excitation on a vicinal surface

— electrostatic, coming from the interaction of electric dipoles present on
  every step.

All these interactions are known to be repulsive and to decrease as r^{-2}. There is
thus a coupling between steps which makes the statistical problem much more
difficult. We are now facing a two- rather than a one-dimensional problem
and from this dimensionality change one can expect that the behaviour of the
correlation function will change from a linear to a logarithmic one. The basis for
a solution has been described by *Villain* et al. [6.16]. The elementary excitation
on a step is a detour of length L limited by two kinks. The energy cost of this
detour is $2\,W_0 + Lw_1$, where w_1 is the energy per unit length spent when a step
is displaced between its two neighbours. The statistical treatment of this model
is carried out by a renormalization approach. The parameters W_0 and w_1 are
involved in two new temperature–dependent renormalized surface tensions, η_x
and η_y, which represent the excess surface tension appearing when the coupled
step array is deformed in the x and y directions respectively. Then it can be
shown that the model undergoes a roughening transition at $T = T_\mathrm{R}$ such that

— for $T < T_\mathrm{R}$ the correlation function of step displacements remains finite at
  large distances. Thus the surface is smooth.
— For $T > T_\mathrm{R}$ the correlation function diverges logarithmically at large distances and takes the form

$$\langle\langle [u(x) - u(0)]^2 \rangle\rangle \;=\; \text{const.}$$

$$+ \; A(\mathrm{T}) \ln[(\eta_x/\eta_y)^{1/2} x^2 \; + \; (\eta_y/\eta_x)^{1/2} y^2]^{1/2} \quad, \tag{6.24}$$

with

$$A(T) = \frac{T}{\pi\sqrt{\eta_x \eta_y}} \;. \tag{6.25}$$

The surface is rough. We note that the distance appearing in the logarithm is renormalized by the energetic anisotropy η_x/η_y.

— For $T = T_{\mathrm{R}}$ one obtains the universal relation

$$A(T_{\mathrm{R}}) \; = \; \frac{2}{\pi^2} \; . \tag{6.26}$$

(b) Scattering from a Disordered Vicinal Surface As a consequence of step interaction, the energy is minimized if the elementary detours form domains as shown in Fig.6.9. Lines of kinks appears which may be considered as forming secondary steps delimiting shifted domains the structure of which is the same as that of the original vicinal surface. Thus the formalism developed previously can be applied to this structure provided that one makes the correspondence:

$$\begin{array}{lcl} \text{terrace} & \rightarrow & \text{vicinal domain,} \\ \text{step edge} & \rightarrow & \text{secondary step.} \end{array}$$

As discussed above, the terrace scattering will be the product of a form factor and a strucure factor. The form factor is completely determined by the corrugation of the vicinal structure. The structure factor is the Fourier transform of a phase factor which depends only upon the statistics of step displacements.

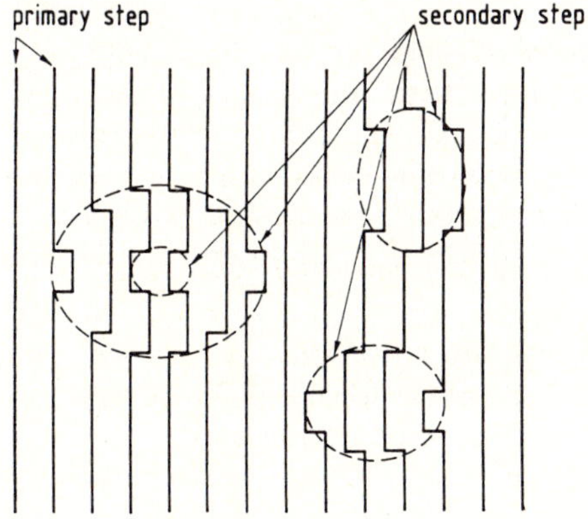

Fig.6.9. Kink arrangement into secondary steps which delimit shifted domains

Above T_{R} the correlation function is logarithmic and thus the structure factor is made of peaks centered on Bragg positions defined by the reciprocal lattice of the vicinal stucture. They decay according to an inverse power law:

$$
I(\boldsymbol{Q}) \approx \left[\left(\frac{\eta_x}{\eta_y}\right)^{1/2}\right) (a_x q_x + 2\pi p_x)^2
$$
$$
+ \left(\frac{\eta_y}{\eta_x}\right)^{1/2} (a_y q_y + 2\pi p_y)^2 \bigr]^{-1 + \frac{\tau}{2}} , \tag{6.27}
$$

where a_x is the lattice constant along the step direction, a_y the lattice constant of the vicinal surface ($a_y = l$ distance between steps), q_x, q_y the momentum exchange expressed in the coordinates of the vicinal terrace, and p_x, p_y integers indexing the diffraction peak. The parameter τ in the exponent is given by

$$
\tau = A(T)W(\varphi) \tag{6.28}
$$

and $W(\varphi)$ has been defined in Sect. 6.3.2.

For in-phase conditions $\tau = 0$. Thus the scattering reduces as discussed above to δ-functions. For anti-phase conditions and $T = T_R$ one gets $\tau \approx 1$. Below T_R the structure factor contains a coherent δ-function contribution superimposed on an incoherent contribution which has a more or less power law decay. As stated previously, the finite instrumental resolution prevents one from characterizing T_R by the vanishing of the coherent term. T_R is better fixed by the observation of the power law decay of the peak shape with the condition $\tau(T_R) \approx 1$. From the measurement of the peak decay and of its anisotropy one obtains the values of η_x, η_y, from which it is possible to deduce the energetic parameters W_0, w_1 [6.26].

(c) Experimental Evidence for the Roughening Transition on Vicinal Surfaces
The first observation of a power law decay of the diffraction peaks on a vicinal surface was made by *den Nijs* et al. [6.17] for Ni(115). They measured the exponent parameter τ versus temperature from which they deduced $T_R = 450K$. As they did not measure the peak anisotropy, a quantitative comparison with the statistical models is not possible for their data. This measurement was achieved by *Fabre* et al. [6.26, 27] on Cu(115). This surface is found to follow Villain's model quite well. A systematic study of the behaviour of vicinal copper surfaces has been carried out in the same laboratory [6.28, 29]. The main results are summarized in Table 6.1.

The following trends are observed:

- At a fixed temperature the roughness increases with the distance between steps as a result of the weakening of the step–step interaction.
- Changing the terrace structure from (100) to (111) does not significantly affect the roughness.
- Steps parallel to the loosely packed direction [100] (Cu(310)) are very much rougher than those parallel to the close-packed direction [110] (Cu(11n)).
- Freezing of the roughness due to low surface mobility always occurs below room temperature.
- The behaviour of the large–terrace surface Cu(1,1,11) deviates from the predictions of the Villain model. The reason is not yet fully understood.

Table 6.1. Roughening temperatures and step energies for vicinal copper surfaces

Surface	Roughening temperature T_R[K]	Energy of a step W_o[K]	Energy between steps w_l[K]
Cu(113)	720	800	560
Cu(115)	380	850	120
Cu(1,1,11)	<300	?	$\approx$5
Cu(331)	650	$\approx$1000	$\approx$500
Cu(310)	<300	$\approx$400	$\approx$300

Ni(113) has also been studied by *Conrad* et al. [6.30]. They have found no energetic anisotropy. This result is hard to understand and it disagrees with a recent X-ray experiment [6.31].

6.6 Conclusion

The scattering of helium atoms from stepped surfaces is now quite well understood. Appropriate models allow quantitative predictions in two cases:

- random steps delimiting large terraces and
- vicinal surfaces with domains limited by kink lines.

Helium scattering is thus a good tool to study step distributions on surfaces. This as been clearly demonstrated for the roughening transition on vicinal surfaces and for the layer-by-layer crystal growth.

Acknowledgments

I am very grateful to all my colleagues G. Armand, L. Barbier, H.-J. Ernst, F. Fabre, B. Salanon and J. Sprösser for the critical reading of the manuscript.

References

6.1   J. Lapujoulade, Y. Lejay: J. de Phys. **38**, L303 (1977)
6.2   Y. Lejay, J. Lapujoulade:*Proc. 7th Int. Vacuum Congr. and 3rd Int. Conf. on Solid Surfaces* ed. by R. Dobrzemski, F. Rüdenauer, F.P. Viebock, A. Breth (Vienna 1977) p. 1373
6.3   J. Lapujoulade: Surf. Sci. **108**, 526 (1981)
6.4   R. Spadacini, C.E. Tommei: Surf. Sci. **133**, 216 (1983)
6.5   A.C. Levi: Surf. Sci. **137**, 305 (1984)
6.6   G. Blatter, T.M. Rice: Phys. Rev. B **27**, 7050 (1983)
6.7   G. Armand, B. Salanon: Surf. Sci. **217**, 317 (1989)
6.8   L.K. Verheij, B. Poelsema, G. Comsa: Surf. Sci. **162**, 858 (1985)
6.9   G. Blatter: Ann. Physics (N.Y.) **162**, 100 (1985)
6.10 A. Lahee, J.R. Manson, J.P. Toennies, Ch. Wöll: Phys. Rev. Lett. **57**, 471 (1986)
6.11 B.J. Hinch: Phys. Rev. B **31**, 2551 (1985)
6.12 J. Lapujoulade, Y. Lejay: Surf. Sci. **69**, 399 (1977)
6.13 J. Lapujoulade, Y. Lejay, N. Papanicolaou: Surf. Sci. **90**, 133 (1979)
6.14 D. Gorse, B. Salanon, F. Fabre, A. Kara, J. Perreau, G. Armand, J. Lapujoulade: Surf. Sci. **147**, 611 (1984)
6.15 J. Lapujoulade, J. Perreau, A. Kara: Surf. Sci. **129**, 59 (1983)
6.16 J. Villain, D.R. Grempel, J. Lapujoulade: J. Phys. F **15**, 809 (1985)
6.17 M. den Nijs, E.K. Riedel, E.H. Conrad, T. Engel: Phys. Rev. Lett. **55**, 1689 (1985) and Erratum **57**, 1279 (1986)
6.18 J. Lapujoulade: Surf. Sci. **178**, 406 (1986)
6.19 A.C. Levi, R. Spadacini, G.E. Tommei: Surf. Sci. **108**, 181 (1981)
6.20 G. Armand, B. Salanon: Surf. Sci. **217**, 341 (1989)
6.21 L. J. Gòmez, S. Bourgeal, J. Ibàñez, M. Salmeròn: Phys. Rev. B **31**, 2551 (1985)
6.22 J.J. de Miguel, A. Sànchez, A. Cebollada, J.M. Gallego, J. Ferròn, S. Ferrer: Surf. Sci. **189/190**, 1062 (1987)
6.23 N. Esbjerg, J.K. Nørskov: Phys. Rev. Lett. **45**, 807 (1980)
6.24 J.F. Annett: Phys. Rev. B **35**, 7826 (1987)
6.25 B. Salanon, F. Fabre, J. Lapujoulade, W. Selke: Phys. Rev. B **38**, 7385 (1988)
6.26 F. Fabre, D. Gorse, J. Lapujoulade, B. Salanon: Europhys. Lett. **3**, 737 (1987)
6.27 F. Fabre, D. Gorse, B. Salanon, J. Lapujoulade: J. de Phys. **48**, 1017 (1987)
6.28 F. Fabre, B. Salanon, J. Lapujoulade: Solid. State Commun. **64**, 1125 (1987)
6.29 B. Loisel: Thesis, Université Paris VII
6.30 E.H. Conrad, L.R. Allen, D.L. Blanchard, T. Engel: Surf. Sci. **187**, 265 (1987)
6.31 I.K. Robinson, E.H. Conrad, D.S. Reed: J. de Phys. **51**, 103 (1990)

7. Resonances in Helium Scattering from Surfaces

H. Hoinkes and H. Wilsch

With the advent of increasingly better methods for the production of UHV, atomic and molecular beams and crystal surfaces including their characterization, studies of gas–solid interactions developed as an important branch of surface physics. Measured in the very first atomic scattering and diffraction experiments and correctly recognized and interpreted from the very beginning, selective *adsorption resonances* now represent a special highlight in atom–surface scattering investigations allowing for a precise determination of *bound state energies* in the physical atom–surface interaction potential and thus providing a unique test for any method of calculating such *interaction potentials*. The theory of the scattering and diffraction of atoms (molecules) from solid (crystalline) surfaces also greatly benefitted from the stringent task of correctly describing the many detailed and characteristic structures caused by selective adsorption resonances. The following sections will give a review of the historical development of selective adsorption studies, the discovery of *diffraction-*, *rotation-* and *phonon-mediated resonances* together with some of their theoretical implications (interaction potentials are discussed in Chap. 3).

7.1 A Historical Overview

7.1.1 First Experimental Evidence of Diffraction and Resonances

Only a few years after the first diffraction experiments with electrons by *Davisson* and *Germer* in 1927 [7.1] the wave nature of atomic particles was verified too by diffraction of atomic beams from single crystal surfaces. This was done by *Estermann* et al. in the early 1930s; they scattered He and H_2 beams of thermal energy from (001)-surfaces of NaCl [7.2,3], LiF [7.3–6], and NaF [7.6] and observed angular scattering distributions which were in complete accordance with the de Broglie relation. Similar results were achieved by *Johnson* [7.7,8] by scattering of atomic hydrogen from LiF(001) and recording the distribution

113

Springer Series in Surface Sciences, Vol. 27   **Helium Atom Scattering from Surfaces**
Editor: E. Hulpke   © Springer-Verlag Berlin, Heidelberg 1992

of diffracted particles with a molybdenum oxide detecting plate. A first review
of all these early diffraction experiments was given by *Frisch* and *Stern* [7.9].

In the first diffraction experiment with a Maxwellian He beam from a
LiF(001)-surface [7.3] *Estermann* and *Stern* already observed a relatively complicated behaviour of the specularly reflected intensity. With constant beam
temperature $T_B = 290$ K and constant angle of incidence $\Theta_i = 78.5°$ (measured from the surface normal) they found a symmetrical minima structure of
the specular intensity when rotating the crystal around an axis parallel to the
surface normal (this rotation is measured by the azimuthal angle ϕ, with $\phi = 0$
corresponding to the $\langle 110 \rangle$ crystal axis in the plane of incidence). This very
first example of resonances in gas atom–crystal surface diffraction is shown in
Fig.7.1 .

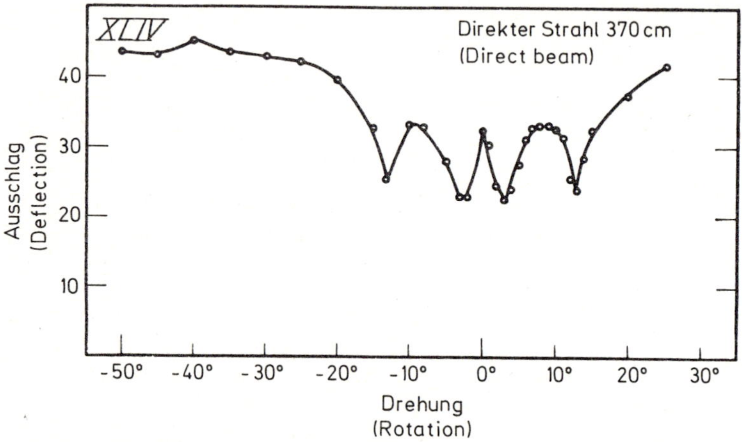

Fig.7.1. First observed resonance minima; specular intensity as a function of azimuthal angle
ϕ for He scattered from LiF(001). From [7.3]

In a subsequent short communication *Frisch* and *Stern* [7.5] showed the
dependence of this minima structure on the angle of incidence with surprisingly
steep and narrow intensity variation at flat angles of incidence ($\Theta_i \geq 85°$). Later
on they also investigated the dependence on beam velocity. This was done
firstly by changing the beam temperature and secondly in a double scattering
experiment, where they used the velocity dispersion in the (0,1) beam of the
first scattering process to get a monoenergetic beam, and then scattered at a
second crystal. In this detailed paper on "anomalies in specular reflection and
diffraction" [7.6] *Frisch* and *Stern* also report on minima observed on NaF and
with H_2 and not only in the specular beam but also in the angular distributions
of the (0, ±1)-diffraction beams. An example of these minima is given in Fig.7.2.

Frisch and Stern did not find a satisfactory interpretation of the observed
minima. They stated that a certain kind of adsorption has to be assumed, with
the probability of adsorption depending in a very definite manner on direction
and velocity of the incident molecules. *Frisch* [7.10] established correlations between the two components of the momentum of particles reflected or diffracted

114

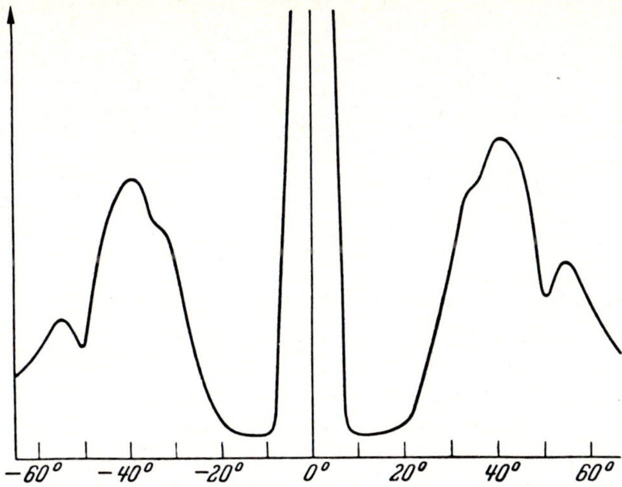

Fig.7.2. He diffracted from LiF(001) with resonance minima in the $(0, \pm 1)$-order diffracted beam distributions. From [7.9]

under minimum conditions and concluded: atoms incident on the crystal or leaving the crystal with certain velocity components relative to the surface lattice are adsorbed exceptionally easily; however, the mechanism behind this special process of adsorption is unknown.

7.1.2 The First Interpretation of the Minima

A mechanism leading to the observed losses in diffracted intensity at certain conditions of incidence was discovered by *Lennard-Jones* and *Devonshire* in 1936 [7.11,12]. They found out that the minima could be explained by transitions of the incident particles into states bound perpendicular to the surface and called this process "selective adsorption" (SA).

In more detail this special scattering process may be described in the following way. An incident atom is transferred to a bound surface state by diffraction; in this surface state the atom is bound normal to the surface in the gas–surface potential with a discrete binding energy $E_n < 0$, but moves parallel to the surface with an energy $E = E_i + |E_n|$ which exceeds the incident energy E_i just by the gained binding energy $|E_n|$. After some time the atom may leave this bound state again by a second diffraction step leading to an outgoing diffracted beam. Since energy conservation and diffraction conditions have to hold simultaneously this process occurs only at certain conditions of incidence. But if these conditions are fulfilled the temporary bound states may act as intermediate states causing a resonance behaviour in the elastic diffraction process. As a result strong changes in the diffracted beam intensities may occur under resonance conditions.

Lennard-Jones and Devonshire discussed this SA process within a first order perturbation theory. They approximated the interaction potential of a gas atom with a cubic crystal surface by the first terms of a Fourier series

$$V(x, y, z) = U_0(z) + 2U_1(z)(\cos\, ax + \cos\, ay) \quad , \tag{7.1}$$

with the coordinates x, y in the surface plane and z perpendicular to the surface. Neglecting $U_1(z)$ they get as zero order solutions free states, but also states bound normal to the surface with a total energy

$$W = \frac{\hbar^2}{2m}(k_x^2 + k_y^2) + E_n \quad , \tag{7.2}$$

where E_n is the negative energy of a state bound in the potential well $U_0(z)$ and $(\hbar^2 / 2m)(k_x^2 + k_y^2)$ is the kinetic energy of free motion parallel to the surface. The perturbation $2U_1(z)(\cos\, ax + \cos\, ay)$ then results in no first order corrections of the energy but in additional terms in the wavefunctions with the wave vector components k_x or k_y changed by a reciprocal lattice vector $\pm a$. These additional terms correspond to four first order diffracted beams $[(0, \pm 1)$ and $(\pm 1, 0)]$. The z-component k_z' of the wave vector of these diffracted beams is given by equations of the type:

$$(k_z')^2 + (k_x + a)^2 + k_y^2 = k_x^2 + k_y^2 + k_z^2 \quad . \tag{7.3}$$

If the value of $k_z'^2$ is negative there is no corresponding diffracted wave, but in the case where the following equation holds:

$$\frac{\hbar^2}{2m}(k_z')^2 = E_n \quad , \tag{7.4}$$

then Lennard-Jones and Devonshire find diffraction into the state bound normal to the surface with the negative binding energy E_n. Using (7.3, 4) they get relations for the components of momentum p_x, p_y, p_z which must hold for the incident particles to be transferred into a bound state.

The appropriate form for fitting the experimental results of *Frisch* and *Stern* [7.10] is the following relation:

$$p_z^2 - 2a\hbar p_y - a^2\hbar^2 = 2m\, E_n \tag{7.5}$$

and the resulting binding energies for ^{4}He–LiF(001) are -2.5 meV and -5.6 meV. The energies extracted for ^{4}He–NaF from less extensive measurements were: -3.5 meV and -8.4 meV, but these could not be confirmed later. Finally they assumed a Morse function for the potential well:

$$U_0(z) = D\exp[-2\kappa(z - b)] - 2\exp[-\kappa(z - b)] \tag{7.6}$$

and determined from the binding energies for He–LiF a well depth $D = 7.6$ meV and a range parameter $\kappa = 1.10 \cdot 10^8$ cm^{-1}. In a subsequent paper [7.13] *Lennard-Jones* and *Devonshire* used a Morse-potential with an exponentially repulsive form of the periodic term

$$\begin{aligned}
V(x, y, z) = \, &D\exp[-2\kappa(z - b)] - 2\exp[-\kappa(z - b)] \\
&+ 2\beta D \exp[-2\kappa(z - b)](\cos\, ax + \cos\, ay)
\end{aligned} \tag{7.7}$$

116

to obtain in first-order perturbation theory formulae for the diffracted beam intensities and also probabilities of selective adsorption and selective evaporation.

For the process of selective evaporation, in which a particle bound normal to the surface and moving parallel to it is diffracted into an unbound state and leaves the surface, they calculated the mean lifetime at the surface to be $\tau \approx 10^{-10}$ s corresponding to a mean free path of about $2 \cdot 10^{-5}$ cm. Viewing this process in connection with SA minima they argued within their first-order perturbation theory: on a perfect crystal surface the adsorbed particle would move parallel to the surface for about 10^{-10} s and would then be diffracted back into the specular beam. So under these conditions SA would not be observed. Since Frisch and Stern observed minima in the specular beam at the critical directions, one has to infer that the atoms selectively adsorbed were lost within their lifetime at the surface due to incoherent scattering from imperfections of the crystal, adsorbed atoms or due to loss of energy to the solid.

Today we know that first-order perturbation theory is inadequate for calculating diffracted beam intensities or intensity changes at bound state resonances, but the main ideas of Lennard-Jones and Devonshire concerning the possibilities of extracting interaction potential parameters from a comparison of measured and calculated scattering distributions are still valid. In particular, the proposed method for determining bound state energies solely from the geometrical conditions where resonant features are observed in diffracted beam intensities is a still frequently applied procedure to get information about the gas-surface interaction potential. After these first very promising results both in experiment and theory, there followed a rather long period with only little work in the field of gas–surface diffraction. Concerning bound state resonances the next experiments were not carried out till the end of the 1960s. *Crews* [7.14] scattered a thermal He beam from cleavage planes of LiF and confirmed the existence of one of the minima in the specular beam previously observed by Frisch and Stern. *O'Keefe* et al. [7.15] investigated the reflection and diffraction of ^{3}He, ^{4}He, H_2 and D_2 from LiF (001) and observed a few relative weakly resolved minima in the specular beam. They extracted from their measurements a new bound state energy level for ^{3}He and three levels for H_2 and D_2 which are questionable since the energies do not differ for the two isotopes.

Gas–surface diffraction and the investigation of resonances received fresh impetus in the early 1970s. Experimental progress came from the application of intense monoenergetic beams and modern quantum theory began to eliminate the shortcomings of the first-order distorted-wave Born approximation (FODWBA).

The new developments in the 1970s, in experiment as well as in theory, will be discussed in the next section.

7.2 Progress in Experiment and Theory in the 1970s

7.2.1 Experimental Advances: New Systems, Higher Precision

Improved vacuum conditions and the development of intense monoenergetic atomic beam sources in the 1960s were the basis of a new impetus in gas–surface scattering. In the early 1970s several groups started new gas atom–crystal surface diffraction experiments. In connection with bound state resonance investigations one should mention the follwing: at Ottawa, *Williams* and coworkers, at the University of Virginia, *Bledsoe* and *Fisher*, at the Pennsylvania State University, *Frankl* and coworkers, and at the University of Genova, *Boato* and coworkers. They all used He beams, whereas the authors' group at the University of Erlangen–Nürnberg initiated new experiments with atomic hydrogen.

The first series of resonance minima, going beyond the results of Estermann, Frisch and Stern, were published in 1971 by *Hoinkes* et al. for the system H–LiF(001) [7.16,17]. They observed minima in the specular intensity as well as in the (1,0)-diffraction distribution of the Maxwellian H beam. An example of these resonance features is shown in Fig.7.3.

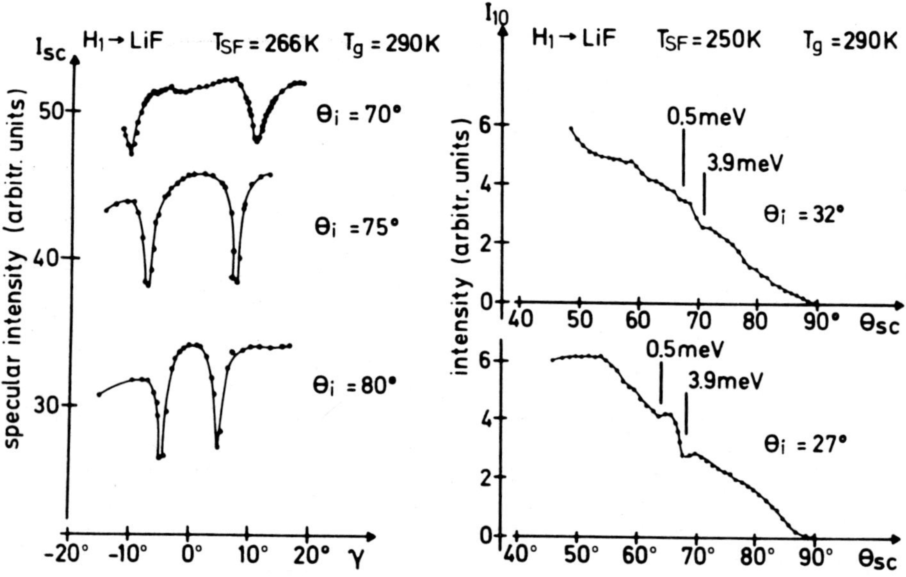

Fig.7.3. Detection of bound surface states for hydrogen atoms on (001)-LiF. Left side: variation of specularly reflected intensity with azimuthal angle ϕ for different incidence angles Θ_i. The dips symmetrically located with respect to $\phi = 0°$ correspond to transitions of atoms into a bound surface state with binding energy $E_n = 12.3$ meV. Less pronounced dips occur in the first order diffraction (right side), also corresponding to transitions of atoms into bound surface states, the angular position of the dips depending on the incidence angle Θ_i and the binding energy of the states. From [7.16]

These first results were complemented by similar observations with atomic deuterium [7.21] so that bound state energies could be determined for both H and D. As expected for isotopes with different mass but the same electron shell, these two different series of energy levels fitted well into one common potential well. This is clearly demonstrated in Fig.7.4 for a Morse potential with depth $D = 17.8$ meV and reciprocal range $\kappa = 1.04 \cdot 10^{10} \mathrm{m}^{-1}$.

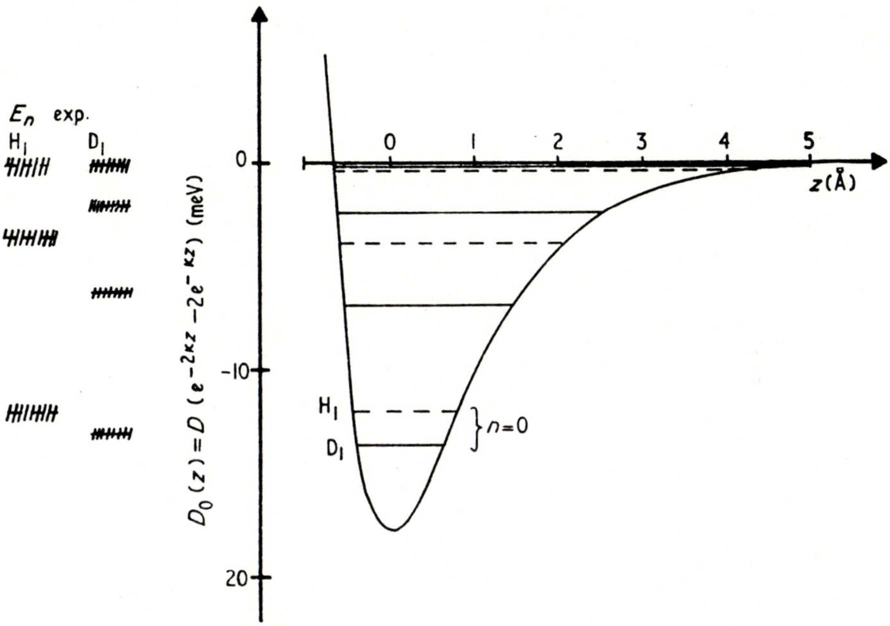

Fig.7.4. Left side: experimentally determined binding energies for the system H$_1$–LiF and D$_1$–LiF. Right side: Morse potential with depth $D = 17.8$ meV and reciprocal range $\kappa = 1.04\cdot10^{10}$ m^{-1} which reproduces all binding energies E_n found according to $E_n = [(2mD)^{1/2}/\kappa h - n - 1/2]^2 \kappa^2 \hbar^2/2m$. From [7.18]

The isotope effect used here for the first time is a good means of confirming the quantum numbers attributed to the determined enery levels E_n. A method which uses the isotope effect to check the labelling of the energy levels was introduced by *LeRoy* [7.88]. From the known asymptotic behaviour of the gas–surface interaction potential $U_0(z) \approx -C_3/\,z^3$ for $z \rightarrow \infty$ he deduced the following formula for the levels near the dissociation limit:

$$|\,E_n\,|^{1/6} = b\,C_3^{-1/3}(\eta_\mathrm{D} - \eta)\quad,\tag{7.8}$$

where $b = 0.2027$ (meV)$^{1/2}$ nm, C_3 is the constant of long range van der Waals attraction, and η, η_D are reduced quantum numbers defined by

$$\eta = (n + 1/2)\left(\frac{m_\mathrm{g}}{m_\mathrm{H}}\right)^{-1/2}\quad,\tag{7.9}$$

119

with η_D corresponding to an effective quantum number of the dissociation limit, which determines the maximum number of bound states in the potential well, and m_g/m_H being the mass ratio of the actual gas particle to the hydrogen atom. Equation (7.8) shows that, if the labelling is correct, a plot of the inverse sixth power of the binding energies E_n of the two isotopes versus η should result in one straight line; the slope and the intercept with the η-axis will give C_3 and η_D respectively. *LeRoy* applied this procedure first to our data for H, D on LiF and NaF [7.25] and also to data of *Meyers* and *Frankl* [7.26] for ^{3}He, ^{4}He on LiF. The plot of these He results is shown in Fig.7.5 .

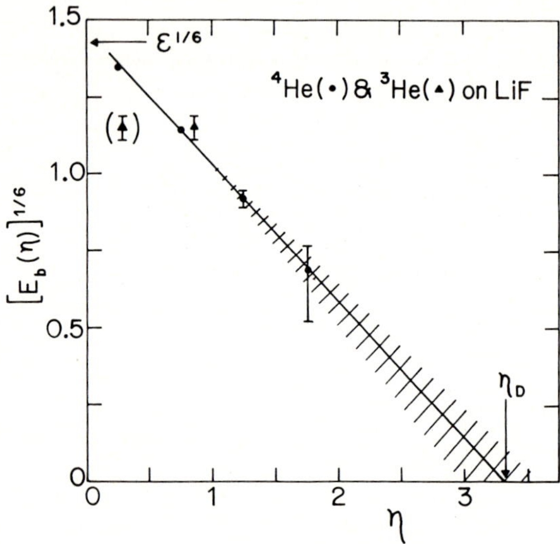

Fig.7.5. Experimental binding energies E_b for ^{3}He, ^{4}He on LiF(001) [7.26] plotted as function of mass-reduced quantum number η defined by (7.9) [7.88]

Very recently *Andersson* et al. [7.129] used this method to check the assignment of energy levels obtained from bound state resonance investigations with H_2 and D_2 on Cu(100). In this connection they also discussed the limitations in uniqueness of an assignment with this kind of analysis. A knowledge of correct level numbers is absolutely necessary for evaluation of the parameters characteristic of the corresponding potential well.

In the following years a series of detailed resonance investigations, now using velocity-selected beams of atomic hydrogen and deuterium, were performed by the Erlangen group. Diffraction from the following surfaces was investigated: LiF(001) and NaF(001) [7.23,25], KCl(001) with an ordered layer of water [7.30], NaCl(001) [7.100] and other alkali halides [7.53]. In all cases resonance minima appeared in the specular beam, but the overall resonance structure in scans of the specular intensity versus azimuthal angle became more and more complicated with increasing corrugation of the systems investigated. For instance with D on NaF they observed for the first time pronounced minima not only by first order transitions to bound states via reciprocal lattice vectors of type G_{10} but also by those of type G_{11} and in some cases (near degeneracy, see later) even via those of type G_{21} [7.25]. Finally on KCl [7.100] and still heavier alkali halides (KBr, RbCl) [7.53] the observed intensity distribution consisted of so many overlapping resonance features that a complete evaluation of contributing G-vectors and bound state levels became impossible.

Returning now to the main aspect of this chapter: "resonances in He atom–crystal surface scattering" we have to mention first the experiment by in the early 1970s *Bledsoe* [7.19] who observed minima on LiF but also some resonance structures on NaCl(001). Significant progress concerning resonances in the standard system He–LiF(001) was contributed by the experiments of Frankl and his group with nozzle beams. The improvement in resolution becomes evident in Fig.7.6 showing pronounced and sharp minima and an additional fine structure in the plots of specular intensity versus azimuthal angle.

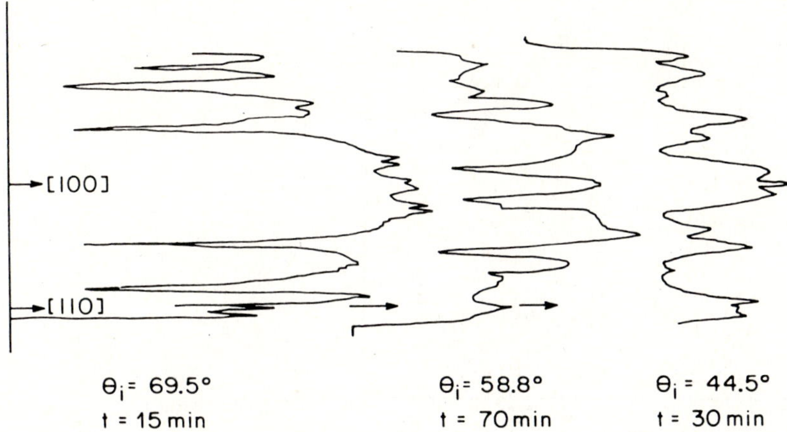

Fig.7.6. He scattered from vacuum-cleaved LiF(001); detector trace of specular beam intensity versus azimuthal angle ϕ at fixed angles of incidence. The age of the sample after cleaving is indicated on each. Maximum intensity is of the order of 30% of the incident beam. From [7.20]

In this paper *Houston* and *Frankl* [7.20] also demonstrated how a geometrical representation of the resonance condition (7.10,11) in K-space may be used to discover or check the indexing of the observed resonance features with the corresponding binding energy E_n and reciprocal lattice vector G_{res}. In generalized form the resonance condition reads:

$$(K_i + G_{\mathrm{res}})^2 = \frac{2m_g}{\hbar^2}(E_i + \mid E_n \mid) \tag{7.10}$$

which obviously shows that for a certain resonance, which is observed by varying the azimuthal angle ϕ_i and the angle of incidence Θ_i but keeping constant the incident energy E_i, the surface component of the incident wave vector K_i must always end on a circle with radius

121

$$K_{\text{res}} = \left[\frac{2m_{\text{g}}}{\hbar^2}(E_{\text{i}} + \mid E_n \mid)\right]^{1/2} \tag{7.11}$$

which is centered at $-G_{\text{res}}$. An example of this representation is given in Fig.7.7. It shows a plot of very recent results obtained by the Penn. State group for He on MgO(001) [7.131]. The circles drawn here belong to transitions by way of different G_{res} vectors which all lead to the same energy level E_0.

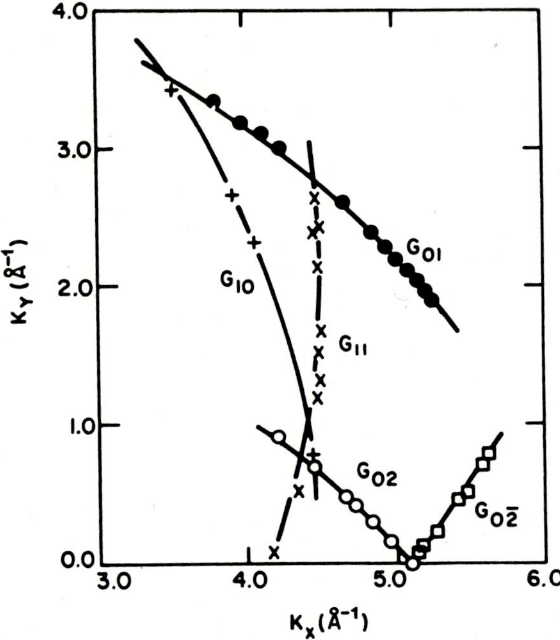

Fig.7.7. Selective adsorption loci in the K-plane for ^{4}He–MgO(001). Lines are circles described by (7.10) with $n = 0$, and centered at G_{mn}. From [7.131]

After these first high resolution results on LiF *Frankl* et al. also reported highly structured resonance behaviour of the specular beam for NaF(001) and NaCl(001) [7.22]. In an experiment with still improved resolution [7.26] they found four bound states for ^{4}He–LiF and observed that in the main minima via $G_{\text{res}} = (0,1)$ and $(0,\bar{1})$ as much as 85% of the specular intensity was removed. Less intense resonance features could be attributed to higher order G vectors.

Another important contribution to the understanding of details of the resonance process was the first observation of intensity maxima under resonance conditions in some higher order diffracted beams by *Wood* et al. [7.24]. These results showed that the minima found in the specular intensity until then need not necessarily be produced by enhanced incoherent or inelastic scattering in the resonance, but can also be caused by a purely elastic effect. At resonance the contribution of the additional intermediate bound state channel may induce strong redistributions in the intensity of the different observed diffracted beams. The Genova group concentrated their first activities on precise measurements of diffracted beam intensities, again first for the standard system He–LiF(001) [7.27]. In runs of specular and $(-1, -1)$-beam intensity versus angle of incidence they observed resonance minima too which could be attributed to transitions via $G_{\mathrm{res}} = (0, 1)$ and $(1,0)$ to energy levels already known.

The new experimental facts etablish in the first half of the 1970s, for instance, relatively strong higher order diffracted beams, bound state resonance transitions via higher order G-vectors, and maxima and minima observed at resonance in different diffracted beams, demonstrated more and more clearly that a theory based on first-order distorted-wave Born approximation (FODWBA) as introduced by Lennard-Jones and Devonshire was by no means suited to describe gas–surface diffraction. Indeed, in about 1970, parallel to the progress in the experimental field, there began a development of improved theoretical methods to overcome previous approximations. This advance in theory will be described in the following section.

7.2.2 Development of the Theory Beyond FODWBA

Around 1970, new, quite general quantum mechanical treatments of elastic diffraction theory with the inclusion of bound state resonances were presented by *Tsuchida* [7.61], by *Cabrera, Celli, Goodman* and *Manson* (CCGM) [7.62] and somewhat later also by *Wolken* [7.65]. In this section on theory, we will give only a brief survey of the basic concepts following a review by *Goodman* [7.54].

The Schrödinger equation for the wavefunction $\psi(\mathbf{r})$ of a gas atom with mass m_{g} and position vector $\mathbf{r}$ is

$$[\frac{\delta^2}{\delta x^2} + \frac{\delta^2}{\delta y^2} + \frac{\delta^2}{\delta z^2} + k_i^2 - 2\frac{m_{\mathrm{g}}}{\hbar^2}V(\mathbf{r})]\psi(\mathbf{r}) = 0 \quad , \tag{7.12}$$

where $\mathbf{k}_i$ is the incident wave vector and $V(\mathbf{r})$ is the gas–surface interaction potential. We assume periodicity parallel to the surface plane, i.e. in the x- and y-directions, which implies that $V(\mathbf{r})$ and $\psi(\mathbf{r})$ may be expanded in the following Fourier series:

$$V(\mathbf{r}) = \sum v_{\mathbf{G}}(z) \exp(\mathrm{i}\mathbf{G} \cdot \mathbf{R}) \tag{7.13}$$

$$\psi(\mathbf{r}) = \sum \psi_{\mathbf{G}}(z) \exp\left[\mathrm{i}(\mathbf{K}_i + \mathbf{G}) \cdot \mathbf{R}\right] \quad , \tag{7.14}$$

where $\mathbf{G}$ are reciprocal lattice vectors of the surface array and vector components parallel to the surface plane are generally denoted by the corresponding capital letter, e.g. $\mathbf{K}$ for a wave vector and $\mathbf{R}$ for the position vector. Substituting (7.13) and (7.14) into (7.12) results in

$$\sum f_{\mathbf{G}}(z) \exp(\mathrm{i}\mathbf{G} \cdot \mathbf{R}) = 0 \quad , \tag{7.15}$$

where

$$f_G(z) = \frac{d^2\psi_G}{dz^2} + k_{Gz}^2\psi_G - \frac{2m_g}{\hbar^2}\sum v_{G-G'}\psi_{G'} \tag{7.16}$$

and

$$k_{Gz}^2 = k_i^2 - (K_i + G)^2 \quad . \tag{7.17}$$

Here k_{Gz} is the wave vector component normal to the surface when the atom is in the diffracted state G. If $k_{Gz}^2 < 0$ then k_{Gz} is imaginary and the state G is a closed (envanescent) channel because an atom in this state cannot reach $z = \infty$; if $k_{Gz}^2 > 0$ then k_{Gz} is real and the state G is an open channel because an atom in this diffracted beam can reach $z = \infty$. In the Fourier sum (7.15) equal to zero each factor f_G must vanish separately, so we get for the partial waves ψ_G an infinite number of coupled differential equations:

$$\frac{d^2\psi_G}{dz^2} + k_{Gz}^2\psi_G = 2\frac{m_g}{\hbar^2}\sum v_{G-G'}\psi_{G'} \quad . \tag{7.18}$$

All information about the diffraction process is contained in $\psi_G(z \to \infty)$. The general form of these asymptotic solutions of (7.18) can be expressed as follows: for open channels $(k_{Gz}^2 > 0)$

$$L^{1/2}\psi_G(z \to \infty) = \exp(-ik_z z)\delta(G, 0) + S_G \exp(ik_{Gz}z + i\xi_G) \tag{7.19}$$

and for closed channels $(k_{Gz}^2 < 0)$

$$L^{1/2}\psi_G(z \to \infty) = 0 \quad , \tag{7.20}$$

where $\delta(G, G')$ is the Kronecker delta function, L is a normalization length and ξ_G is a phase. The interesting quantity is the amplitude S_G which is related to the observable diffraction probability P_G into the beam of order G by:

$$P_G = \frac{k_{Gz}}{k_{0z}} \mid S_G \mid^2 \quad . \tag{7.21}$$

In purely elastic scattering, particle conservation requires the sum over all open diffracted channels to be unity:

$$\sum P_G = 1 \quad . \tag{7.22}$$

In this chapter on resonances in He–surface diffraction we are interested in solutions of the system of coupled differential equations (7.18) for a gas–surface interaction potential with bound states; especially solutions for those cases where the conditions of incidence fulfil the following resonance condition:

$$\frac{\hbar^2 k_i^2}{2m_g} - E_n^{\parallel}(K_i + G) = E_n < 0 \quad . \tag{7.23}$$

124

In this equation E_n is the binding energy of the atom in one of the states bound normal to the surface in the potential well $v_{00}(z)$ which represents the gas–surface interaction potential averaged parallel to the surface and $E_n^{\parallel}(\boldsymbol{K}_i + \boldsymbol{G})$ is the energy of an atom moving parallel to the surface in the periodic potential with the wave vector $\boldsymbol{K_G} = \boldsymbol{K} + \boldsymbol{G}$, when bound normal to the surface in the state with quantum number n. If (7.23) is fulfilled the incident beam is energetically in resonance with a closed channel in which the atom is bound normal to the surface and moves parallel to the surface with the corresponding higher energy. This resonant intermediate channel may cause drastic changes in the intensities of the observed diffracted beams.

The first attempts, around 1970, to solve the coupled equations (7.18) with inclusion of bound state resonances suffered from the limited computer capacities at that time, so they had to restrict the infinite number of channels in the system to just a few, or had to use other approximations. *Cabrera* et al [7.62] applied the so-called CCGM approximation (CCGMA) and found: "when exact resonance occurs, the intensity of the specular beam rises sharply, and that of each of the other beams falls". The minima generally observed in the specular beam were explained by enhanced inelastic scattering at resonance. CCGMA is essentially a partial summation of the Born series and gives unitary results; a discussion of this approximation can be found for instance in [7.54–56].

A first application of the CCGM theory was the calculations by *Goodman* [7.63] for the elastic scattering of ^{3}He and ^{4}He from LiF(001). In a more recent paper concerning CCGMA *Goodman* [7.64] pointed out that the statement of Ref. [7.62] cited above is misleading if not incorrect and he showed that with a suitable basis of reciprocal lattice vectors CCGMA may result in minima as well as in maxima in the intensity of the specular beam depending on details of coupling between the diffracted channels. *Tsuchida* [7.61] and somewhat later *Wolken* [7.65] were the first to apply the close-coupling formalism (CCF) to gas–surface diffraction with resonances. CCF consists, essentially, of a numerical integration of the set of coupled equations (7.18) using the boundary conditions (7.19, 20) together with

$$L^{1/2}\psi_G(z \to -\infty) = 0 \quad . \tag{7.24}$$

Details of the close-coupling method can be found for instance in [7.55, 65]. *Tsuchida* [7.61] included in his calculations a maximum of four open and no closed channels and got, out of resonance, at most qualitative agreement with the experimental intensities of *Estermann* et al. [7.4]. He discussed the resonance phenomenon with inclusion of inelastic effects, but did not calculate the behaviour of intensity when the conditions of incidence pass through a resonance.

More detailed calculations for He–LiF diffraction with resonances were performed by *Wolken* [7.65]. He used a Morse potential with first order, exponentially repulsive corrugation, the potential form used by Lennard-Jones and Devonshire (7.7). The potential parameters were those of *Goodman* [7.63]. Only the coupling parameter β was varied, but in most cases it was fixed to $\beta = 0.10$.

To investigate the resonance structure in the specular beam intensity he solved the system of coupled equations (7.18) for a finite basis set of 29 channels. As an example of his results Fig.7.8 shows the specular intensity as a function of incident energy, with the main bound state resonances indicated. In most cases he found intensity maxima in the specular beam. The maxima appeared still more pronounced if he restricted the basis set to the first five channels (specular and first order diffraction only). Though no quantitative comparison of calculated and experimental results was possible, the essential outcome of Wolken's work was that an adequate choice of the basis set of included channels (open and closed) is the prerequisite to get results which can be compared with experiments.

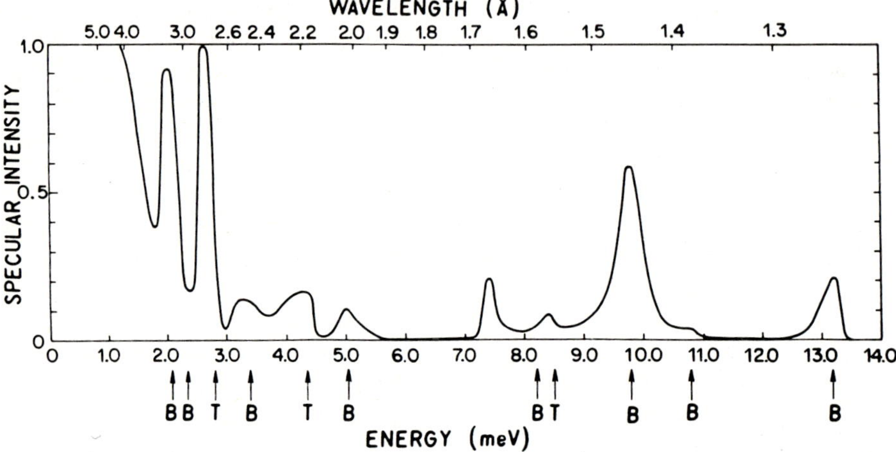

Fig.7.8. Calculated specular intensity vs incident energy: $\beta = 0.10$, $\Theta_i = 40°$, $\phi_i = 0°$, ^{4}He and using a 29 channel basis set. Some (but not all) of the bound state resonances (B) and thresholds (T) are indicated. From [7.65]

After this first series of calculations [7.61–63, 65] it seemed that elastic diffraction theory always results in resonance maxima in the specular beam. The minima observed in experiment were assumed to be caused by enhanced inelastic scattering under resonance conditions (the particle stays for a longer time near the surface). So it was an important advance when *Chow* and *Thompson* [7.66, 67] showed in 1976 that minima in the specular intensity could also be produced by purely elastic scattering calculations. To get this result it was necessary to include all "important" open and closed channels in the basis set chosen to truncate the infinite set of coupled equations (7.18). Important closed channels for instance are those that are strongly coupled by large Fourier components of the potential to the incident beam or to the resonant bound state. Because of particle conservation a resonance minimum in the specular beam has to be accompanied by a maximum in some diffracted beam, and, indeed, Chow

126

and Thompson found maxima in those diffracted beams which are strongly coupled to the resonant state.

As already mentioned, maxima of this kind were first observed by *Wood* et al. [7.24] for He–NaF(001). A more detailed experimental verification of the resonance behaviour of several diffracted beams by *Liva* and *Frankl* [7.29] again for He–NaF followed soon after the predictions of *Chow* and *Thompson* [7.66]. Later the maxima–minima correlation was confirmed for other systems too, in our group for instance for H–KCl(H_2O) [7.31] and D–LiF [7.46], but in the latter case much less intensity appeared in the maxima than disappeared in the minima. As discussed later, this is caused by relatively strong inelastic effects.

The clarifying results of Chow and Thompson led to further efforts to give a complete description of experimentally observed resonance structures, and indeed almost quantitative fits were obtained for the specular intensity of the standard system He–LiF(001). These high resolution measurements, compared with theory here, were accomplished by *Frankl* et al. [7.37] at a rather low surface temperature $T_S = 125K$ so that inelastic contributions are really small. The experimental scan of specular intensity versus azimuthal angle ϕ is compared to resonance structures calculated by two different methods. In Fig.7.9 results of *Harvie* and *Weare* [7.71] obtained with an expansion method and a corrugated wall model with attractive well are shown, and Fig.7.10 compares the experimental features with results of an exact non-perturbative calculation by *Garcia* et al. [7.75] again with the interaction potential approximated by a corrugated hard wall with an attractive well with flat bottom [7.76].

From calculations of the type cited above, general rules governing the intensity features near selective adsorption resonances were derived. *Wolfe* and *Weare* [7.72, 73] proposed rules which correlate features in the specular intensity with the strength of Fourier components of the interaction potential:

(1) A minimum will be observed when the resonant channel couples directly and strongly to the specular channel and to at least one other open channel.
(2) Mixed extrema (up and down wiggle) will be observed when the only open channel to which the resonant state couples strongly and directly is the specular.
(3) Maxima will be observed for resonant channels that couple only directly to the specular.

Examples of rule (1) and (3) can be found in Fig.7.9 pronounced minima are induced by the strong Fourier components v_{01}, whereas resonances with higher order G-vectors of type (1,1) and (0,2) result in maxima.

On the basis of the flat-bottom hard-wall potential [7.76] *Celli* et al.[7.77] developed a relatively simple formula for the lineshape of an isolated resonance (transition via $G = N$ to bound state level n) appearing in any diffracted beam of order F.

For the resonant dependence of diffraction probability P_F on the energy ε_N normal to the surface

$$P_F(\varepsilon_N) = \left| S(F,0) + \frac{iS(F,N)R_N S(N,0)}{(d\delta/d\varepsilon)(\varepsilon_N - \varepsilon_n + i\Gamma/2)} \right|^2 . \tag{7.25}$$

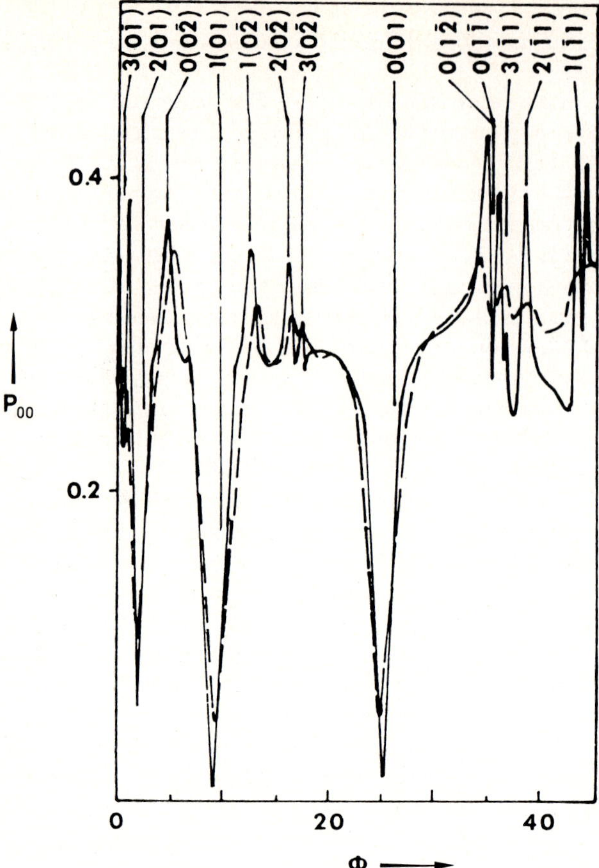

Fig.7.9. Resonance structure in the specular intensity P_{00} as a function of azimuthal angle ϕ for He diffracted from LiF(001): comparison of attractive-well, corrugated-wall theory (solid line) versus experiment (dashed line) for $\lambda = 1.03$ Å, $\Theta = 70°$. From [7.71]

The z-component p_G of the wave vector and the corresponding energy ε_G are measured from the bottom of the well; they are given by

$$\varepsilon_G = \frac{\hbar^2}{2m_{\mathrm{g}}}p_G^2 = E_i + D - \frac{\hbar^2}{2m_{\mathrm{g}}}(\boldsymbol{K}_i + \boldsymbol{G})^2 \quad , \tag{7.26}$$

$S(\boldsymbol{G}', \boldsymbol{G})$ is the amplitude for hard-wall diffraction from channel $\boldsymbol{G}$ to channel $\boldsymbol{G}'$, and R_G is the coefficient of reflection of a wave propagating in the z-direction against the long-range attraction. The phase shift of the perpendicular motion of the particle in the well can be generally expressed by $\delta(\varepsilon) = 2\pi n(\varepsilon)$ or expanded near resonance as

$$\delta(\varepsilon) = 2\pi n + \frac{d\delta}{d\varepsilon}(\varepsilon - \varepsilon_n), \quad n \text{ integer} > 0 \quad , \tag{7.27}$$

128

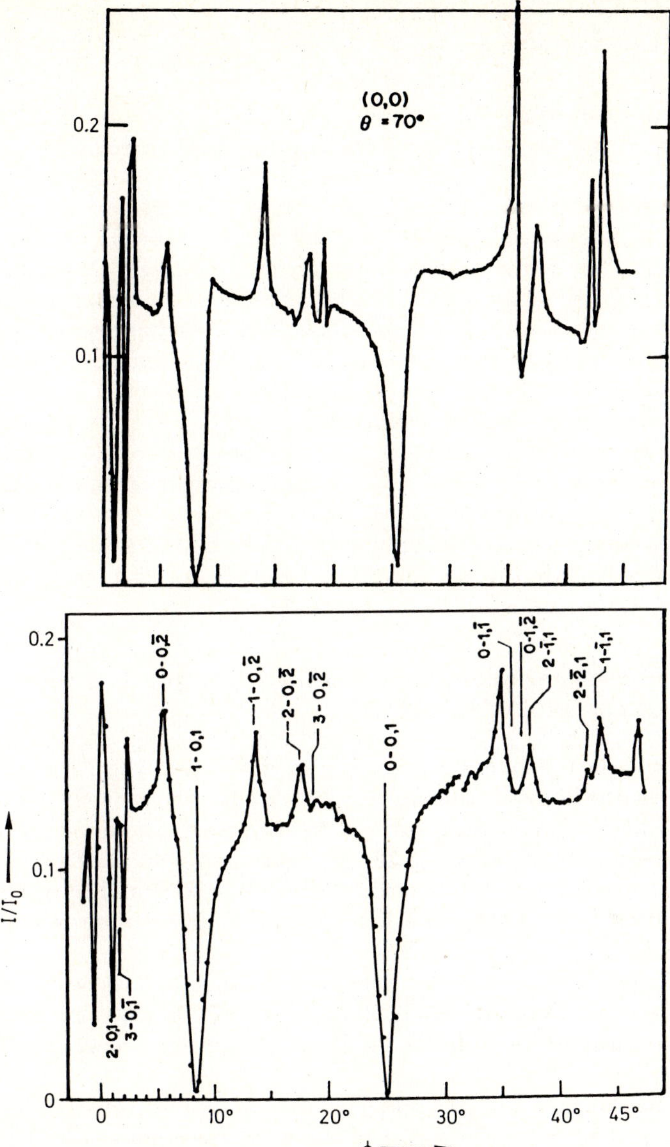

Fig.7.10. Theory (above) and experiments from [7.37] (below) for azimuthal (ϕ_i) dependence of the specular intensity at $\Theta_i = 70°$. The theoretical plot gives the elastic intensity multiplied by a factor of 0.43, accounting for the Debye-Waller factor and other possible corrections. Straight line segments join points determined at $\Delta\phi_i = 0.3°$ in the theory and $0.333°$ in the experiment. From [7.75]

with $\varepsilon_n = D + E_n$ the binding energy measured from the bottom of the well, and finally the width Γ of the resonance is given by

$$\Gamma = 2 \left(\frac{d\delta}{d\varepsilon}\right)^{-1} [1 - \mid S(\boldsymbol{N}, \boldsymbol{N}) \mid] \quad . \tag{7.28}$$

Sufficiently out of resonance, i.e. for $\mid \varepsilon - \varepsilon_n \mid > \Gamma$, only the term $S(\boldsymbol{F}, \boldsymbol{0})$ contributes and the scattering is that produced by the hard corrugated wall only.

The results of this formalism and also the rules of Wolfe and Weare where checked in detail in an extensive experimental study for ^{4}He–LiF(001) by *Wesner* and *Frankl* [7.91]. They found practically complete agreement. Deviations from these theoretical predictions have to be expected only in systems with high corrugation, i.e. surfaces with adsorbates for instance, or in systems with strong inelastic contributions. As will be discussed later, the resonance formula (7.25) of *Celli* et al. [7.77] was also used to introduce the effect of inelastic scattering on the resonance line shapes [7.46, 80].

Close coupling calculations as introduced in the 1970s are now frequently used to compare experiment and theory and to extract information on the gas–surface interaction potential. Especially for highly corrugated systems with many interacting resonances, this is the only way to determine the interaction potential from observed scattering distributions. Examples of such cases will be discussed in Sect. 7.3.

7.2.3 High Precision Experiments and Detailed Comparison with Theory up to the End of the 1970s

(a) New systems and exact bound state levels At the end of the 1970s the investigation of He-scattering with resonances was extended to new classes of surfaces. But the known alkali halide surfaces continued to be investigated with improved precision. *Derry* et al. [7.32, 35] presented results of precise measurements with both ^{3}He and ^{4}He on both LiF and NaF and used the isotope effect to obtain from determined bound state energies more reliable gas–surface potentials.

Selective adsorption from a material other than alkali halides was first observed by *Cantini* et al. with He [7.33, 39] and H_2 [7.33] scattered from NiO(001). They found three energy levels for He–NiO and five deeper energy levels for H_2–NiO.

Graphite(0001) has become a second standard surface for resonance studies. On this surface detailed bound state resonance investigations have been done with ^{3}He and ^{4}He but also with other gases such as H, D, H_2, D_2. The resulting well depths and periodic terms of the gas–surface interaction potential were of particular interest in comparison to results of thermodynamic measurements (for a detailed discussion see *Cole* et al. [7.57]) and also to test potential functions obtained from summation of pair potentials (see for instance [7.81, 82, 84]). The first resonances for ^{4}He–graphite were reported by *Boato* et al. [7.36]. In scans of the specular intensity versus azimuthal angle ϕ_i and versus angle of incidence Θ_i they observed minima corresponding to four bound state levels. More precise investigations of the energy levels of this system followed by *Boato* et al. [7.40, 41] and for the two isotopes ^{3}He and ^{4}He by *Derry* et al. [7.42]. Bound state energies on graphite were also determined for atomic hydrogen and deuterium by *Ghio* et al. [7.49] and for molecular hydrogen and deuterium by *Mattera* et al. [7.45].

In 1977 *Cantini* et al. [7.34] reported the first observation of resonances for Ne–LiF (a detailed analysis of selective adsorption of this system was given by *Semerad* et al. [7.125] ten years later; see Sect.7.3.1) and in 1980 *Iannotta* and *Valbusa* [7.51] observed resonances for H–NaCl (the bound state levels determined in this paper had to be corrected later, see [7.100, 119]).

Very important with regard to further investigations in the future was the first observation of resonances on a metal surface. *Lapujoulade* et al. investigated the diffraction of He on Cu(117) [7.43] and found resonance features in several diffracted beams. An example of these results is shown in Fig.7.11.

These first resonances on metal surfaces were the starting point for a series of investigations on metals in the 1980s which will be discussed in Sect.7.3.2.

A detailed compilation of all the bound state energy levels determined from resonance investigations of several gas–surface systems until 1980 can be found in [7.58] together with a review of model potentials of the attractive well $v_{00}(z)$ used in order to fit the experimental binding energies. Selective adsorption, however, yields information not only on the main term $v_{00}(z)$ of the interaction potential but observed splitting of degenerate bound states also allows one to evaluate the strength of the periodic terms $v_G(z)$; this will be discussed in the next section.

(b) Band structure effects in resonant scattering Though band structure effects for the motion of a He-atom parallel to the surface of an alkali halide crystal had already been discussed by *Lennard–Jones* and *Devonshire* [7.59], the free atom approximation was usually used in the analysis of bound state resonances. This means for an atom bound normal to the surface in state n the energy $E_n''(K)$ for the motion parallel to the surface in the periodic potential is assumed to be given by:

$$E_n''(\boldsymbol{K}) = \frac{\hbar^2 \boldsymbol{K}^2}{2m_{\mathrm{g}}} \quad , \qquad (7.29)$$

an assumption which generally worked well. But in 1976 *Chow* and *Thompson* [7.67] showed that this approximation may no longer hold for degenerate bound state resonances. If the degenerate states are coupled by strong periodic potential terms, split resonances can occur at the point of degeneracy. This splitting of bound state energies by the periodic potential terms is quite analogous to the energy gaps occurring at the Brillouin zone boundaries and may thus be classified as a band structure effect.

Soon after the predictions by Chow and Thompson, split specular minima corresponding to degenerate (1,0) and (1,1) resonances of He–NaF were observed by *Liva* et al. [7.28]. A resonance circle representation (according to (7.10)) of these results is shown in Fig.7.12.

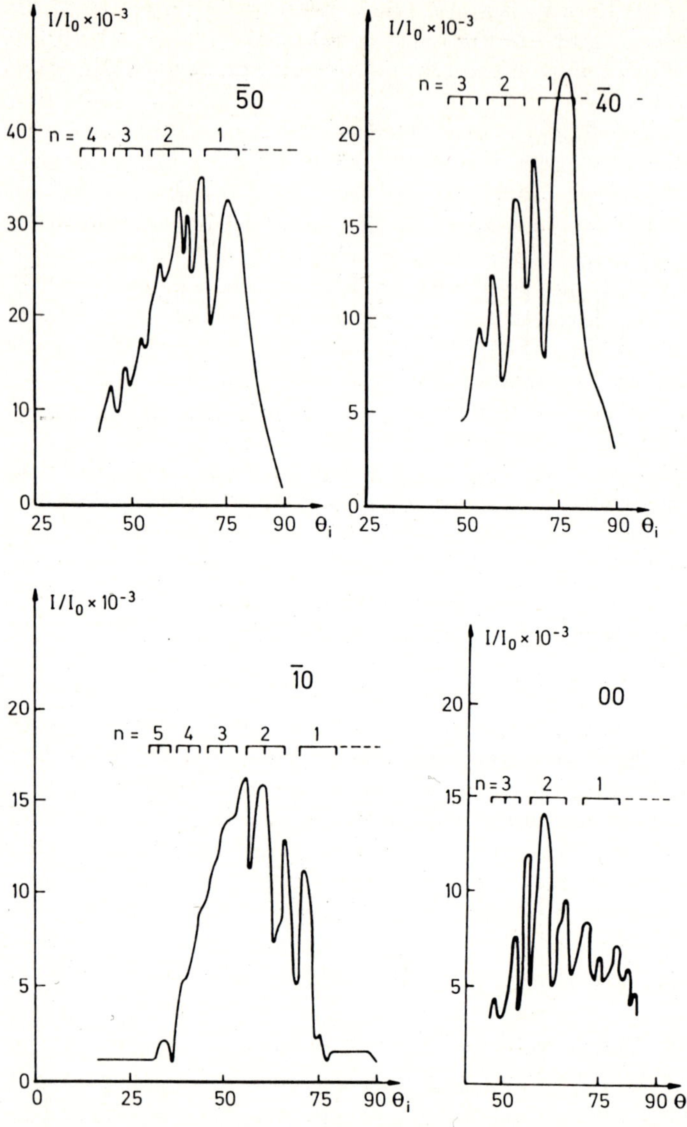

Fig.7.11. Resonance features in the scattering of He from Cu(117). Plot of the intensity ratios I/I_0 (corrected for instrumental broadening) of diffracted peaks versus incidence angles at incident wave vector $k_i = 6.3$ Å^{-1} and surface temperature $T_S = 70$K. From [7.43]

As mentioned by *Chow* [7.68], experimental evidence of band structure effects in resonant scattering can also be found in earlier work. For example *Finzel* et al. [7.25] have observed in the system D–NaF a (1,2) specular minimum close to a (0,1) minimum, both belonging to transitions to the deepest

132

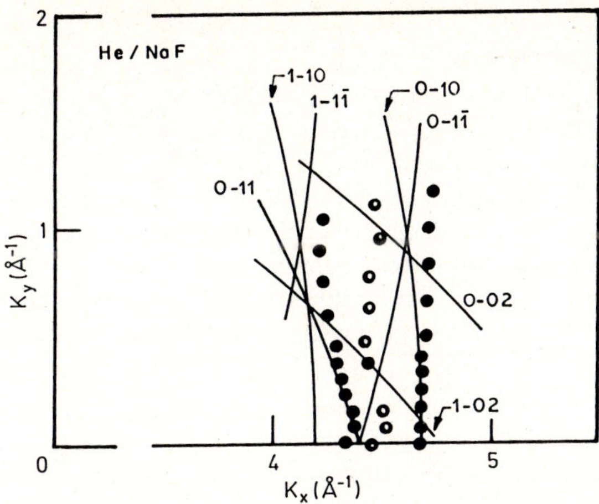

Fig.7.12. Wave-vector components for minima observed with He diffracted from NaF. Solid lines are circles corresponding to zeroth-order energy levels according to (7.10). Dot size indicates estimated error. From [7.28]

level ($n = 0$). Since the (1,2) resonance is normally too weak for observation, this observed (1,2) minimum can be attributed to an admixture of the (1,2) and the (0,1) resonance by v_{11} which is almost as strong as v_{10} for D–NaF. A more detailed investigation of the minimum splitting of this D–NaF system was made by *Hoinkes* et al. [7.31]. They used a perturbation method similar to that proposed by *Chow* [7.68] to get a relation between the observed splitting and the strength of the periodic potential term admixing the two degenerate bound states. From perturbation theory the energy eigenvalues $E_{a,b}$ of the admixed states are known to be

$$
\begin{aligned}
E_{a,b} = \frac{1}{2}\{ &(E(0;\ 0,1) + E(0;\ 1,2) \\
&\pm \left[(E(0;\ 0,1) - E(0;\ 1,2))^2 + 4H^2_{G-G'} \right]^{1/2} \}
\end{aligned}
\tag{7.30}
$$

which at the point of degeneracy, where $E(n = 0;\ \boldsymbol{G} = 0,1) = E(n' = 0;\ \boldsymbol{G'} = 1,2)$, results in an energy splitting of

$$
\begin{aligned}
\Delta E_{a,b} = E_a - E_b &= 2\, H_{\boldsymbol{G}-\boldsymbol{G'}} \\
&= 2\langle 0;\ 0,1 \mid v_{11}(z)\exp(\mathrm{i}\boldsymbol{G}_{11}\cdot\boldsymbol{R}) \mid 0;\ 1,2\rangle \\
&= 2\langle 0 \mid v_{11}(z) \mid 0\rangle \quad .
\end{aligned}
\tag{7.31}
$$

Thus from experimentally observed splitting the corresponding matrix element may be determined, which in the special case described above turned out to be $H_{11} = 0.48$ meV. Further, by comparing the experimental matrix element with that calculated with the eigenfunctions of $v_{00}(z)$ and an appropriate form of $v_{11}(z)$, the strength β_{11} of this periodic potential term can be evaluated. A detailed band structure calculation for the D–NaF system discussed above was performed by *Garcia-Sanz* and *Garcia* [7.85].

For alkali halide surfaces an additional example of band structure splitting for He on NaF was shown by *Derry* et al. [7.35] and an extensive study of band structure effects in selective adsorption of ^{4}He on LiF was presented by *Carlos* et al. [7.44]. In this work splittings of ground state ($n = 0$) and first-excited-state ($n = 1$) resonances by (0,1) and (1,1) Fourier components of the potential are investigated and experimental matrix elements of v_{01} and v_{11} are compared with those calculated from pairwise sum potentials. They found that the pairwise sum potentials overestimate the strength of v_{11} compared with v_{01}.

For He on graphite(0001) band structure effects taking place at the crossing of resonances were studied for a variety of experimental conditions by *Boato* et al. [7.40, 41]. From the observed splitting, it is confirmed that only the first Fourier component v_{10} gives relevant contributions to the periodic part of the potential and matrix elements belonging to v_{10} are evaluated. Figure 7.13 shows, as an example of these investigations [7.41], polar scans of the specular intensity for different azimuthal angles around $\phi_i = 30°$. Admixing of (10) and (01) resonances and splitting is observed for both transitions to $n = 2$ and $n = 3$, but is not measurable for transitions to $n = 4$. The disappearance of one of the split minima can clearly be seen at the point of degeneracy near $\phi = 30°$. As predicted by *Chow* [7.68], at the crossing of the two symmetric states only one resonance appears, but this is shifted.

A second example which clearly demonstrates the splitting is given in Fig.7.14. The experimentally observed positions of the minima are well described by best-fit curves calculated by a perturbation formula like (7.30). The full lines represent resonance positions as expected from free-atom approximation. The matrix element determined from this fit is 0.185 meV.

Level crossings in the selective adsorption of ^{4}He on graphite (0001) were also investigated by *Derry* et al. [7.47]. Weak second-order resonances were seen by means of their admixture with stronger resonances, as predicted by *Chow* and *Thompson* [7.67] and *Harvie* and *Weare* [7.71], and matrix elements measured at incident energies as low as 4.7 meV were found to agree with those determined by *Boato* et al. [7.41] at higher incident energies.

Besides bound state levels E_n, the experimentally observed resonance splittings served as an additional test of model potentials developed for He–graphite. *Hutchison* and *Celli* [7.79] obtained expressions for surface band splittings for the model of a hard corrugated surface with a well (HCSW) [7.76] and made an explicit connection between the HCSW model and perturbation theory. They got good agreement with experimental matrix elements of *Boato* et al. [7.41] if they included a periodicity of the well bottom in the HCSW model potential.

Carlos and *Cole* assumed that the He–graphite surface potential $V(r)$ can be represented as a sum of He–carbon pair potentials. In a first paper [7.81] they used a Lennard–Jones 6–12 pair potential, with carbon atoms at r_i, to get:

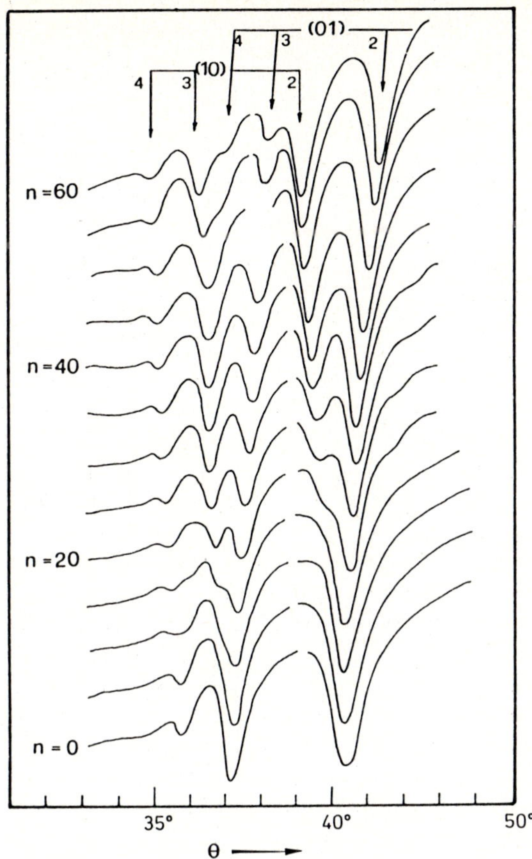

Fig.7.13. Polar scans of specular He intensity diffracted from graphite near the azimuthal angle $\phi = 30°$; n is the number of turns of the azimuthal driver as measured from the symmetry direction corresponding to $\phi = 30°$ ($n = 100$ corresponds to $\Delta\phi = -10.43°$). From [7.41]

$$V(\boldsymbol{r}) = 4\varepsilon \sum \left[\left(\frac{\sigma}{|\,\boldsymbol{r} - \boldsymbol{r}_i\,|} \right)^{12} - \left(\frac{\sigma}{|\,\boldsymbol{r} - \boldsymbol{r}_i\,|} \right)^{6} \right] \quad . \tag{7.32}$$

They fit the parameters ε and σ to the bound state spectrum of *Boato* et al. [7.36] and predicted matrix elements $\langle n' \,|\, v_G \,|\, n \rangle$ between eigenstates of $v_{00}(z)$ for the smallest reciprocal lattice vector of type (0,1).

Using more precise bound state energies E_n for both ^{3}He and ^{4}He determined by *Derry* et al. [7.42] and matrix elements by *Boato* et al. [7.41] they found [7.82,84] none of a variety of central potentials $U(|\,\boldsymbol{r} - \boldsymbol{r}_i\,|)$ to be consistent with the data. In contrast, an anisotropic pair potential agreed with scattering and thermodynamic data, and with estimates derived from electronic properties of graphite. The energy band-structure and thermodynamic

135

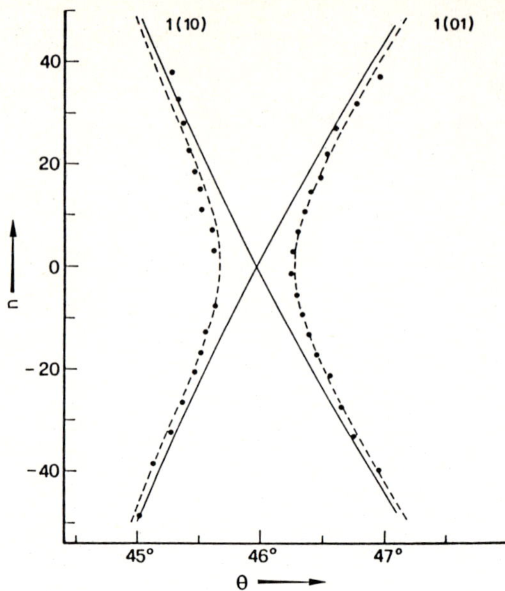

Fig.7.14. Plot showing how the evaluation of the matrix element V_{10} was carried out for He on graphite (0001). Points represent positions of experimental minima. Full lines correspond to the free-atom approximation, while dashed lines represent curves described by (7.30) with $V_{10}^{11} = 0.185$ meV. From [7.41]

properties (in the noninteracting limit) of He atoms on graphite were calculated by *Carlos* and *Cole* using this anisotropic pair potential [7.83].

(c) Inelastic effects in resonance line-shapes In Sect.7.2.2 we have shown (Figs.7.9, 10) that the resonance structure observed at relatively low surface temperature $T_S = 125$K and small incident wave vector $k_i = 57.6$ nm^{-1} in the specular intensity of He–LiF(001) can be described almost quantitatively by elastic scattering theory. Resonance-independent inelastic effects can be taken into account by an overall Debye-Waller factor.

Similar behaviour seems to hold for He–NaF since *Krishnaswamy* et al. [7.38] find almost no temperature dependence of specular resonance features in the range of 90K $\leq T_S \leq$ 300K and *Wolfe* and *Weare* [7.74] find agreement of experimental and calculated resonance widths without including any inelastic broadening. But a purely elastic description did not work for several other gas-surface systems, as shown in detail for atomic hydrogen on alkali halides and He on graphite. For the specular bound-state resonance features observed for H, D on LiF(001) and NaF(001) [7.25] at $T_S = 240$K and k_i in the range $30 - 70$ nm^{-1}, *Chow* and *Thompson* [7.70] accomplished coupled channels calculations and found a quantitative fitting of some of the resonance features to be possible only if the effect of inelastic contributions is taken into account by a complex interaction potential.

136

$$v(\boldsymbol{r}) = v_{\mathrm{R}}(\boldsymbol{r}) - iv_{\mathrm{I}}(\boldsymbol{r}) \quad . \tag{7.33}$$

A comparison of calculations and experiment for the specular intensity versus incident energy of atomic hydrogen on LiF(001) is shown in Fig.7.15.

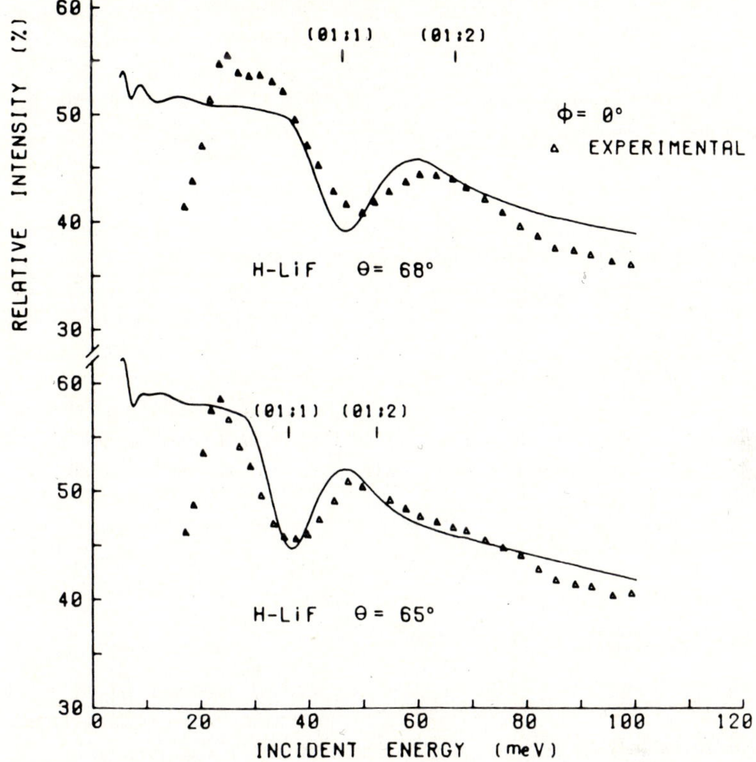

Fig.7.15. Calculated and experimental specular intensity for H–LiF(001). The experimental data (Δ) were taken from [7.25]. The calculations were done with $v_{00}^{\exp-3}$ ($D = 19.7$ meV; $\alpha = 36.7\mathrm{nm}^{-1}$; $z_{\mathrm{e}} = 0.24$ nm) and the strength of periodic and imaginary terms $\beta_{10} = 0.06$; $\beta_{11} = 0.03$, and $\gamma = 0.04$ resulting from a fit of the experimental data (12% FWHM velocity distribution is included). From [7.70]

There an exp-3 potential [7.58] was used for $v_{00}(z)$ together with exponentially repulsive corrugation terms $v_G(z)$ of strength β_G. Unknown was the form of the imaginary part v_{I}, but since it should account here in an empirical way for the inelastic scattering from interaction with phonons, v_{I} should be large in the region where inelastic scattering occurs, that is near the surface. So v_{I} was assumed to be of the form:

$$v_{\mathrm{I}}(z) = \begin{cases} \gamma D; & z \le z_{\mathrm{e}} \\ 0; & z > z_{\mathrm{e}} \end{cases}, \tag{7.34}$$

137

where D is the well depth, z_e the position of the potential minimum, and γ the strength parameter. The strength parameters β_{10} and β_{11} of the periodic terms and γ of the imaginary term were varied to get the best fit. The most appropriate values obtained were: $\beta_{10} = 0.06$, $\beta_{11} = 0.03$ and $\gamma = 0.04$.

The effect of inelastic contributions on resonance line shapes was also described by *Hamauzu* [7.86,87]. He derived a formula for the diffracted intensities near resonance, where the inelastic effects are taken into account by two empirical parameters: the shift of the resonance energy and a resonance width Γ. For specular resonances observed with H–NaF(001) by *Finzel* et al. [7.25] this method worked well with reasonable parameters ($\Gamma = 1.0$ meV for a $[n = 0, G = (0,1)]$-resonance).

Another example, where purely elastic calculations yielded resonances that were too narrow and too deep (or high) compared to experimental data and where much more intensity disappeared in specular resonance minima than appeared in corresponding resonance maxima in other beams, was studied by *Greiner* et al. [7.46]. To improve the agreement between experiment and calculated resonances they introduced a multiple Debye-Waller (DW) effect in the resonance formula (7.25) developed by *Celli* et al. [7.77].

A similar DW-correction procedure was used by *Hutchison* [7.80] to describe the effect of inelastic scattering on line-shapes of He–graphite. As in Ref. [7.46] the following replacement was made in formula (7.25) for all hard-wall diffraction amplitudes

$$S(\boldsymbol{G}, \boldsymbol{G}') \rightarrow \exp[-W(\boldsymbol{G}, \boldsymbol{G}')]S(\boldsymbol{G}, \boldsymbol{G}') \quad , \tag{7.35}$$

where the DW-exponent

$$W(\boldsymbol{G}, \boldsymbol{G}') = (1/2)\langle u_z^2 \rangle (\boldsymbol{p}_G + \boldsymbol{p}_{G'})^2/\hbar^2 \tag{7.36}$$

contains the Beeby correction [7.60], i.e. the momentum transfer $(\boldsymbol{p}_G + \boldsymbol{p}_{G'})$ corresponds to the energy computed from the bottom of the potential well, and $\langle u_z^2 \rangle$ represents the thermally averaged displacement of the surface atoms. Experiment and curves calculated with this method are compared in Fig.7.16.

Part (a) of Fig.7.16 gives the specular intensity versus angle of incidence observed by *Boato* et al. [7.41]; part (b) results from purely elastic calculations, part (c) from calculations with DW-correction as described above, and finally part (d) from calculations with DW-correction and convolution with the energy dispersion of the experimental beam. The resonance marked by an asterisk is an example where a small elastic maximum is changed to a strong minimum due to the inelasticity.

An appreciable amount of inelastic scattering for the system He–graphite was also found in experiments by *Wesner* et al. [7.48]; this is evidenced by deviations from predictions of elastic scattering theory. Calculations for He–graphite where inelastic effects were accounted for by use of optical potentials to get better agreement with experimental resonance structure can be found in [7.69,74,78,177].

More recently *Mantovani* et al. [7.89] proposed a method for handling thermal effects in resonant atom–surface scattering in the t-matrix formalism. This

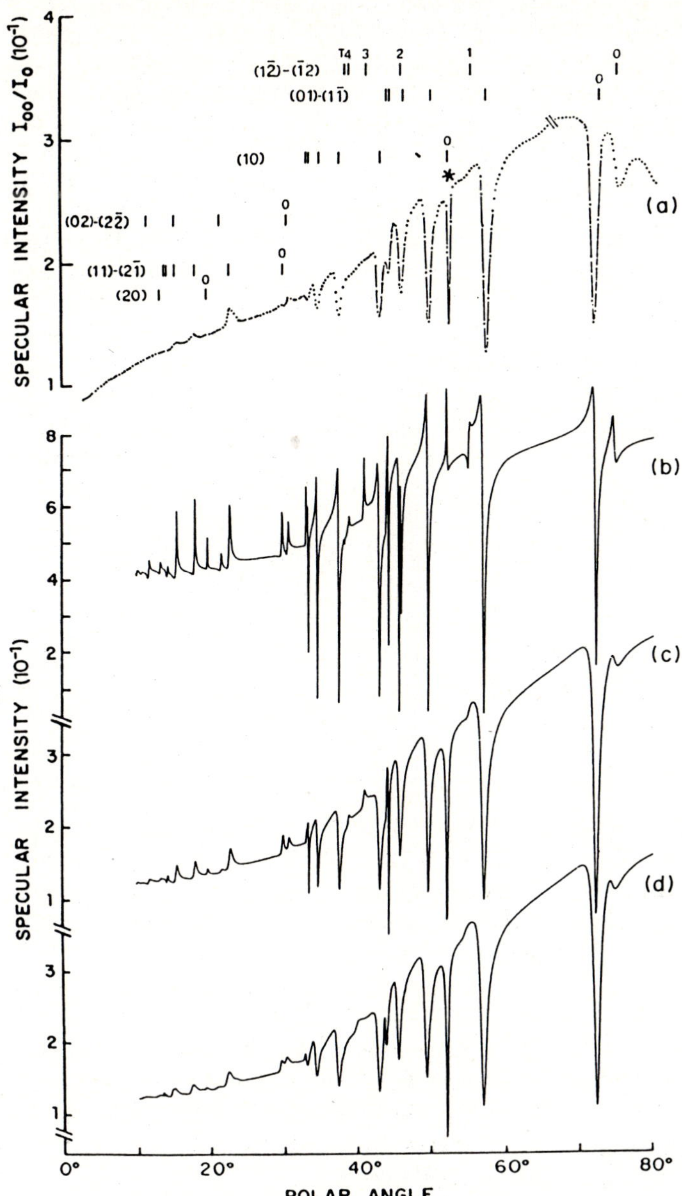

Fig.7.16. Specular scattering of 22-meV helium atoms from the basal surface of graphite. The intensity observed by Boato et al. [7.41] is reproduced in panel (a) with the other panels calculated as follows: (b) elastic theory, (c) Debye-Waller attenuation, and (d) Debye-Waller attenuation and convolution with the energy dispersion of the experimental beam. The resonance features are labeled according to the diffraction order and the energy level of the laterally averaged potential. The asterisk indicates a channel strongly coupled only to the specular beam. From [7.80]

formalism is particularly well suited to handle more realistic soft potentials and was applied by *Mantovani* et al. to the case of helium scattered from Cu(113) by the use of a corrugated Morse potential. Figure 7.17 shows a detailed comparison of calculated curves with the experimental specular intensity around the $(n = 0; \boldsymbol{G} = 1, 0)$ resonance.

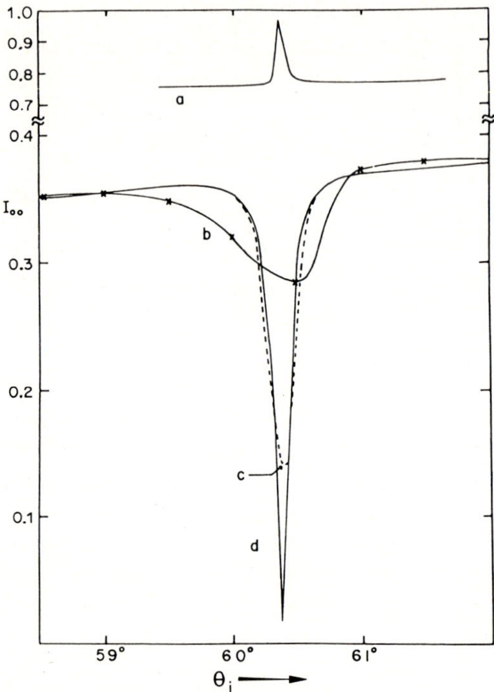

Fig.7.17. Details of the $(\boldsymbol{G} = (1, 0), n = 0)$-resonance for 21-meV-He atoms incident upon a Cu(113) surface: (a) elastic calculation, (b) experiment, (c) present theory with a simple correction for the 0.2° angular dispersion, and (d) present theory without angular dispersion. From [7.89]

The elastic calculation (a) shows a maximum which is changed to a deep and narrow minimum (d) after inclusion of the thermal effects. The experimental minimum (b) is wider and not as deep as the calculated one. Even after a 0.2° angular dispersion in the incident beam is taken into account the calculated minimum (c) is still narrower than in experiment.

To conclude this section we can make the following statements: at the end of the 1970s most of the experimentally observed resonance features were understood theoretically. Investigation of resonances in gas–surface scattering has become a well established method to determine physical gas–surface interaction potentials for light gases such as H, D, H_2, D_2, ^{3}He, ^{4}He and perhaps also Ne. Parameters of model potentials can be obtained from bound state energies and matrix elements extracted from experimental curves or a complete comparison of experimentally observed resonance structures with close-coupling calculations can serve as a test of potential functions obtained for instance from summation of pair potentials. So this experience gained mainly on surfaces of alkali halides and graphite can now be applied to new classes of surfaces. Some examples of new trends have already emerged. *Cantini* et al. [7.39] investigated an oxide surface, *Lapujoulade* et al. [7.43] and *Rieder* et al. [7.50] showed that resonances may be observed from metal surfaces too, and the Saclay group [7.52] reported resonances from the strongly corrugated adsorbate system Cu(110)–O(2 × 1). Results for such new systems will be discussed in detail in the following section in terms of the progress in resonance investigations in the 1980s.

7.3 Since 1981 : Still More Precision in Experiment and Theory for Many More Gas–Surface Systems

7.3.1. Selective Adsorption on Insulators and Graphite

The famous LiF(001) surface still provided the basis for some new investigations and results. *Wesner* and *Frankl* [7.91] studied selective adsorption line shapes in 17 meV He-atom scattering and showed that specular line shapes obey the three rules of *Wolfe* and *Weare* [7.72] and that specular and diffracted beam line shapes obey the rules of *Celli* et al. [7.77] in all cases except for one, where the mixed-extrema Wolfe-Weare structure is seen. This line shape was very sensitive to surface temperature and confirmed the predicted dependence of SA line shapes on inelastic effects.

Scoles and coworkers [7.120] performed very low energy (i.e. down to 4 meV) He-atom scattering experiments offering the advantage of being very sensitive to small changes in the attractive part of the interaction potential and the advantage of reduced computer time for close-coupling calculations. The scattering intensities are dominated by SA effects and are rapidly varying functions of both polar and azimuthal angles. Nevertheless comparing the results with close-coupling calculations on the potential of *Celli* et al. [7.166], very good agreement was found, the quality of which is demonstrated in Fig.7.18 .

Brusdeylins et al. [7.92] measured angular distributions for scattering of very monoenergetic low energy ($\simeq$ 20 meV) He-beams and found a large number of mostly small maxima and minima between the specular and diffraction peaks. Time-of-flight spectra of the scattered atoms showed that the atoms in most of the maxima had suffered a two step process in which the atoms undergo a DMSA (diffraction mediated selective adsorption) and subsequently desorb as a result of a single phonon inelastic process. This will be discussed in more detail later.

Other interesting results concerning the SA in the specular diffraction of ^{20}Ne from LiF(100) were obtained by *Semerad* et al. [7.125]. Using a time-of-flight method it was possible to resolve the high density of resonant structures

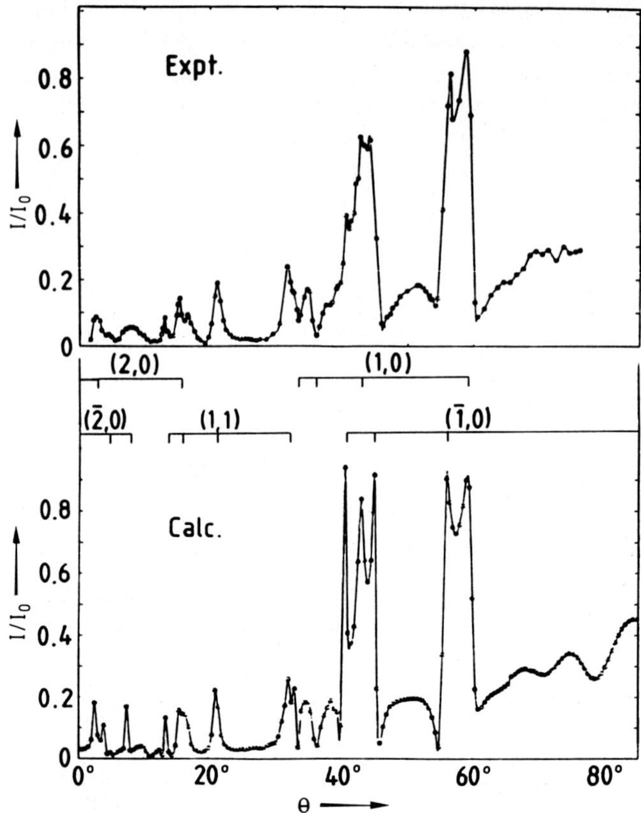

Fig.7.18. He-scattering from LiF(001). Comparison of scattering intensities (upper part) with close-coupling calculations (lower part) for $k = 4.01$ Å^{-1} and azimuth $\phi = 45°$. The measured and calculated points are denoted by dots; the lines are simply interpolating cubic splines. The calculated positions of zeroth order resonances are shown above the calculated spectrum, labelled with the resonant G vector. From [7.120]

over coherent areas in K-space, thus applying a TOF technique in the investigation of DMSA for the first time. Imagine an azimuthal scan of the specular intensity at fixed angle of incidence: this will show resonant structures. Selecting e.g. a certain minimum therein, a TOF-measurement of the scattered flux will decompose the energy spread in the beam used, thus providing additional details of resonant features in K-space which is explored as usual by variation of k_i and ϕ. After changing the TOF-intensity scans into a bright/dark modulation, the data can be plotted in the $K_x - K_y$ plane and resonance circles [derived by a straightforward transformation from (7.10)] can be identified. An example is shown in Fig.7.19. Here the transformed data of a TOF-scan are shown (one transformation is indicated together with resonance circles involving $G = (3, -4)$ and $n = 1, \dots 6$. A corresponding plot of specular intensity data calculated by the HCSW model after *Celli* et al. [7.77] is also shown for

142

comparison. This "best-fit" was obtained for $\xi_0 = 0.297$ Å, $D = 13.45$ meV and $\sigma = 1.9$ Å as parameters of a (9-3) potential, these being in good agreement with the literature [7.58].

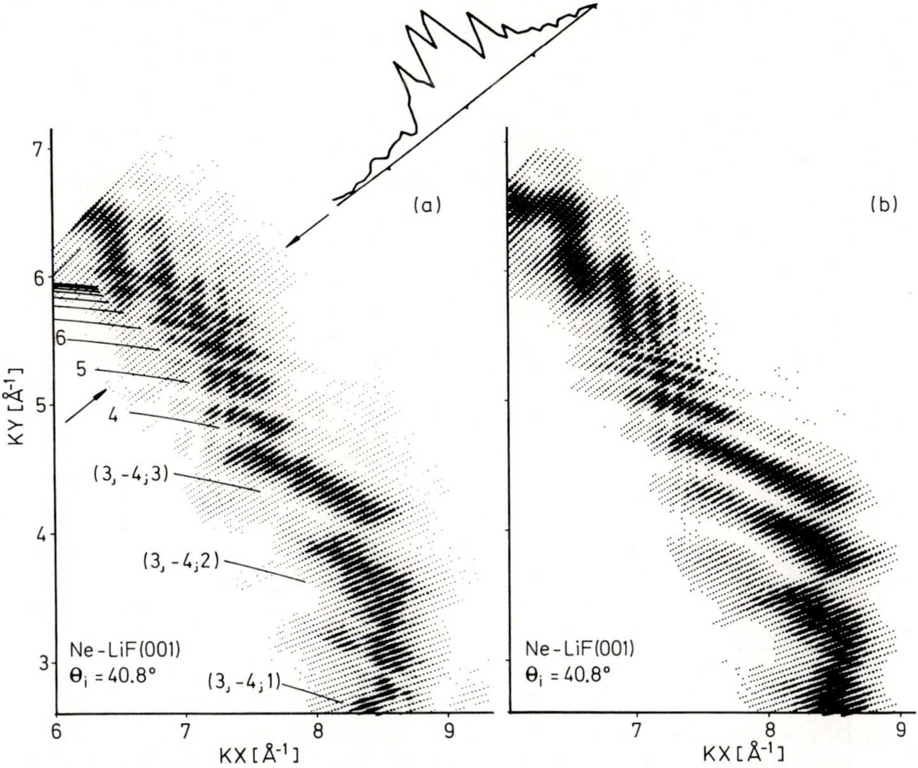

Fig.7.19. Specular intensities of Ne scattered from LiF(001) at $\Theta_i = 40.8°$. Normalized intensity in bright/dark presentation (a) from experimental TOF-scans for many azimuthal directions, (b) from HCSW-calculations. For description of resonant circles see text and original source [7.125]

NaCl(001) is another alkali halide surface which can be prepared without water layer [7.167]. So *Scoles* and coworkers [7.90] have performed selective adsorption measurements with H and D atomic beams, identifying 6 and 3 bound states respectively, the exact energies of which had to be revised later on [7.119] because of a mistake in the E_i-calibration. In 1982 *Baetz* et al. [7.100] published SAR-measurements for H-atoms on NaCl and deduced 3 bound states which in 1985 were confirmed by *Scoles* and coworkers [7.119] and supplemented by 5 bound states for D. It could be shown that these latter data agreed well with the universality law for physisorption potentials as proposed by *Vidali* et al. [7.168].

Helium time-of-flight spectroscopy with high resolution together with angular distribution measurements $I_{sc}(\Theta_i)$ at different azimuthal angles was applied

143

in experiments by *Toennies* and coworkers [7.105] not only in order to explore the Rayleigh phonon dispersion curve for NaCl but also to demonstrate the existence of kinematic focussing giving rise to structure in the angular distributions. Other structures were identified as DMSA followed by single-phonon-mediated desorption (called "selective desorption", DMSAPMD) and 4 bound states (BS) were deduced for the He–NaCl system, the deepest BS being assigned $E_0 = -(7.2 \pm 0.1)$ meV. Intensive investigations for the He/NaCl-system with DMSA however by *Frankl* and coworkers — one with 63 meV [7.111], the other with 17.6 meV [7.112] incident atom energy — gave no evidence for this low lying level whereas the three remaining levels were always reproduced in very good agreement.

MgO(001) is another ionic crystal which was used in He-scattering studies. *Cantini* and *Cevasco* [7.116] quote a preliminary analysis of BS resonances together with a deduced well depth of approximately 8 meV. *Brusdeylins* et al. [7.147] reported BS energies of 10.2, 6.0, 2.6 and 1.2 meV and a well depth of 12.5 meV based on DMSAPMD ("selective desorption") features in polar angle scans of near specular intensity. Again, in a DMSA-investigation by *Frankl* et al. [7.131] using a vacuum-cleaved, oxygen-treated MgO surface, the deepest lying level could not be found but the other three were identified together with two weakly bound states at 0.54 and 0.26 meV. The corresponding depth of a (9-3) potential was $D = (7.5 \pm 0.2)$ meV and it was proposed to give the BS at -10.2 meV of [7.147] the alternate assignment of the -1.2 meV BS.

The experiments discussed up to now used insulators of the ionic crystal type, but there is also work of *Vidali* and *Frankl* [7.102] which was done with a covalently bonded insulator, i.e. with the unreconstructed (111) face of diamond. By observation of DMSA three BS energy levels were identified: 6.4 meV, 3.0 meV and 1.1 meV but most probably the diamond surface was covered with chemisorbed hydrogen.

The other crystalline form of carbon also was investigated, here we only concentrate on the clean basal plane whereas adsorbed layers on that surface will be discussed later. *Cantini* et al. [7.93] investigated the inelastic effect on the resonance line shapes in He–graphite scattering. From a comparison with calculations it was concluded that the multiple Debye-Waller attenuation [7.46, 80] appears inadequate to describe the inelastic effect on the resonance line shapes.

In 1986 the group of *Frankl* [7.121] published a redetermination of BSR energies in He–graphite scattering. They found a total of 6 BS from -12.42 meV up to -0.16 meV, all values somewhat deeper in energy than previously published and in excellent agreement with theoretical calculations following the method of *Toigo* and *Cole* [7.169].

Finally some rather unique work on He-scattering from a semiconductor surface will be mentioned. *Cardillo* et al. [7.95] presented an extensive set of in-plane diffraction data for the scattering of He from the GaAs(110) surface. This reconstructed surface has a very strong corrugation ($\simeq 1.1$ Å going across the surface troughs and still $\simeq 0.3$ Å perpendicular to that direction) and scans of the specular intensity exhibit many high frequency extrema characteristic of BSR together with broad modulations believed to be characteristic of interference in Δk; however no fitting of BS is reported.

7.3.2 Selective Adsorption Resonances on Clean Metals

Selective adsorption resonances (SAR) on clean metals have been studied by several groups. Many efforts concentrated on Cu surfaces, mainly stepped ones, and on the two silver faces Ag(110) and Ag(111), but several other metals were also investigated. Let us start with the work of Lapujoulade and his group on SAR of He, H_2 and Ne on Cu(11n) with $n = 0, 3, 5, 7$. Rather detailed results from [7.98, 101] showed up to four bound states of He on Cu(113), Cu(115) and Cu(117) their energy practically independent of surface crystallography. For H_2 up to six BS on Cu(110) and Cu(115) were found. Both He and H_2 BS levels could be well fitted by (9-3)-potentials whose van der Waals constants C_3 correctly scaled with the polarizabilities of He und H_2 and could even be deduced quantitatively from the semi-empirical formula of *Hoinkes* [7.58] with good agreement. *Gorse* et al. [7.114] then used the He BS levels together with many data on diffracted intensities from the different faces of copper to carefully analyse theoretical models based on the superposition of charge densities and atomic densities. First, the zero order Fourier term $v_{00}(z)$ of the potential was determined making use of the energy levels as deduced from the SAR on the (113) and (115) faces. Then the higher order Fourier components were adjusted to match the scattering intensities. The resulting realistic soft potential could also be deduced from an electron density profile obtained by atomic density superposition (with conveniently adjusted proportionality constant). In [7.138] *Armand* et al. investigated the dependence of the line shape of a resonance in $I_{\mathrm{spec}}(\theta_i)$ with the deepest BS and compared the measurements to their theory for inelastic effects under resonant conditions. The thermal attenuation at resonance was substantially stronger than off resonance; single phonon processes seemed to prevail and it was concluded that the width and signature of resonances at (very) low surface temperature are strongly affected by disorder on the surface and much less so by inelastic effects.

Salanon and *Lapujoulade* [7.106] also reported on Ne scattering from Cu(110). BSR were observed on the specular peak and 6 BS energy levels were identified ranging from -8.25 meV to -2.09 meV. However when fitting a "modified corrugated Morse potential" to the diffraction intensities, the existence of a still deeper level at -10.66 meV was indicated which would have been kinematically unobservable as a resonance with the $G = (1, 0)$ reciprocal lattice vector.

Andersson et al. used a Cu(100) surface in their investigations of SAR and of sticking of H_2 and D_2 in the quantum regime beginning with [7.135]. In [7.129] extensive data on DMSA of H_2 and D_2 on this copper surface at $T_S = 80$ K with incident energies and angles in the ranges (5–50) meV and 40°–80° were reported. The data analysis was compatible with a single gas–surface potential curve for H_2 and D_2 and a well depth of 30.9 meV, which was substantially larger than the hitherto accepted value ($\simeq 22$ meV) but surprisingly close to the values 31.5–31.7 meV that were reported for the H_2 Ag system. Finally, in [7.134] the role of resonance phenomena in promoting sticking of H_2 and D_2 molecular beams in the energy range (8–45) meV and for angles of incidence between 0° and 60° on a Cu(100) surface at $T_S \simeq 10$ K has been broadly explored. The sticking coefficient at zero coverage, S_0, was shown to consist of a background that falls off with increasing energy (based mainly on surface phonon excitations) and superposed well-defined peaks at characteristic energies that coincide with features in the specular reflectivity and with SA conditions. The peaks are due to the formation of quasi-bound states that are entered via DMSA, RMSA or mixed DRMSA (thus involving an initial elastic collision into a resonant state as predicted theoretically by *Böheim* [7.170] and decay via sticking and backscattering channels.

Our next topic will be work on the Ag(110) surface. In 1982, *Luntz* et al. [7.99] presented the first SAR measurements for He on this surface, yielding 4 BS and a potential depth of 6.0 meV. Shortly afterwords *Schinke* and *Luntz* [7.103] derived an empirical He–Ag(110) interaction potential by fitting simultaneously these SAR and all diffracted intensities using close-coupling calculations and compared their results to theoretical and other He–metal interaction potentials. Improved experimental resolution in SAR measurements in 1986 led to 5 new slightly different BS [7.123] and the conclusion that the theoretical potential of *Nordlander* and *Harris* [7.171] based on the sum of a Hartree-Fock term, essentially proportional to the unperturbed electron density at the metal surface and a damped Van der Waals term, provides quite an accurate description of the He–Ag(110) interaction. These aspects were investigated in more detail in 1988 when, in addition to the known BS, high-resolution He-diffraction measurements were included into the analysis [7.127].

Another interesting experiment was reported by *Chiesa* et al. [7.117] in 1985. Here DMSA measurements for n-H_2, p-H_2 and n-D_2 molecules scattered by Ag(110) yielded experimental evidence of the anisotropy of the molecule–surface potential by measuring for the first time the sub-level splitting due to the different values of the magnetic quantum number of n-H_2-molecules. The many BS identified from DMSA for n-H_2, p-H_2 and n-D_2 were used to fit a single phenomenological potential leading to a depth $D = 31.7 \pm 0.2$ meV which compares quite well with $D = 31.5$ meV for the H_2–Ag(111) potential measured by *Yu* et al. [7.109].

This work reported the first clear observation of diffractively mediated selective adsorption DMSA on a close-packed low corrugation metallic surface Ag(111), (together with the first detection of RMSA for a homonuclear diatomic on a solid surface, see below). The DMSA results have been discussed in more detail in [7.162] and led to a comprehensive study of the spatially isotropic component of the laterally averaged molecular hydrogen/Ag(111) physisorption potential. As mentioned before the well depth was determined to be 32 meV consistent with a reduced universal potential as formulated by *Vidali* et al. [7.168] for closed-shell atoms/noble metals physisorption systems and also in excellent agreement with available ab initio predictions for H_2/Ag(111) by *Nordlander* and *Holmberg* [7.172].

146

Some other clean metal surfaces have also been studied with respect to DMSA. In 1979 *Lapujoulade* et al. [7.43] reported DMSA measurements with He on a stepped Cu(117) and in 1980 *Rieder* et al. [7.50] presented results for the Au(110)-(1 × 2) reconstructed surface with 4 BS and a potential depth of $D = 7.15$ meV for He. Also 4 BS and a rather similar potential depth $D = -8.05$ meV were found for the He/Pd(110) system by *Rieder* and *Stocker* [7.107] as well as by *Kaufmann* et al. [7.124] for the He/Ni(115) system. In 1987 *Salanon* and *Lapujoulade* [7.126] again took up He diffraction experiments from W(112) as first reported in 1971 by *Tendulkar* and *Stickney* [7.173] and found 3 BS which could be fitted by a potential well depth of $D \simeq 6.07$ meV. Selective adsorption structures of He on Rh(110) were traced with low-temperature beams of different wavelengths by *Parschau* et al. [7.132] and allowed the determination of 4 BS levels. The measured energies were fitted with different analytical potential forms and yielded a potential well depth of $D \simeq 8.2$ meV. In subsequent work *Kirsten* et al. [7.133,139] investigated He diffraction from the missing row reconstructed (and thus highly corrugated) Pt(110)-(1x2) surface. After analyzing a large number of He-diffraction intensities for deriving an interaction potential, close coupled channels calculations concerning the angular dependence of the specular beam were performed in excellent agreement with experimental data thus confirming the potential form and depth $D \simeq 8.5 \pm 0.2$ meV and the proper assignment of resonant scattering features as well as of 4 BS energies. Figure 7.20 shows an example from this work: rather pronounced regular undulations in intensity (due to constructive and destructive interference between the top and the bottom parts of the corrugation) and superimposed resonant scattering structures labeled with the corresponding energy levels and compared to the calculations.

It should be noted that the resonant structures at large angles of incidence Θ_i (all coupled with the (01)-beam) clearly appear as maxima, this being in contrast to the usual observations of minima for less corrugated metal surfaces which when calculated purely elastically, are also maxima, but turn into minima by intensity losses due to inelastic effects because of the longer lifetimes in the bound states as compared to the enhanced probability for back-diffraction from BS at highly corrugated surfaces. This work also presents a comparison of the hitherto known potentials for He atoms on (110)-surfaces of fcc metals, i.e. Ag, Cu, Pd, Rh and missing row Pt, for more details see Chap. 4 by Rieder.

7.3.3 Selective Adsorption Resonances on Adsorbate Systems

SAR measurements have also made rapid progress for adsorbate systems which, in general, represent strongly corrugated surfaces. A widely used substrate was the basal plane of graphite (0001), in many cases with Xe as the adsorbate. In 1982 *Ellis* et al. [7.97] reported the first measurements of BSR in the scattering of an atomic beam (H) from an ordered layer of gas (Xe) physisorbed on a crystal surface (graphite (0001)). In a later paper [7.113], however, thresholds were found to strongly affect the resonance data leading to a new interpretation

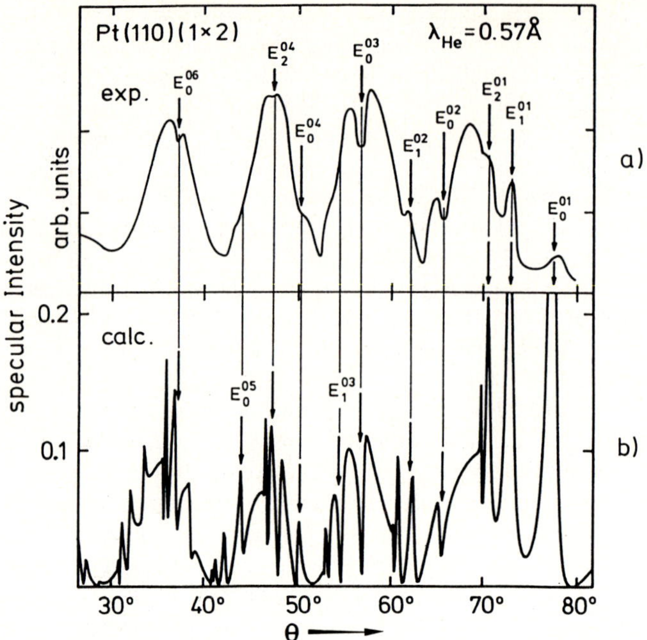

Fig.7.20. (a) Intensity of the specular beam as a function of angle of incidence measured with a He-wavelength of 0.57 Å. (b) Intensity of the specular beam from purely elastic close coupled channels calculations. The most prominent resonant structures are indicated by arrows and labeled with the corresponding energy levels. The rather pronounced regular undulations in intensity (with periodicity of 10°) are due to interference between the top and the bottom parts of the corrugation. From [7.139]

of the resonance features and demonstrating that the behaviour of resonances at some crossings is different from that seen in weakly corrugated systems.

The first scattering measurements carried out with a He-atom supersonic beam on a xenon layer adsorbed on the basal plane of graphite were reported in 1983 by *Bracco* et al. [7.108]. In the same year *Hutson* and *Schwartz* [7.174] used still earlier results of *Bracco* [7.175] in order to demonstrate that only accurate close-coupling calculations can reproduce the observed BSR structures in this strongly corrugated system and that the calculations are significantly affected by small changes in the corrugation strength as well as by other fine details of the potential. The high quality of the calculations may be recognized from Fig.7.21

In 1984 further results of SA measurements for the He–Xe/graphite system were published by *Bracco* et al. [7.110] identifying 3 BS at 4.83 meV, 1.99 meV and 0.66 meV. The derived (laterally averaged) helium-rare gas overlayers potential was found to be well represented by a sum of pair interactions between the He atom and the Xe atoms plus the long range interaction of the He atom with the graphite substrate. Very similar BS energy levels were found by *Larese*

148

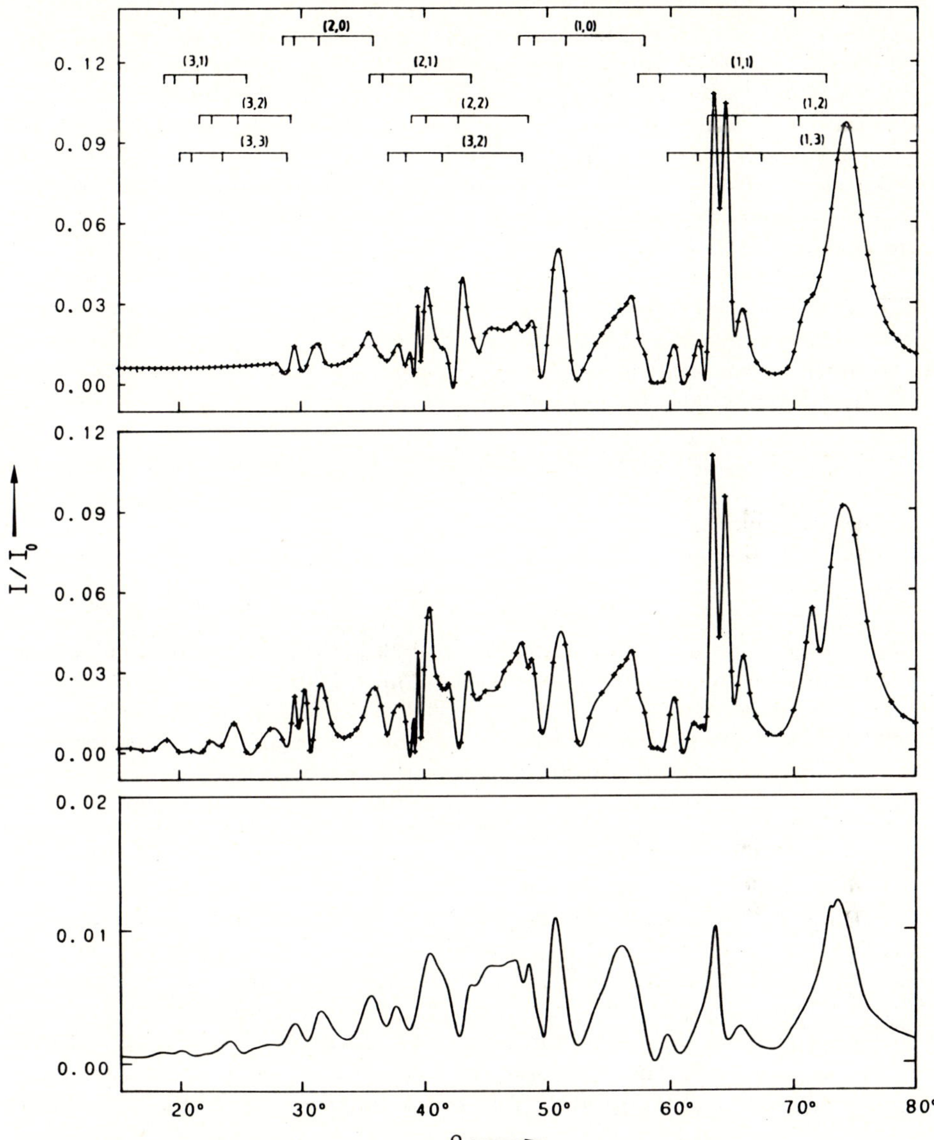

Fig.7.21. SAR in the calculated and experimental specular intensity for He–Xe/C with $k = 6.5$ Å^{-1} at azimuth 0°. Top: minimal basis calculation for a certain potential; center: full basis calculation for the same potential; bottom: experimental data from [7.175]). Calculated points are denoted by the symbol + and a cubic spline has been used to draw the solid lines. The resonance energies are given in the top panel for the important G vectors; for each vector, the angles shown are, from left to right, those corresponding to threshold and the three deepest bound states ($n = 2, 1, 0$). From [7.174]

et al. [7.118] including a weakly bound fourth level for commensurate Kr films adsorbed on graphite whereas for an incommensurate Kr layer only the lowest level at somewhat different energy could be identified. All binding energies agreed with calculations incorporating three-body interactions (i.e. He–Kr–Kr and He–Kr–C(0001) interactions in addition to the He–Kr and He–graphite two-body interactions). Extensive investigations on the problem of three-body interactions in gas–surface scattering of He-atoms from graphite (0001) coated by a monolayer of Xe were finally presented by *Aziz* et al. [7.136] in 1989 starting with the precise determination of two-body He–Xe interactions from crossed-beam gas phase experiments. Then a He–surface potential was constructed by adding together the He–Xe pairwise sum, the long-range He–graphite interaction, the three-body He–Xe–Xe triple dipole interactions and the surface-mediated three-body interaction He–Xe–C(0001). Using this potential together with a small correction for thermal motion at the surface, good agreement in a comparison of close-coupling scattering calculations and experimental results on He-diffraction and $I_{\mathrm{spec}}(\Theta_i)$ dependencies including SAR features could be obtained.

Before discussing other results on He scattering from rare-gas overlayers on the metal surface Ag(111), let us summarize other earlier investigations of adsorbate/metal substrate systems.

In 1981 *Soler* et al. [7.94] measured asymmetric resonance line shapes in the specular intensity for He diffraction from Ni(110)-(1 × 2)H and through calculations using a hard corrugated surface model with an attractive well they were able to resolve the ambiguity as to whether the corrugation function $+D(x, y)$ or $-D(x, y)$ correctly describes the surface profile. *Lapujoulade* et al.[7.96] studied oxygen adsorption on Cu(110) which for saturation results in a (2 × 1) structure and found many He-diffraction beams $(0,0; \pm 1, 0; \pm 1/2, 0)$ all of them with pronounced resonance structures when measured as a function of ϕ_i. These data could be explained by 4 BS and a (9-3)-potential of depth $D \simeq$ 8.2 meV replacing an earlier incorrect calculation [7.52]. The Ni(110)-(2 × 1)O system was investigated using He diffraction by *Engel* et al. [7.115]. Whereas resonant effects were negligible at small wavelengths (0.57 Å) thus enabling a hard-wall model analysis with respect to corrugations, measurements at longer wavelength (1.09 Å) clearly yielded SAR structures both in the $I_{00}(\Theta_i)$ and $I_{20}(\Theta_i)$ scans. For values of Θ_i at which the (0,0) intensity shows minima, the (2,0) intensity shows maxima and vice versa. Although this is consistent with the essentially elastic nature of resonant scattering, quantitative comparison of intensity losses to corresponding intensity gains always indicated missing particles which are due to the enhanced probability for inelastic scattering of a selectively adsorbed atom. BS energy levels were not deduced.

An analysis of the resonance scattering of He from the Ag(001)c(2 × 2)Cl surface by *Becker* et al. [7.104], however, concentrated on BS and the attractive well depth. Five energy levels were identified corresponding to a one-dimensional well depth of $D = 17.5$ meV. By studying the (considerable) momentum-transfer dependence of the Debye-Waller factor the authors were able to confirm that there is no level deeper than -13.7 meV.

150

The adsorption of hydrogen on Ni(115) was the subject of a He diffraction study by *Kaufmann* et al. [7.122] in 1986. Resonant scattering from the saturated Ni(115)-(1 × 1)H surface was evaluated to yield 4 BS levels which turned out to be not much different from those at the clean Ni(115) surface.

Cantini et al. in 1988 [7.130] chemisorbed atomic deuterium on Ag(110) at surface temperatures below 170K. An Ag(110)-c(4 × 4)D phase was observed at saturation coverage and studied by He diffraction. SAR structures in $I_{00}(\Theta_i)$-scans (both maxima and minima) enabled a set of 4 BS energies to be assigned and a potential depth of $D \simeq 10.1$ meV which can be compared to $D \simeq 6.0$ meV for the He–Ag(110) potential.

Extensive investigations of elastic helium scattering from ordered overlayers of Ar, Kr and Xe physisorbed on Ag(111) were presented by *Gibson* et al. [7.128] in 1988 done on a layer-by-layer basis for 1, 2, 3 and $\simeq 25$ overlayers. Besides diffraction, SAR were measured in a very large number of $I_{00}(\Theta_i)$ scans and compared to accurate close-coupling calculations, as exemplified by the results for a Xe monolayer and a Xe trilayer shown in Fig.7.22.

Such measurements proved to be extremely sensitive to the He–surface potential and observable changes in the calculated SA-scans appeared for different He–rare gas potentials or when various nonadditive terms were included in the potential; for details the reader is referred to the original paper.

A very recent investigation will end this discussion on SAR of adsorbate systems: the work of *Jung* et al. [7.137] on He scattering from an adsorbed monolayer of CH_4 on a MgO(001) surface at $T_S = 41$ K. The experimental BSR positions were compatible with predictions based on either a free rotor or fixed dipole-oriented CH_4 layer. The line shapes, however, did not conform to the rules of *Celli* et al. [7.77] or *Wolfe* and *Weare* [7.72] but only to close-coupled channel calculations, which, however, were difficult to perform because inelastic effects had to be included.

7.4 Bound State Resonances Coupled to Phonon and Rotational Energy Exchange Processes

Up to now we have concentrated on processes in which atoms (or molecules) make a purely elastic transition into (or out of) physically bound states at a surface, the changes in momentum being mediated by a reciprocal lattice vector $\mathbf{G}$, i.e. by diffraction.

Since diffraction depends on the corrugation of the atom–crystal surface interaction potential, these types of resonant transition are called diffraction- or corrugation-mediated selective adsorption resonances DMSAR or CMSAR. However inelastic processes also may be involved in the population and depletion of (or interchange between) bound states in several ways and indeed these processes have been identified in experiments and proved to be important in many respects. For the sake of completeness the discussion will not be restricted

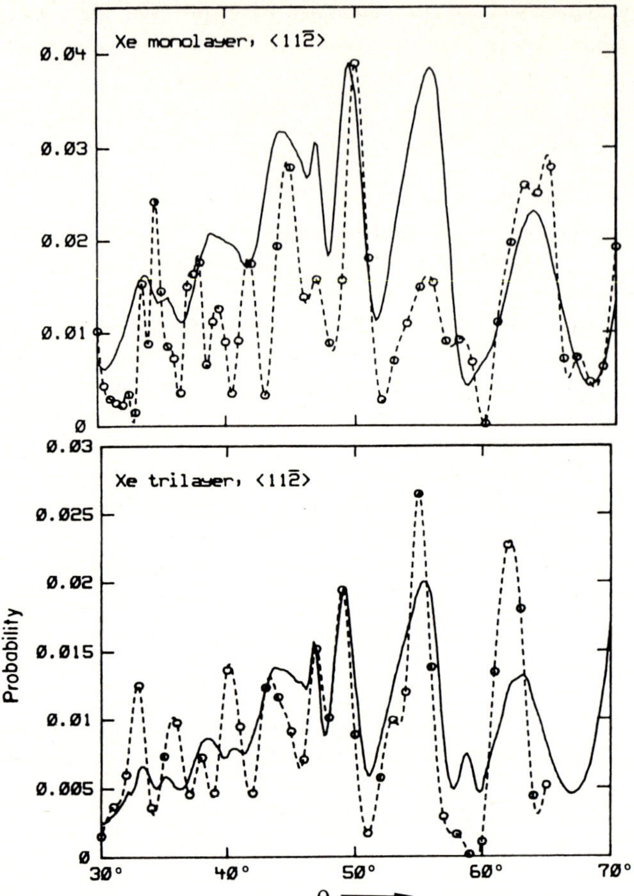

Fig.7.22. Comparision between calculated (dashed lines) and experimental (full lines) results for He scattering from a Xe monolayer and Xe trilayer along the ⟨112⟩ azimuth at incident energy of $E = 17.978$ meV and $E = 17.928$ meV respectively. The accurate close-coupling calculations used a MS pair potential plus V_s plus V_3^{a-a}. From [7.128] (see original paper for details)

to He atom scattering but will also include hydrogen molecule scattering where appropriate.

7.4.1 Phonon-Mediated Selective Adsorption and Desorption: PMSA and PMSD

It has been well known for long time that the intensity in angular atomic scattering distributions between the Bragg diffracted beams is largely due to scattering processes including one-phonon creation and annihilation, and observed structures in the scattered distribution were directly correlated with special features of the phonon spectrum.

In 1976, however, *Cantini* et al. [7.140] presented scattering measurements for He (and Ne) on LiF(001) clearly demonstrating that the inelastic reflection probability can be strongly affected by the presence of bound state resonances and even used the selective character of the inelastic resonances to analyze the phonon spectrum at the surface. For this purpose the authors first associated bound-state resonance conditions in the inelastic scattering with minima in the reflection probability and, second, assumed that the scattered intensity is mainly due to Rayleigh waves. Figure 7.23 shows some examples of the measurements. In part (a) the structure in the inelastic scattering intensities in the tails of specular beams can be clearly seen, the position of these structures being nearly independent of the angle of incidence. From comparison to the SAR in a $I_{\mathrm{spec}}(\Theta_\mathrm{i} = \Theta_\mathrm{f})$ scan (b) the correlation becomes evident.

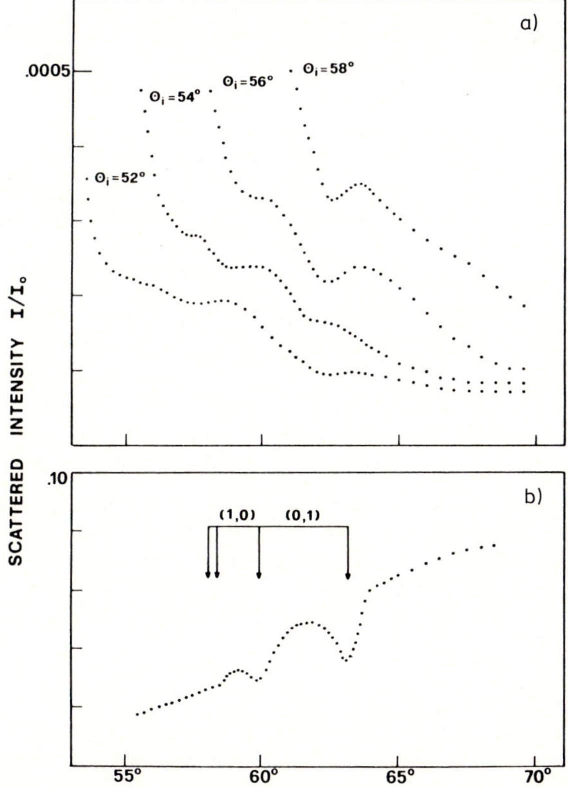

Fig.7.23. He scattering from LiF(001); $\langle 100 \rangle$ azimuth, $k_\mathrm{i} = 10.95$ Å^{-1}. (a) Angular distribution of the inelastically scattered intensity at different angles of incidence. (b) Specular intensity at different angles of incidence ($\Theta_\mathrm{i} = \Theta_\mathrm{f}$). Indicated are the positions of the elastic resonances. From [7.140]

Cantini and *Tatarek* were able to confirm their findings and interpretations also in the scattering of He from graphite (0001) during the following years in [7.141, 142, 144] and together with other coworkers in [7.145]. Again angular distributions were studied and four types of inelastic transitions from an inital state (incident atom) to a free final state (outgoing atom with a particular k_f) were discussed:

(1) a direct inelastic transition by creation or annihilation of an appropriate phonon;
(2) elastic DMSA into a bound state ε_j and subsequent reemission by an appropriate phonon;
(3) inelastic PMSA into a bound state by an appropriate phonon and subsequent reemission by diffraction (now also known as "selective desorption");
(4) elastic DMSA into a bound state ε_i followed by an inelastic transition by phonon annihilation/creation into another bound state ε_j and finally elastic DM reemission from this state.

For a schematic representation of processes (2) and (3), see Fig.7.27.

The basic kinematic conditions for any scattering process involving the exchange of a single phonon are

$$k^2 = k_0^2 \pm 2M\omega_q/\hbar \quad , \tag{7.37}$$

$$\boldsymbol{K}_G = \boldsymbol{K}_0 + \boldsymbol{G} \pm \boldsymbol{Q} \quad , \tag{7.38}$$

where $\boldsymbol{k} = (\boldsymbol{K}, k_z)$ is the wave vector of the gas atom of mass M, while $\hbar\omega_q$ and $\boldsymbol{Q}$ are the energy and parallel momentum of the exchanged phonon; $\boldsymbol{G}$ is a surface reciprocal lattice vector. The gas atom may undergo a phonon-assisted resonant transition into a bound state ("inelastic selective adsorption") if its perpendicular energy corresponds to an energy level ε_j of the laterally-averaged gas–surface potential $v_{00}(z)$, while the atom travels in a nearly-free particle state parallel to the surface after exchange of a phonon. The "selectively adsorbed" atom may ultimately leave the surface via elastic diffraction into an allowed final channel, or via some further inelastic process. Due to the discrete nature of the bound state eigenvalues ε_j, the phonon-assisted resonance selects, at any incident angle and energy, a well defined family of phonons. Their frequencies $\omega_{Nj}(Q)$ are given, through the kinematic conditions (7.37) and (7.38), by

$$k_0^2 \pm 2M\omega_{Nj}(Q)/\hbar - (\boldsymbol{K}_0 + \boldsymbol{N} \pm \boldsymbol{Q})^2 = 2M\varepsilon_j/\hbar^2 \quad , \tag{7.39}$$

where $\boldsymbol{N}$ is a closed-channel reciprocal lattice vector. On the other hand, when a scan of the inelastic angular distribution (on both sides of elastic peaks) is taken, each final scattering angle Θ_f corresponds to inelastic processes which select a family of phonons $\omega_f(Q)$ given, for in-plane scattering, by the parabolic equation

$$k_0^2 \pm 2\,M\omega_f(Q)/\hbar - \frac{(\boldsymbol{K}_0 + \boldsymbol{F} \pm \boldsymbol{Q})^2}{\sin^2\Theta_f} = 0 \quad , \tag{7.40}$$

where $\boldsymbol{F}$ is the reciprocal lattice vector of the final channel.

In general, the family of phonons yielding a resonant contribution, (7.39), and that selected by the second kinematic condition (7.40) do not coincide; therefore, the position (Θ_0, Θ_f^*) of a resonant structure observed in the tail of a diffraction peak will fix, through (7.39) and (7.40) both the energy $\hbar\omega_q^*$ and the parallel momentum $\boldsymbol{Q}^*$ of the phonon involved. This procedure, repeated for a set of incident angles Θ_0, allows a phonon dispersion curve to be obtained without energy analysis of the scattered particles.

In [7.142] the double-resonance transition process (4) was observed for the first time and a first attempt to apply true inelastic calculations to selective adsorption processes (2), (3) and (4) was presented. The new theory for the single-phonon resonant scattering used ideas from the work of *Celli* et al. [7.77] on DMSA and yielded explicit formulas for the above inelastic processes in the (semiclassical) eikonal approximation with a time-dependent hard corrugated wall HCW.

In [7.144] *Cantini* and *Tatarek* concentrated on a special case of PMSA where the bound state is reached without the exchange of a reciprocal lattice vector [$N = 0$ in (7.39)] this process was termed "specular inelastic SA" and for the first time investigated both experimentally and theoretically. This SISA gives a very large contribution to the inelastic scattering of He from graphite near grazing incidence and might contribute to the sizeable drop in the elastic intensity often observed at such angles.

Finally in [7.145] *Boato* et al. used PMSAR in He scattering from graphite in order to obtain the dispersion curve for the phonons involved (corresponding to transverse acoustic modes of bulk graphite). An anomalous temperature dependence of the Debye-Waller factor at low temperature was also found due to the layered structure and explained in terms of the known low frequency phonon spectrum of graphite.

All the results described so far were obtained without energy analysis of the scattered particles. Although there was strong evidence that the interpretations given were correct more direct proofs were desirable and indeed possible by complementing angular distribution measurements with momentum dispersive time-of-flight (TOF) measurements.

In 1981 *Brusdeylins* et al. [7.143] used a fixed geometry $\Theta_i + \Theta_f = 90°$ and a highly monoenergetic ($\Delta v/v = 0.8\%$) low energy ($\simeq 20$ meV) He beam to measure first angular scattering distributions from LiF(001) and NaF(001) along the $\langle 001 \rangle$ and $\langle 110 \rangle$ directions (for LiF only) and resolved a large number of maxima and minima between the specular and neighbouring diffraction peaks with low intensities ($\sim 10^{-3}$ of the specular peak), see Fig.7.24.

Subsequent TOF measurements of scattered atoms at the incident angles corresonding to the maxima clearly revealed that the atoms were inelastically scattered by single phonons and from the known BS energies it was possible to explain most of the maxima by process (2), i.e. elastic diffraction into a BS with subsequent inelastic reemission via the appropriate phonon ("selective desorption"). Evidence for minima involving DMSA into out-of-plane states was also found. Furthermore the possible appearance of maxima by kinematic fo-

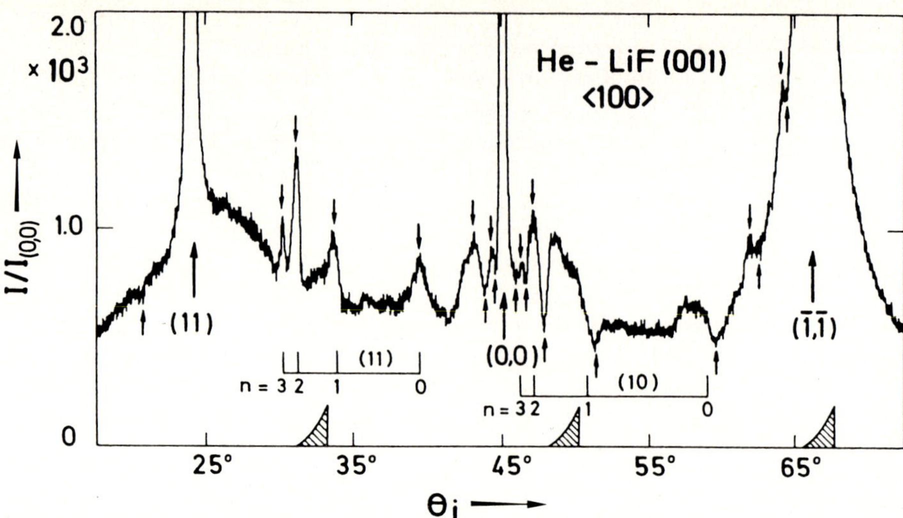

Fig.7.24. Recorder trace showing the intensity as a function of incident angle ($\Theta_i + \Theta_f = 90°$) measured for He ($k = 6.174$ Å^{-1}) scattered from LiF(001)⟨100⟩ at a target temperature of $T_S = 293$K. The small arrows above the trace mark secondary maxima and minima, the bars below the trace mark the diffraction peaks and the predicted angles at which selective adsorption with the indicated momentum transfer is expected. The "wedges" at the bottom show the angular regions and relative intensity enhancement expected from kinematical focussing (KF). Near $\Theta_i = 32.7°$ a selective desorption peak coincides with a KF-singularity. From [7.143]

cussing KF was considered; this is caused through the tangency of the Rayleigh surface-mode dispersion curve to the parabolic scan curve resulting from condition (7.40) and representing the locus of possible inelastic events at a given final scattering angle. Figure 7.25 shows an example of TOF measurements for the large selective desorption peak at $\Theta_i = 32.7°$ (see also Fig.7.24).

Three annihilation peaks are resolved for $\Theta_i = 30.7°$ which is on the bright side of the kinematical focussing singularity. The two leftmost of the three annihilation peaks belong to the same branch of the surface phonon dispersion curve for which the tangency point for KF is very close to $32.7°$. These peaks are indeed strongly diminished at the dark side of the KF singularity ($\Theta_i = 34.7°$), whereas the third peak belonging to another branch of the dispersion curve is not affected very much.

At $\Theta_i = 32.7°$, however, all peaks are enhanced because of the selective adsorption resonance. Thus this work was the first direct observation of particles inelastically ejected from a selectively adsorbed state. In addition to that, the lifetimes of He atoms in the BS were deduced from the half-width of the resonances, varying from $\tau \simeq 7 \cdot 10^{-12}$ s for the deepest level at -5.9 meV to $\tau \simeq 60 \cdot 10^{-12}$ s for the weakest bound level at -0.21 meV.

In a subsequent paper [7.146] *Lilienkamp* and *Toennies* identified an anomalously large selective resonance maximum in the angular distribution of He scattered from LiF(001) as being due to the 3-resonant-step process (4): DMSA into

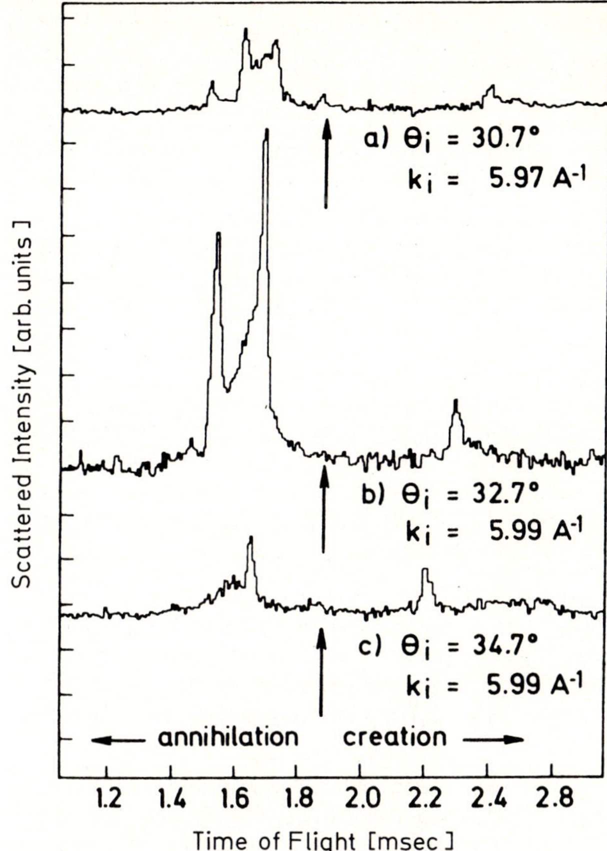

Fig.7.25. The time of flight spectrum along the $\langle 100 \rangle$ direction in (b) was measured at an incident angle at which a selective desorption peak is observed. The spectra in (a) and (c) were taken at two angles 2° removed. The relative intensities can be compared. The arrow shows the expected location of the elastic scattering, which is not observed. Structures at smaller time of flight are due to annihilation and structures at larger time of flight are due to creation of single phonons. From [7.143]

the $n = 2$ BS level followed by a transition $n = 2 \rightarrow n = 0$ through phonon creation and completed by ejection from the $n = 0$ state into the continuum again via elastic diffraction.

Selective desorption structures at angles close to the specular peak were also investigated by TOF methods for He scattering from 750K MgO(001), published by *Brusdeylins* et al. in 1983 [7.147].

In [7.148] *Lilienkamp* and *Toennies* presented an extensive investigation of all types of phonon mediated inelastic processes (1) – (4) again in the He/LiF(001) system and discussed the results in terms of a simple quantum resonance theory. Angular distributions and TOF spectra for He scattering from NaCl (001) revealed three BS in the He–surface potential by selective desorption features unambiguously separable from KF features also present [7.105]; surface phonon dispersion relations could also be determined. Similar investigations were made by *Benedek* et al. [7.152] concerning the scattering of He from a NaF (001) surface. Here DMSA, PMSA and KF were found and it was shown that the dispersion relation of Rayleigh phonons can be obtained directly from the positions of such KF singularities without the use of TOF measurements.

However in the atom scattering spectrometry of surface phonons the KF singularities must really be distinguished from resonant features since the interference between the directly scattered beam and the beam passing through a BS can give sharp maxima and minima (corresponding to the particular phonon momentum and energy required to accommodate the atom in the bound state) thus producing dramatic effects on the TOF spectra of the He atoms. This was pointed out in [7.149] by *Evans* et al. and resonant intensities were calculated in order to explain the observed sharp enhancement in the response of the surface phonon spectrum wherever inelastic resonances occur. The calculations generalized the formalism of *Celli* et al. [7.77] to include inelastic processes and took into account temperature effects through a Debye-Waller factor corresponding to $T_S = 300$ K.

An example from this work is shown in Fig.7.26 where for $\Theta_i = 64.2°$ the intense resonance $N = (1,1)$, $n = 1$ accounts for the observed enhancement in the Rayleigh wave peak R. The comparison of the experimental data to the calculations with and without inclusion of resonant effects clearly demonstrates the order of magnitude of amplification of the R-phonon peak and the rather good general agreement.

The theoretical treatment of resonant and kinematic enhancement in inelastic scattering was further advanced in 1986. In [7.151] *Nichols* and *Weare* calculated one-phonon inelastic-scattering intensities with the Rayleigh solution to the hard-wall atom–surface scattering problem which included inelasticity by allowing the infinite hard-wall potential of the Rayleigh expansion to include the amplitude of motion of the surface due to thermal excitation in addition to describing the static corrugation. A somewhat similar HCW-type theory had been applied by *Evans* et al. [7.149], see above.

However in [7.150] *Eichenauer* and *Toennies* used a different approach to a fully quantum mechanical theoretical description of resonant inelastic one-phonon scattering processes in the framework of the distorted wave Born approximation. First the elastic diffractive scattering problem was solved exactly using close coupling calculations for a pairwise additive semi ab initio potential and a rigid lattice containing both the attractive well and the corrugation in a realistic manner, thus taking BSR fully into account. Next the inelastic coupling due to one-phonon processes was treated as a perturbation of the elastic wavefunctions obtained from the close coupling solutions and the LiF surface lattice dynamics were described by realistic Green's function calculations. The results provided a good description of most of the structures seen in experimental angular distributions and time of flight spectra. The resonant features could be interpreted in terms of one-phonon assisted adsorption into and (rep-

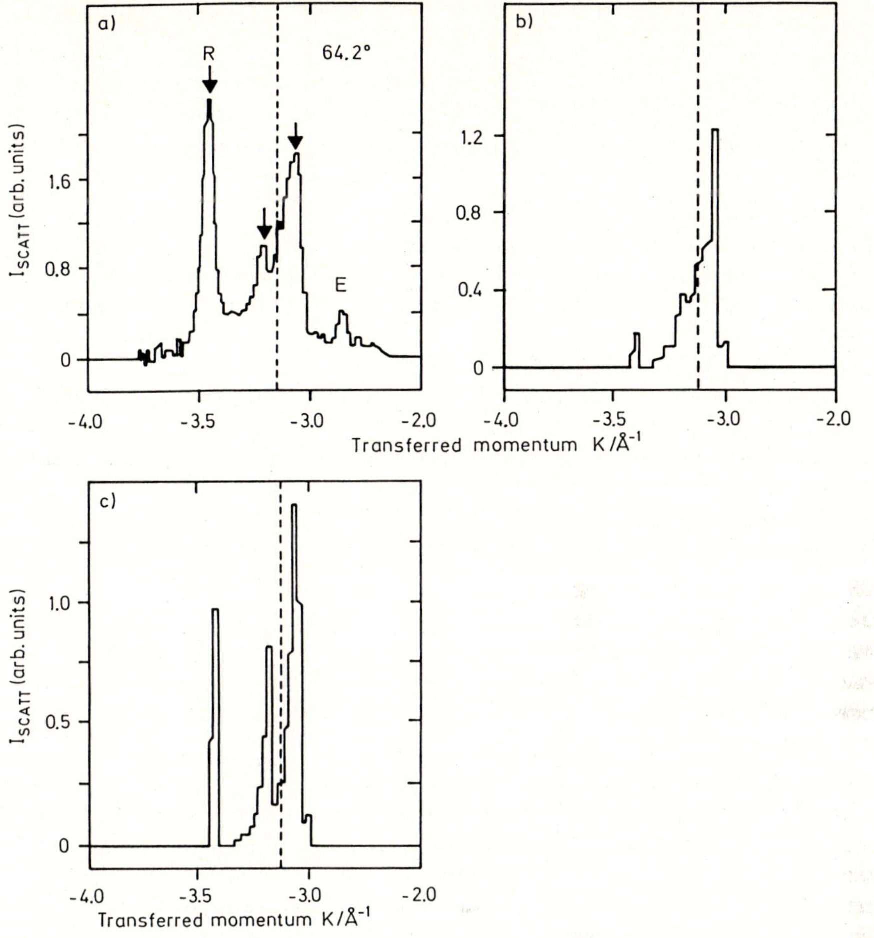

Fig.7.26. TOF spectra for He scattering from LiF(001) along $\langle 100 \rangle$ for $k_i = 6.06$ $\text{\AA}^{-1}$ and $\Theta_i = 64.2°$. The experimental data (a) are compared to the calculated nonresonant spectra (b) and to the results of the full calculation (c). The maxima [marked by arrows in (a)] correspond to the inelastic resonances for $N = (1,1)$ and $n = 1, 2, 3$ (from left to right). From [7.149]

resenting the time-reversed process) desorption out of specified BS of the atom surface potential. In Fig.7.27 these processes are sketched schematically.

In Fig.7.28, a comparison of experimental and calculated intensities of a resonance maximum near $\Theta_f \simeq 50°$ is shown ($\Theta_i = \Theta_{SD} - \Theta_f$ with $\Theta_{SD} = 105°\text{--}135°$). From this and other similar results it can be concluded that the Rayleigh mode is the most important one for the inelastic He–LiF (001) one-phonon scattering and that a realistic theoretical description of the resonance effects in inelastic He-atom scattering (treating the static and dynamic interactions for the first time consistently within one model) is available.

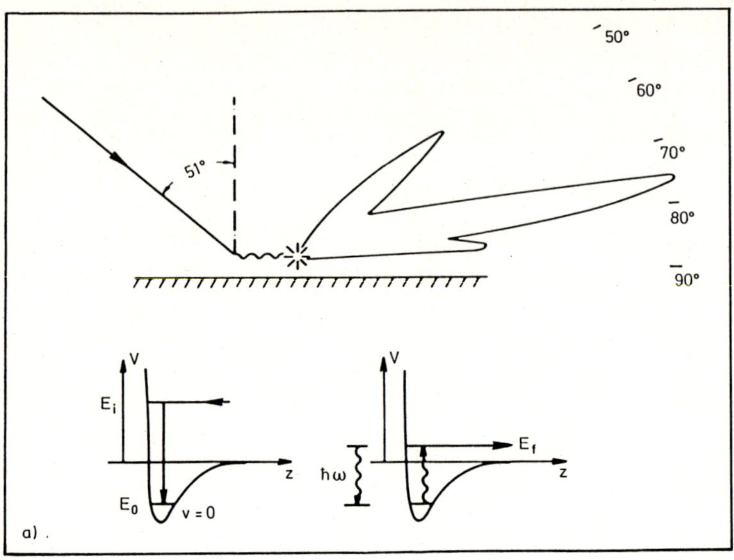

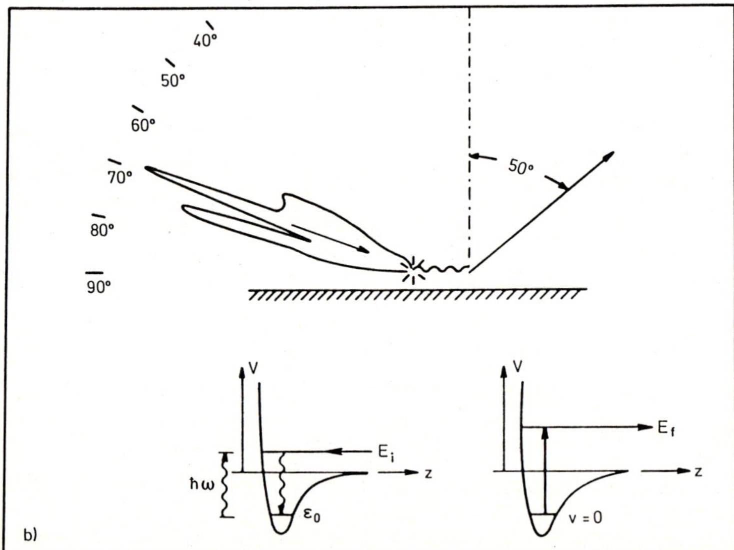

Fig.7.27. Schematic diagram showing the interpretation of the experimental results for inelastic scattering of monoenergetic He atoms with 17.75 meV scattered from LiF along the ⟨110⟩ azimuth. In (a) the atoms with incident angle 51° are trapped by diffraction via a (10) reciprocal lattice vector into the $v = 0$ bound state. The final angular distribution of atoms leaving the surface after annihilating a single phonon on the surface is shown on the right. In (b) the time-reversed process is shown. Because of special kinematic conditions atoms in the $v = 0$ vibrational bound state leave the surface at $\Theta_f = 50°$ for a wide range of energies. Thus from the time-of-flight spectra it can be concluded that the particles are trapped by phonon creation. The incident angle distribution for the trapped atoms is shown at the left. From [7.150]

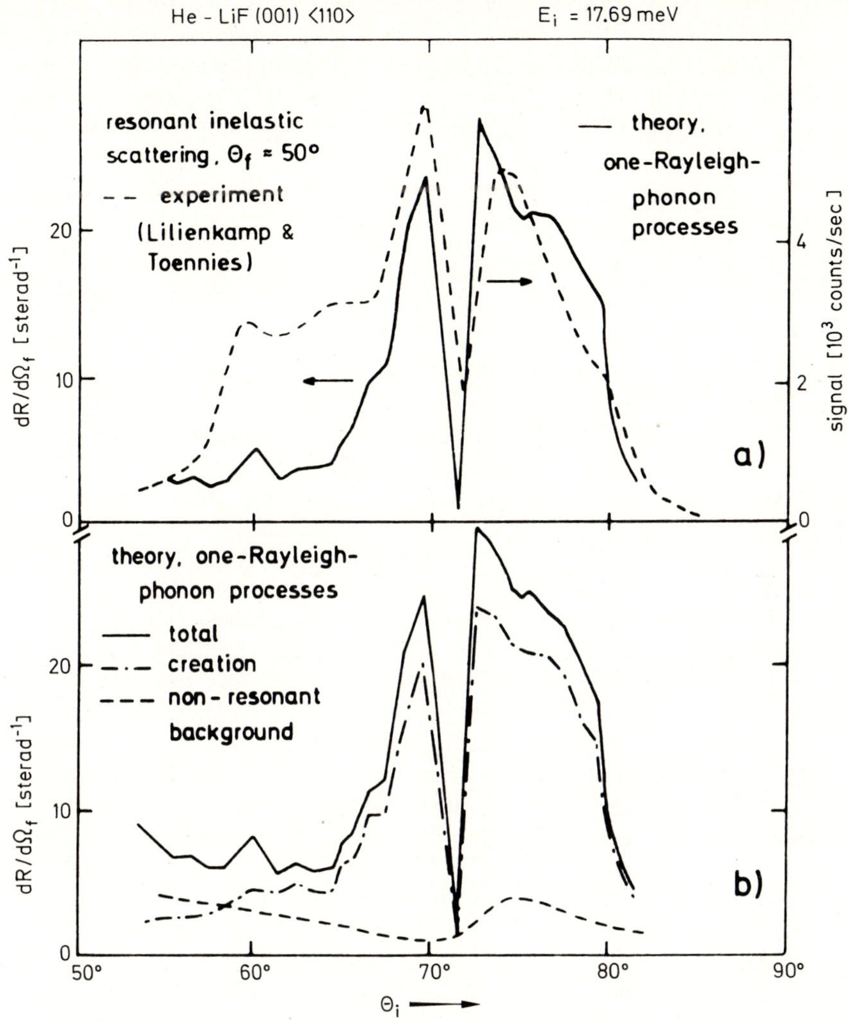

Fig.7.28. Angular dispersion for He–LiF(001) scattering along the $\langle 110 \rangle$ direction as function of the initial angle Θ_i with $E_i = 17.69$ meV. The final angles Θ_f were chosen as the positions of resonance maxima near $\Theta_f = 50°$. (a) comparison of experiment and theory for the resonant (background subtracted) intensity, (b) contributions of the phonon creation processes and the nonresonant scattering to the total (resonant plus background) scattered intensity. From [7.150]

7.4.2 Rotation-Mediated Selective Adsorption: RMSA, or Rotational Feshbach Resonances

Whilst for an atom such as He only diffraction and phonons can mediate selective adsorption (DMSA and PMSA, see above), the internal degrees of freedom in molecular scattering can also mediate SA from surfaces. Whereas for most molecules vibrational excitation energies are too large in comparison to thermal energies in scattering (at least for H_2 and its isotopic variants) translational energy may often be transferred into rotational energy in the molecules during scattering and rotation-mediated selective adsorption RMSA, the kinematic condition for resonance (without phonon interaction) may now be written as

$$E_i = \frac{\hbar^2}{2m}(\boldsymbol{K}_i + \boldsymbol{G})^2 + \frac{\hbar^2}{2I}J'(J'+1) + \varepsilon_n + \varepsilon_{n,Jm}^{(1)} \quad . \tag{7.41}$$

Here E_i is the translational energy of the incident molecules of mass m and moment of inertia I, ε_n is the BS energy of quantum number n, J' is the final state rotational quantum number, $\varepsilon_{n,Jm}^{(1)}$ is a first order perturbation theory correction to the energy for small interaction potential anisotropy, and $\boldsymbol{K}_i$ and $\boldsymbol{G}$ have their usual meanings. Thus, even for a perfectly isotropic potential (i.e. $\varepsilon_{n,Jm}^{(1)} = 0$), mixed DMSA-RMSA resonances are possible (e.g. for the asymmetric HD-molecule) or pure RMSA or pure DMSA resonances are described by setting $\boldsymbol{G} = (0,0)$ or $J' = 0$, respectively, if the initial state was $J = 0$, otherwise $\Delta J = 0$ must hold.

For $\varepsilon_n = 0$ (and $\varepsilon_{n,Jm}^{(1)} = 0$) one has collision–induced rotationally inelastic scattering which leads to additional peaks in the angular distribution of scattered molecules because energy losses due to transitions $J = 0 \rightarrow J' = 1, 2, 3, \ldots$ decrease the z-component of the He wavevector. Thus diffracted beams for all $\boldsymbol{G}$ vectors are displaced by discrete amounts in the angular distributions.

First experimental evidence for RMSA was published in 1981 when *Cowin* et al. [7.153] reported on the collision-induced translation-to-rotation energy transfer of HD $(J = 0)$ scattering from Pt(111), and on the observation of sharp modulations of these $J = 0 \rightarrow J' = 0, 1, 2, \ldots$ rotational transition probabilities as a function of incident angle, see Fig.7.29. These modulations were ascribed to RMSA where the final HD rotational state was $J_n = (J' + 1)$ of a (nearly) free rotor in the n-th BS in the physisorption potential. Analysis of these resonances determined for the first time BS energies of a molecule on a *smooth metallic surface*.

Stimulated by the work of *Cowin* et al. [7.153] *Schinke* in 1982 performed model calculations which computationally verified the existence of SAR in rotationally elastic and inelastic molecule–surface scattering [7.155]. The various rotational transition probabilities within the specular diffraction channel $(\boldsymbol{G} = (0,0))$ were calculated within an approximation to the exact close-coupling treatment of diatom–rigid–surface scattering. The rotational degree of freedom was treated exactly whereas the diffraction channels were decoupled in the sudden approximation established by *Gerber* et al. [7.176].

Shortly after that RMSA of HD on Pt(111) was also examined theoretically by *Whaley* et al. [7.154] using Wigner R-matrix scattering techniques (treating the molecule as a rigid rotor and neglecting diffraction). With a laterally averaged surface–molecule Morse potential interaction together with an anisotropic potential term (transformed from H_2), excellent agreement was obtained between the resonances and the (first-order perturbed) bound state–free rigid rotor energies.

Schinke continued his theoretical studies on rotationally inelastic scattering and RMSA and in [7.156] presented detailed calculations performed at two levels of accuracy:

(i) exact close-coupling calculations including both rotational and diffractional states in the expansion of the total scattering wavefunction, and

(ii) diffractionally sudden calculations, which however treated the rotational degree of freedom exactly.

All calculations assumed a rigid surface (i.e. no interactions with phonons and no thermal motion of surface atoms) and treated the scattered HD molecules as rigid rotors. As an example of Schinke's results we reproduce from [7.156] a calculation for a flat surface varying the incidence angle and including averaging over the energy spread in the inital beam. These curves can be compared to experimental data for HD on Pt(111) of *Cowin* et al. [7.153]. All results together are shown in Fig.7.29 and one observes good agreement between experiment and calculations for several rotationally inelastic transition probabilities including the appearance of RMSA resonances.

Next *Yu* et al. [7.109] measured DMSA and RMSA resonances for the scattering of n-H_2, p-H_2, n-D_2 and o-D_2 on Ag(111). This was the first clear observation of DMSA on a close-packed (low corrugation) metallic surface, Ag(111) (cf. Sect. 7.4.1 above), and the first detection of RMSA for homonuclear diatomics, the potential well depth for the system H_2/Ag(111) was deduced as $D \simeq 32$ meV and small energy shifts and linewidth differences were observed between n-H_2 and p-H_2-SAR which were attributed to a weak anisotropy of the molecule–surface interaction.

In [7.157] *Whaley* and *Light* concentrated on a nonpropagative Wigner R-matrix theory for rotationally inelastic scattering of HD from smooth metal surfaces. The model again assumed scattering of rigid rotors from stationary surfaces but additionally used an optical potential to describe energy dissipation by phonon effects. The agreement between calculated rotational transition probabilities including RMSA resonances and the experimental results of Cowin et al. was of similar quality as that of *Schinke*'s results [7.156] which are shown in Fig.7.29

In 1985 *Cowin* et al. presented a full report [7.158] on the HD-scattering experiment on Pt(111) in which RMSA was first discovered [7.153]. Seven BS energy levels were now located through RMSA resonances in accordance with the assignment given by *Schinke* [7.156], inferring a well depth of 55 meV for this system. With an experimental resolution by far exceeding that of electron or neutron scattering experiments, it could be shown that the rotational hindering of HD on Pt is small ($\leq$ 1–2 meV) compared to BS energy level spacings (10 meV) or rotational energy (11 meV for $J = 1$).

Another point of view was adopted by *Stiles* and *Wilkins* [7.160] when they theoretically investigated the sticking probability of H_2 and HD on noble metal surfaces. One-phonon distorted-wave Born-approximation calculations of the

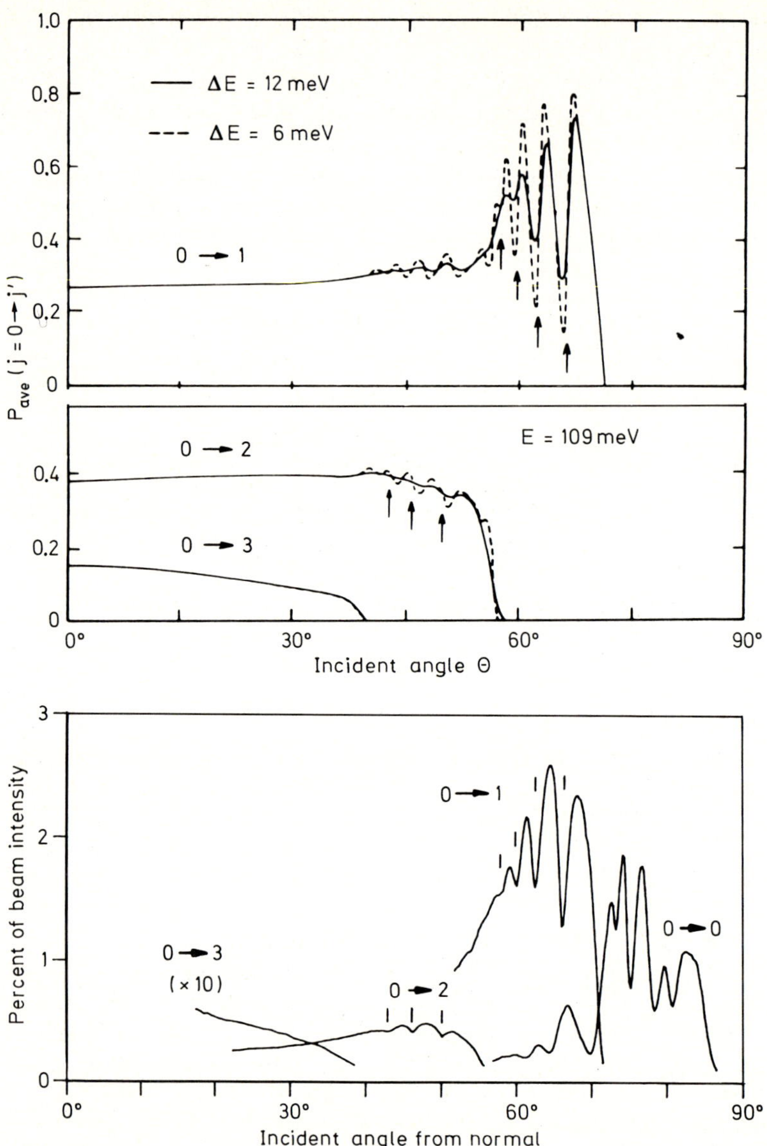

Fig.7.29. Rotational transition probabilities ($J \rightarrow J'$) in the scattering of HD ($j = 0$) from Pt(111). Lower part: experimental peak height versus incident angle for several rotationally inelastic transitions ($E_i = 109$ meV, $\langle 101 \rangle$ azimuth) (from [7.153]). Vertical lines indicate resonance dips used to estimate BS energies. Upper part: calculated rotational transition probabilities versus incident angle for $E_i = 109$ meV, energy averaged for two FWHM, $\Delta E = 6$ meV and $\Delta E = 12$ meV, energy spread of the incident beam (from [7.156]). The arrows mark the experimental resonance angles from [7.153]

164

phonon-assisted sticking probabilities yielded peaks as a function of incident energy due to RMSA resonances.

The strong phonon scattering between a rotationally excited trapped state and a rotationally deexcited bound final state was found to lead to a large enhancement of sticking by these resonances.

In a subsequent paper [7.159] *Rettner* et al. then addressed the question of whether RMSA could also provide a route to dissociative chemisorption and thus make important contributions to its probability. Experimental studies of the fate of selectively adorbed HD molecules on W(110) — in particular through RMSA — were presented. However, no correlation was found between chemisorption and SAR although molecules which underwent RMSA were apparently trapped in a physisorbed state at the surface. Most probably these molecules were released from that BS by diffuse scattering before chemisorption occurred.

Also in 1985 the group at Chicago published comprehensive studies of the scattering of rotationally state-selected H_2, D_2 and HD from Ag(111) and investigated both the spatially isotropic and anisotropic component of the laterally averaged molecular hydrogen/Ag(111) physisorption potential.

In two papers [7.162, 163] DMSA and RMSA resonances were mapped out and compared to Wigner R-matrix calculations as well as exact close-coupling quantum scattering calculations.

Several model potentials were checked and a well depth of $D = 32$ meV was derived from the data. Both the attractive and repulsive parts of the anisotropic potential exhibited only a weak orientation dependence but the RMSA-resonances in particular showed well resolved J- and m-dependent energy shifts resulting in part from the oriential anisotropy of the potential. These two papers [7.162, 163] provided the most comprehensive studies of a low energy molecule–surface potential available to that date and demonstrated the capabilities of molecule–surface scattering experiments and corresponding calculations.

Similar to *Stiles* and *Wilkins* in [7.160], *Andersson* et al. [7.161] concentrated on the initial sticking probability of molecular hydrogen on a cold (15K) Cu(100) surface, using n-H_2, p-H_2, n-D_2 and o-D_2 nozzle beams with incident energies of 20–75 meV and observed maxima occurring when the incident energy equalled a rotational excitation energy. A later interpretation assumed the presence of DMSA and RMSA (or both mixed) and the possible decay of the molecules trapped in BS also via sticking; see [7.134].

A theoretical study in connection with the microscopic dynamical description of sticking probabilities by *Kaufhold* and *Toennies* followed in 1986 [7.165] with emphasis on contributions through RMSA resonances. These authors derived an expression for the strength of the optical potential for describing the attenuation of HD molecules undergoing RMSA on smooth metal surfaces by equating a special optical potential ansatz to the Debye-Waller factor. Close coupling calculations with this optical potential for RMSA yielded a good fit of the experimental results of *Cowin* et al. [7.153] (in a modified representation) for HD–Ag(111). This success indicated that phonon interactions —as opposed to electron–hole pair interactions— indeed provide an important process leading to trapping of molecules as proposed earlier by *Stiles* and *Wilkins* [7.160] (see above) assuming a single-phonon coupling model. Now, however, the important role of multi-phonon processes was elucidated and realistically described.

A new method for measuring RMSA (rotational Feshbach) resonances was then published by *Harten* et al. [7.164]. The new technique involves the measurement of the velocity dependence of the reflection coefficient of the specularly scattered HD beam with an initially broad distribution by a TOF method over a wide range of beam energies (a very similar principle was applied for DMSA resonances by *Semerad* et al. shortly after [7.125], see Sect. 7.4.1).

At the resonance energy, molecules may be trapped and delayed resulting in strong interference effects producing sharp dips or peaks in the velocitiy dependence of the scattered intensity. Applying a high resolution TOF technique the energy resolution in the RMSA measurements is not limited by the relative energy spread of the primary beam as in the method of *Cowin* et al. [7.153]. The efficiency of the new technique was demonstrated for the first time by measurements with HD and D_2 on Cu(111) and Au(111). An example is shown in Fig.7.30 (for more details see the original paper).

The results were analyzed by comparison with a close-coupling calculation to provide potential well parameters (for a model potential of type exponential repulsive plus z^{-3}-attractive, both parts with a $P_2(\cos\Theta)$ anisotropy term). The deduced well depths were 22.2 ± 0.1 meV for HD–Cu(111) in perfect agreement with previous results and 40.7 ± 0.2 meV for the new system HD–Au(111) which therefore exceeds that of HD–Cu(111) and that of HD–Ag(111) with $D \simeq 32$ meV [7.162].

Very recently *Andersson* et al. [7.134] published comprehensive experimental and theoretical studies on the sticking probability of H_2 and D_2 on a cold (10K) Cu(100) surface, thus extending preceding work [7.135, 161] and providing new interpretations and conclusions. Experimentally the initial sticking coefficient, S_0, determined using partial monolayer desorption, was measured for H_2 and D_2 molecular beams in the energy range $8-45$ meV and for angles of incidence between 0° and 60°.

S_0 displayed a background that fell off with increasing energy on which well-defined peaks are superposed at characteristic energies that all coincided with features in the specular reflectivity and with one of the conditions for DMSA, RMSA or DRMSA (i.e. mixed resonances). These BS were found to decay via backscattering and sticking channels, the latter giving important contributions to the resonance width.

Comparison was made to calculations using a single-phonon distorted-wave Born-approximation and a multiphonon forced oscillator model yielding satisfactory account for most of the data including the strengths of DMSA resonances in S_0 and the specular reflectivity, whereas RMSA (DRMSA) resonances tended to be weaker (stronger) than expected. From the observed large background sticking it was concluded that positive-energy trapping on initial collision prevails at large incident angles, but that many of the trapped par-

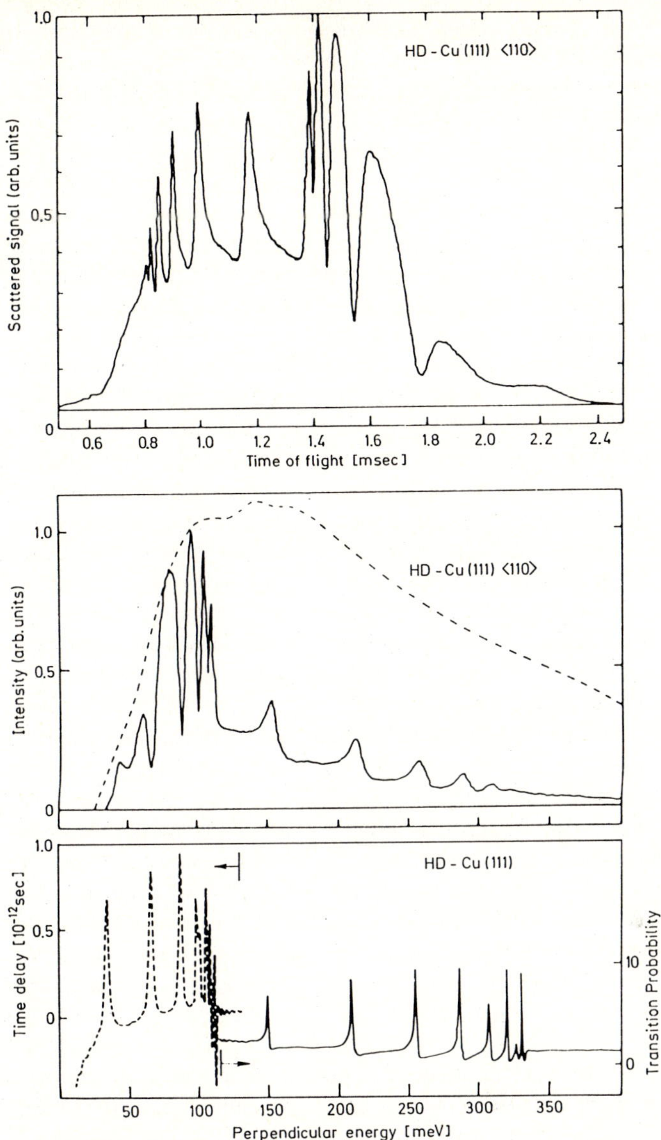

Fig.7.30. Measured time-of-flight spectra and perpendicular energy distributions for HD–Cu(110) and D_2–Cu(111) both along the ⟨110⟩ azimuth. The experimental time-of-flight curve in (a) has been transformed into a distribution over the perpendicular energy in (b). Measurements for D_2–Cu(111) are shown as dashed line and here only very weak resonances are seen, thus this curve essentially shows the distributions without resonances. (c): calculated relative time delays (left part) and transition probabilities (right part) for HD–Cu(111). (The experimental time delays were measured with respect to that of the nonresonant background which has been arbitrarily set to zero). Note that the maxima in the calculated time delays are expected to correlate with minima in the measured distribution, whereas maxima in the transition probability are expected to correlate with the measured maxima. From [7.164]

ticles are re-released into the gas phase without contributing to the sticking coefficient.

7.5 Conclusions

In the very first gas–surface diffraction experiments, conducted by Estermann, Frisch and Stern in the early 1930s to demonstrate the wave nature of atoms and molecules, pronounced intensity minima were already observed.

A few years later in 1936, Lennard-Jones and Devonshire explained the observed features in terms of resonant, diffraction-mediated transitions to bound surface states. They called this phenomenom "selective adsorption" (SA) and showed the potential of SA-investigations to yield information on the physical gas–surface interaction potential.

But not until the 1970s did the investigation of resonances develop into a frequently applied method to determine binding energies in the gas–surface potential. The gas–surface systems first investigated in detail were He, but also H, D, H_2 and D_2 on relatively easy to handle alkali halide surfaces and on graphite (0001). The binding energies could be used to determine the parameters of appropriate model potentials or to check the quality of potentials obtained from summation of pair potentials between the gas atom and the atoms or ions forming the crystal surface.

The resonant intensity variation in complete angular distributions observed for the above systems was used to test methods of gas–surface diffraction theory, and by the end of the 1970s close-coupling calculations were able to describe almost all features observed for systems without too much inelastic scattering.

This advance in the theory led, in the 1980s, to studies of diffraction from strongly corrugated adsorbate-systems with many overlapping and interacting resonances. On the other hand, the experimental advances in resolution and sensitivity of detection made it possible to observe resonances even on weakly corrugated metal surfaces. Thus it is now possible, for a relatively wide range of surfaces, to investigate the physical interaction with light atoms or molecules by studying SA resonances.

The discovery of two additional channels which, besides diffraction, also lead to bound state resonances, may further extend the field of accessible gas–surface systems. Rotationally mediated selective adsorption (RMSA) was successfully applied to find bound state energies for HD molecules interacting with flat metal surfaces. Phonon-mediated selective adsorption (PMSA) together with time of flight (TOF) analysis of scattered atoms was used to complement the information on bound state levels from diffraction–mediated selective adsorption (DMSA) but can also supply information on gas–atom–phonon interaction without TOF analysis.

Thus it is clear that selective adsorption resonances, although first discovered 60 years ago, have developed only in the past 20 years into a valuable source of information about physical gas–surface interactions.

168

References

7.1   C. Davisson, L.H. Germer: Phys. Rev. **30**, 705 (1927)

7.2   O. Stern: Naturwissenschaften **17**, 391 (1929)

7.3   I. Estermann, O. Stern: Z. Phys. **61**, 95 (1930)

7.4   I. Estermann, R. Frisch, O. Stern: Z. Phys. **73**, 348 (1931)

7.5   R. Frisch, O. Stern: Naturwissenschaften **20**, 721 (1932)

7.6   R. Frisch, O. Stern: Z. Phys. **84**, 430 (1933)

7.7   T.H. Johnson: Phys. Rev. **35**, 1299 (1930)

7.8   T.H. Johnson: Phys. Rev. **37**, 847 (1931)

7.9   R. Frisch, O. Stern: Handbuch der Physik XXII/2, 337 (1933)

7.10   R. Frisch: Z. Phys. **84**, 443 (1933)

7.11   J.E. Lennard-Jones, A.F. Devonshire: Nature **137**, 1069 (1936)

7.12   A.F. Devonshire: Proc. Roy. Soc. London **A156**, 37 (1936)

7.13   J.E. Lennard-Jones, A.F. Devonshire: Proc. Roy. Soc. London **A158**, 253 (1937)

7.14   J.C. Crews: In *Fundamentals of Gas-Surface Interaction*, ed. by H. Saltsburg, J.N. Smith, M. Rogers (Academic, New York 1967) p. 480

7.15   D.R. O'Keefe, J.N. Smith, Jr., R.L. Palmer, H. Saltsburg: J. Chem. Phys. **52**, 4447 (1970)

7.16   H. Hoinkes, H. Nahr, H. Wilsch: Entropie **42**, 111 (1971)

7.17   H. Hoinkes. H. Nahr, H. Wilsch: Surf.Sci. **30**, 363 (1972)

7.18   H. Hoinkes, H. Nahr, H. Wilsch: J. Phys. C **5**, L 143 (1972)

7.19   J.R. Bledsoe: Ph.D. thesis, University of Virginia (1972)

7.20   D.E. Houston, D.R. Frankl: Phys. Rev. Lett. **31**, 298 (1973)

7.21   H. Hoinkes, H. Nahr, H. Wilsch: Proc. 8th Intern. Symp. Rarefied Gas Dynamics, Stanford 1972; Rarefied Gas Dynamics (Academic, New York 1974) p.421

7.22   J.A. Meyers, D.E. Houston, D.R. Frankl: Proc. 6th Intern. Vacuum Congr. and 2nd Intern. Conf. on Solid Surfaces, Kyoto 1974; Jap. J. Appl. Phys. suppl.2, Part 2 (1974) p. 563

7.23   H. Wilsch, H.-U. Finzel, H. Frank, H. Hoinkes, H. Nahr: Proc. 6th Intern. Vacuum Congr. and 2nd Intern. Conf. on Solid Surfaces, Kyoto 1974; Jap. J. Appl. Phys. suppl. 2, Part 2 (1974) p. 567

7.24   B. Wood, B.F. Mason, B.R. Williams: J. Chem. Phys. **61**, 1435 (1974)

7.25   H.-U. Finzel, H. Frank, H. Hoinkes, M. Luschka, H. Nahr, H. Wilsch, U. Wonka: Surf. Sci. **49**, 577 (1975)

7.26   J.A. Meyers, D.R. Frankl: Surf. Sci. **51**, 61 (1975)

7.27   G. Boato, P. Cantini, L. Mattera: Surf. Sci. **55**, 141 (1976)

7.28   M.P. Liva, G. Derry, D.R. Frankl: Phys. Rev. Lett. **37**, 1413 (1976)

7.29   M.P. Liva, D.R. Frankl: Surf. Sci. **59**, 643 (1976)

7.30   H. Frank, H. Hoinkes, H. Wilsch: Surf. Sci. **63**, 121 (1977)

7.31   H. Hoinkes, L. Greiner, H. Wilsch: Proc. 7th Intern. Vacuum Congr. and 3rd Intern. Conf. on Solid Surfaces, Vienna 1977 edited by R. Dobrozemsky et al. (Berger, Vienna 1977) p. 1349

7.32   G. Derry, D. Wesner, S.V. Krishnaswamy, M.W. Cole, D.R. Frankl: Proc. 7th Intern. Vacuum Congr. and 3rd Intern. Conf. on Solid Surfaces, Vienna 1977 edited by R. Dobrozemsky et al. (Berger, Vienna 1977) p. 1353

7.33   P. Cantini, G.P. Felcher, R. Tatarek: Proc. 7th Intern. Vacuum Congr. and 3rd Intern. Conf. on Solid Surfaces, Vienna 1977 edited by R. Dobrozemsky et al. (Berger, Vienna 1977) p. 1357

7.34   P. Cantini, R. Tatarek, G. Felcher: Surf. Sci. **63**, 104 (1977)

7.35   G. Derry, D. Wesner, S.V. Krishnaswamy, D.R. Frankl: Surf. Sci. **74**, 245 (1978)

7.36   G. Boato, P. Cantini, R. Tatarek: Phys. Rev. Lett. **40**, 887 (1978)

7.37   D.R. Frankl, D. Wesner, S.V. Krishnaswamy, G. Derry, T.J. O'Gorman: Phys. Rev. Lett. **41**, 60 (1978)

7.38 S.V. Krishnaswamy, G. Derry, D. Wesner, T.J. O'Gorman, D.R. Frankl: Surf. Sci. **77**, 493 (1978)

7.39 P. Cantini, R. Tatarek, G. Felcher: Phys. Rev. B **19**, 1161 (1979)

7.40 G. Boato, P. Cantini, R. Tatarek, G.P. Felcher: Surf. Sci. **80**, 518 (1979)

7.41 G. Boato, P. Cantini, C. Guidi, R. Tatarek, G.P. Felcher: Phys. Rev. B **20**, 3957 (1979)

7.42 G. Derry, D. Wesner, W. Carlos, D.R. Frankl: Surf. Sci. **87**, 629 (1979)

7.43 J. Lapujoulade, Y. Lejay, N. Papanicolaou: Surf. Sci **90**, 133 (1979)

7.44 W.E. Carlos, G. Derry, D.R. Frankl: Phys. Rev. B **19**, 3258 (1979)

7.45 L. Mattera, F. Rosatelli, C. Salvo, F. Tommasini, U. Valbusa, G. Vidali: Surf. Sci. **93**, 515 (1980)

7.46 L. Greiner, H. Hoinkes, H. Kaarmann, H. Wilsch, N. Garcia: Surf. Sci. **94**, L195 (1980)

7.47 G. Derry, D. Wesner, G. Vidali, T. Thwaites, D.R. Frankl: Surf. Sci. **94**, 221 (1980)

7.48 D. Wesner, G. Derry, G. Vidali, T. Thwaites, D.R. Frankl: Surf. Sci. **95**, 367 (1980)

7.49 E. Ghio, L. Mattera, C. Salvo, F. Tommasini, U. Valbusa: J. Chem. Phys. **73**, 556 (1980)

7.50 K.H. Rieder, T. Engel, N. Garcia: Proc. 4th Intern. Conf. on Solid Surfaces, Cannes 1980, Supplément à la Revue "Le Vide, les Couches Minces" **201**, 861 (1980)

7.51 S. Iannotta, U. Valbusa: Surf. Sci. **100**, 28 (1980)

7.52 J. Lapujoulade, Y. Le Cruër, M. Lefort, Y. Lejay, E. Maurel: Phys. Rev. B **22**, 5740 (1980) and B **24**, 3667 (1981)

7.53 L. Greiner, H. Hoinkes, H. Wilsch: unpublished results

7.54 F.O. Goodman: CRC Crit. Rev. Solid State Mater. Sci. **7**, 33 (1977)

7.55 F.O. Goodmanm, H.Y. Wachman: *Dynamics of Gas–Surface Scattering* (Academic, New York 1976)

7.56 K.L. Wolfe, D.J. Malik, J.H. Weare: J. Chem. Phys. **67**, 1031 (1977)

7.57 M.W. Cole, D.R. Frankl, D.L. Goodstein: Rev. Mod. Phys. **53**, 199 (1981)

7.58 H. Hoinkes: Rev. Mod. Phys. **52**, 933 (1980)

7.59 J.E. Lennard-Jones, A.F. Devonshire: Proc. Roy. Soc. London **A158**, 242 (1937)

7.60 J.L. Beeby: J. Phys. C **4**, 359 (1971)

7.61 A. Tsuchida: Surf. Sci. **14**, 375 (1969)

7.62 N. Cabrera, V. Celli, F.O. Goodman, R. Manson: Surf. Sci. **19**, 67 (1970)

7.63 F.O. Goodman: Surf. Sci. **19**, 93 (1970)

7.64 F.O. Goodman: Surf. Sci. **94**, 507 (1980)

7.65 G. Wolken, Jr.: J. Chem. Phys. **58**, 3047 (1973)

7.66 H. Chow, E.D. Thompson: Surf. Sci. **54**, 269 (1976)

7.67 H. Chow, E.D. Thompson: Surf. Sci. **59**, 225 (1976)

7.68 H. Chow: Surf. Sci. **66**, 221 (1977)

7.69 H. Chow: Surf. Sci. **79**, 157 (1979)

7.70 H. Chow, E.D. Thompson: Surf. Sci. **82**, 1 (1979)

7.71 C.E. Harvie, J.H. Weare: Phys. Rev. Lett. **40**, 187 (1978)

7.72 K.L. Wolfe, J.H. Weare: Phys. Rev. Lett. **41**, 1663 (1978)

7.73 K.L. Wolfe, C.E. Harvie, J.H. Weare: Sol. State Commun. **27**, 1293 (1978)

7.74 K.L. Wolfe, J.H. Weare: Surf. Sci. **94**, 581 (1980)

7.75 N. Garcia, V. Celli, F.O. Goodman: Phys. Rev. B **19**, 634 (1979)

7.76 F.O. Goodman, N.Garcia, V. Celli: Surf. Sci. **85**, 317 (1979)

7.77 V. Celli, N. Garcia, J. Hutchison: Surf. Sci. **87**, 112 (1979)

7.78 N. Garcia, W.E. Carlos, M.W. Cole, V. Celli: Phys. Rev. B **21**, 1636 (1980)

7.79 J. Hutchison, V. Celli: Surf. Sci. **93**, 263 (1980)

7.80 J.S. Hutchison: Phys. Rev. B **22**, 5671 (1980)

7.81 W.E. Carlos, M.W. Cole: Surf. Sci. **77**, L173 (1978)

7.82 W.E. Carlos, M.W. Cole: Phys. Rev. Lett. **43**, 697 (1979)

7.83 W.E. Carlos, M.W. Cole: Phys. Rev. B **21**, 3713 (1980)

7.84 W.E. Carlos, M.W. Cole: Surf. Sci. **91**, 339 (1980)

7.85 J. Garcia-Sanz, N. Garcia: Surf. Sci. **79**, 125 (1979)

7.86 Y. Hamauzu: Phys. Lett. **57A**, 275 (1976)

7.87 Y. Hamauzu: J. Phys. Soc. Jap. **42**, 961 (1977)

7.88 R.J. LeRoy: Surf. Sci. **59**, 541 (1976)

7.89 J.G. Mantovani, J.R. Manson, G. Armand: Surf. Sci. **143**, 536 (1984)

7.90 T.H. Ellis, S. Iannotta, G. Scoles, U. Valbusa: J. Vac. Sci. Technol. **18**, 488 (1981)

7.91 D.A. Wesner, D.R. Frankl: Phys. Rev. B **24**, 1798 (1981)

7.92 G. Brusdeylins, R.B. Doak, J.P. Toennies: J. Chem. Phys. **75**, 1784 (1981)

7.93 P. Cantini, S. Terreni, C. Salvo: Surf. Sci. **109**, L491 (1981)

7.94 J.M. Soler, V. Celli, N. Garcia, K.H. Rieder, T. Engel: Surf. Sci. **108**, 1 (1981)

7.95 M.J. Cardillo, G.E. Becker, S.J. Sibener, D.R. Miller: Surf. Sci. **107**, 469 (1981)

7.96 J. Lapujoulade, Y. Le Cruer, M. Lefort, Y. Lejay, E. Maurel: Surf. Sci. **118**, 103 (1982)

7.97 T.H. Ellis, G. Scoles, U. Valbusa: Surf. Sci. **118**, L251 (1982)

7.98 J. Perreau, J. Lapujoulade: Surf. Sci. **119**, L292 (1982)

7.99 A. Luntz, L. Mattera, M. Rocca, F. Tommasini, U. Valbusa: Surf. Sci. **120**, L447 (1982)

7.100 L. Baetz, H. Hoinkes, H. Wilsch: Surf. Sci. **120**, L417 (1982)

7.101 J. Perreau, J. Lapujoulade: Surf. Sci. **122**, 341 (1982)

7.102 G. Vidali, D.R. Frankl: Phys. Rev. B **27**, 2480 (1983)

7.103 R. Schinke, A.C. Luntz: Surf. Sci. **124**, L60 (1983)

7.104 G.E. Becker, M.J. Cardillo, J.A. Serri, D.R. Hamann: Phys. Rev. B **28**, 504 (1983)

7.105 G. Benedek, G. Brusdeylins, R.B. Doak, J.G. Skofronik, J.P. Toennies: Phys. Rev. B **28**, 2104 (1983)

7.106 B. Salanon, J. Lapujoulade: Proc. IX IVC – V ICSS, Madrid (1983) Ext. Abstracts p. 23

7.107 K.H. Rieder, W.Stocker, J. Phys. C.: Solid State Phys. **16**, L783 (1983)

7.108 G. Bracco, P. Cantini, A. Glachant, R. Tatarek: Surf. Sci. **125**, L81 (1983)

7.109 C.F. Yu, K.B. Whaley, C.S. Hogg, S.J. Sibener: Phys. Rev. Lett. **51**, 2210 (1983)

7.110 G. Bracco, P. Cantini, E. Cavanna, R. Tatarek, A. Glachant: Surf. Sci. **136**, 169 (1984)

7.111 W.Y. Leung, J.Z. Larese, D.R. Frankl: Surf. Sci. **136**, 649 (1984)

7.112 W.Y. Leung, J.Z. Larese, D.R. Frankl: Surf. Sci. **143**, L398 (1984)

7.113 H. Jónsson, J.H. Weare: Phys. Rev. B **30**, 4203 (1984)

7.114 D. Gorse, B. Salanon, F. Fabre, A. Kara, J. Perreau, G. Armand, J. Lapujoulade: Surf. Sci. **147**, 611 (1984)

7.115 T. Engel, K.H. Rieder, I.P. Batra: Surf. Sci. **148**, 321 (1984)

7.116 P. Cantini, E. Cevasco: Surf. Sci. **148**, 37 (1984)

7.117 M. Chiesa, L. Mattera, R. Musenich, C. Salvo: Surf. Sci. **151**, L145 (1985)

7.118 J.Z. Larese, W.Y. Leung, D.R. Frankl, N. Holter, S. Chung, M.W. Cole: Phys. Rev. Lett. **54**, 2533 (1985)

7.119 S. Iannotta, G. Scoles, U. Valbusa: Surf. Sci. **161**, 429 (1985)

7.120 L. Danielson, J.-C. Ruiz, C. Schwartz, G. Scoles, J.M. Hutson: paper prepared for the Faraday Discussion, McMaster University, 23–25 July 1985

7.121 S. Chung, A. Kara, D.R. Frankl: Surf. Sci. **171**, 45 (1986)

7.122 D.S. Kaufman, L.R. Allen, E.H. Conrad, R.M. Aten, T. Engel: Surf. Sci. **173**, 517 (1986)

7.123 M.G. Dondi, L. Mattera, S. Terreni, F. Tommasini, U. Linke: Phys. Rev. B **34**, 5897 (1986)

7.124 D.S. Kaufman, R.M. Aten, E.H. Conrad, L.R. Allen, T. Engel: J. Chem. Phys. **86**, 3682 (1987)

7.125 S. Semerad, P. Sequard-Base, E.M. Hörl: Surf. Sci. **189/190**, 975 (1987)

7.126 B. Salanon, J. Lapujoulade: Phys. Rev. B **36**, 4428 (1987)

7.127 M.G. Dondi, S. Terreni, F. Tommasini, U. Linke: Phys. Rev. B **37**, 8034 (1988)

7.128 K.D. Gibson, C.Cerjan, J.C. Light, S.J. Sibener: J. Chem. Phys. **88**, 7911 (1988)

7.129 S. Andersson, L. Wilzén, M. Persson: Phys. Rev. B **38**, 2967 (1988)

7.130 P. Cantini, L. Mattera, M.F.M. De Kieviet, K. Jalink, C. Tassistro, S. Terreni, U. Linke: Surf. Sci. **211/212**, 872 (1989)

7.131 M. Mahgerefteh, D.R. Jung, D.R. Frankl: Phys. Rev. B **39**, 3900 (1989)

7.132 G. Parschau, E. Kirsten, A. Bischof, K.H. Rieder: Phys. Rev. B **40**, 6012 (1989)

7.133 E. Kirsten, G. Parschau, K.H. Rieder: Phys. Rev. B41, 5392 (1990)

7.134 S. Andersson, L. Wilzén, P. Persson, J. Harris: Phys. Rev. B 40, 8146 (1989)

7.135 S. Andersson, L. Wilzén, J. Harris: Phys. Rev. Lett. 57, 1603 (1986)

7.136 R.A. Aziz, U. Buck, H. Jónsson, J.C. Ruiz-Suárez, B. Schmidt, G. Scoles, M.J. Slaman, J. Xu: J. Chem. Phys. 91, 6477 (1989)

7.137 D.R. Jung, J. Cui, D.R. Frankl, G.Ihm, H.-Y. Kim, M.W. Cole: Phys. Rev. B 40, 11893 (1989)

7.138 G. Armand, J. Lapujoulade, J.R. Manson: Phys. Rev. B 39, 10514 (1989)

7.139 E. Kirsten, G. Parschau, K.H. Rieder: Surf. Sci. 236, L365 (1990)

7.140 P. Cantini, G.P. Felcher, R. Tatarek: Phys. Rev. Lett. 37, 606 (1976)

7.141 P. Cantini, R. Tatarek: Proc. 4th ICSS and 3rd ECOSS, Vol. II, p.830 Cannes 1980

7.142 P. Cantini, R. Tatarek: Phys. Rev. B 23, 3030 (1981)

7.143 G. Brusdeylins, R.B. Doak, J.P. Toennies: J. Chem. Phys. 75, 1784 (1981)

7.144 P. Cantini, R. Tatarek: Surf. Sci. 114, 471 (1982)

7.145 G. Boato, P. Cantini, C. Salvo, R. Tatarek, S. Terreni: Surf. Sci. 114, 485 (1982)

7.146 G. Lilienkamp, J.P. Toennies: Phys. Rev. B 26, 4752 (1982)

7.147 G. Brusdeylins, R.B. Doak, J.G. Skofronick, J.P. Toennies: Surf. Sci. 128, 191 (1983)

7.148 G. Lilienkamp and J.P. Toennies: J. Chem. Phys. 78, 5210 (1983)

7.149 D. Evans, V. Celli, G. Benedek, J.P. Toennies, R.B. Doak: Phys. Rev. Lett. 50, 1854 (1983)

7.150 D. Eichenauer, J.P. Toennies: J. Chem. Phys. 85, 532 (1986)

7.151 W.L. Nichols, J.H. Weare: Phys. Rev. Lett. 56, 753 (1986)

7.152 G. Benedek, G. Brusdeylins, J.P. Toennies, R.B. Doak: Phys. Rev. B 27, 2488 (1983)

7.153 J.P. Cowin, C.F. Yu, S.J. Sibener, J.E. Hurst: J. Chem. Phys. 75, 1033 (1981)

7.154 K.B. Whaley, J.C. Light, J.P. Cowin, S.J. Sibener: Chem. Phys. Lett. 89, 89 (1982)

7.155 R. Schinke: Chem. Phys. Lett. 87, 438 (1982)

7.156 R. Schinke: Surf. Sci. 127, 283 (1983)

7.157 K.B. Whaley, J.C. Light: J. Chem. Phys. 81, 2144 (1984)

7.158 J.P. Cowin, C.F. Yu, L. Wharton: Surf. Sci. 161, 221 (1985)

7.159 C.T. Rettner, L.A. Delouise, J.P. Cowin, D.J. Auerbach: Chem. Phys. Lett. 118, 355 (1985)

7.160 M.D. Stiles, J.W. Wilkins: Phys. Rev. Lett. 54, 595 (1985)

7.161 S. Andersson, L. Wilzen, J. Harris: Phys. Rev. Lett. 55, 2591 (1985)

7.162 C.F. Yu, K.B. Whaley, C.S. Hogg, S.J. Sibener: J. Chem. Phys. 83, 4217 (1985)

7.163 K.B. Whaley, C.F. Yu, C.S. Hogg, J.C. Light, S.J. Sibener: J. Chem. Phys. 83, 4235 (1985)

7.164 U. Harten, J.P. Toennies, Ch. Wöll: J. Chem. Phys. 85, 2249 (1986)

7.165 A. Kaufhold, J.P. Toennies: Surf. Sci. 173, 320 (1986)

7.166 V. Celli, D. Eichenauer, A. Kaufhold, J.P. Toennies: J. Chem. Phys. 83, 2504 (1985)

7.167 J. Estel, H. Hoinkes, H. Kaarmann, H. Nahr, H. Wilsch: Surf. Sci. 54, 393 (1976)

7.168 G. Vidali, M.W. Cole, J.R. Klein: Phys. Rev. B 28, 3064 (1983)

7.169 F. Toigo, M.W. Cole: Phys. Rev. B 32, 6989 (1985) and Phys. Rev. B 33, 4330 (1986)

7.170 J. Böheim: Surf. Sci. 148, 463 (1984)

7.171 P. Nordlander, J. Harris: J. Phys. C17, 114 (1984)

7.172 P. Nordlander, C. Holmberg, J. Harris: Surf. Sci. 152/153, 702 (1985)

7.173 D.V. Tendulkar, R.E. Stickney: Surf. Sci. 27, 516 (1971)

7.174 J.M. Hutson, C. Schwartz: J. Chem. Phys. 79, 5179 (1983)

7.175 G. Bracco: M. Sc. Thesis, University of Genova (1981)

7.176 R.B. Gerber, A.T. Yinnon, J.N. Murrell: Chem. Phys. 31, 1 (1978)

7.177 K. Wolfe-Brannon, J.H. Weare: Phys. Rev. B 24, 5753 (1981)

8. Theoretical Aspects of Atom–Surface Scattering

J.R. Manson

Atom scattering at thermal energies has proven to be one of the most sensitive experimental methods for obtaining detailed microscopic information on surfaces [8.1]. In some cases the necessary theory is very simple, as for example in identifying surface structures from the positions of diffraction peaks, or obtaining surface phonon dispersion relations from the positions of the phonon peaks in an inelastic experiment. In most other cases, however, sophisticated theory which often involves intensive numerical calculations is necessary in order to fully exploit the extreme sensitivity of the method to surface structure, disorder and surface vibrations. As with thermal neutrons, the wavelengths of small mass and low energy atoms such as He are comparable to interparticle spacings in solids, and the energies are comparable to maximum crystal phonon energies. Thus such particles are ideally suited for studies of both surface structure and surface vibrations. The theory of scattering of atoms from an extended target such as a surface has similarities with many of the highly developed techniques used to interpret scattering from bulk solids or liquids, as for example neutron, X-ray or electron scattering. The major difference from bulk scattering is that the presence of the surface breaks the translational symmetry normal to the surface, hence momentum is no longer conserved in that direction. One immediate consequence of this is that diffraction peaks from ordered surfaces are two-dimensional in character and can be observed for all incident beam conditions.

The potential describing the interaction of an atom with a surface has a long range attractive part coming from van der Waals forces and a short range, strongly repulsive part arising from exchange forces when the atomic electron cloud overlaps the surface electronic distribution. In many cases a model potential is chosen in which the parameters can be varied to suit the characteristics of the surface under investigation. More detailed potentials are developed from pairwise summations of atomic and molecular interactions, but even this is not completely adequate as many-body forces are non-negligible especially at close distances. The detailed nature of the potential and how to develop it is the subject of another chapter in this volume (Chap.3). Although one of the more important tasks of scattering theory is to aid in determining the potential, we

173

Springer Series in Surface Sciences, Vol. 27   Helium Atom Scattering from Surfaces
Editor: E. Hulpke   © Springer-Verlag Berlin, Heidelberg 1992

will assume here that the potential is understood and set about the task of evaluating the scattering amplitudes and intensities.

Once the potential is established, the problem reduces to finding solutions of Schrödinger's equation. Even in the case of elastic scattering from a rigid, ordered surface, which is a straightforward single-body problem, the lack of symmetry and the extended nature of the potential means that solution methods will not always be simple. In the case of inelastic scattering from a vibrating surface, we are faced with a many-body problem which must be treated with a variety of approximate methods. Disorder on the surface can contribute to both elastic and inelastic scattering and manifests itself in the diffuse intensity as well as by reducing the overall coherent intensity. In the following we discuss elastic, inelastic and disorder scattering and develop some of the primary theoretical methods used for each.

8.1 The Wavefunction and Transition Matrix

The objective of a scattering calculation is to solve the Schrödinger equation, or equivalently to obtain the transition matrix which contains the same information in a different form. In either case we start from the Hamiltonian

$$H = H_0 + H_c + V \quad , \tag{8.1}$$

in which H_0 is the Hamiltonian of the incident free particle, H_c is the many-body Hamiltonian of the unperturbed semi-infinite solid, and V is the interaction which depends on both the variables of the particle and the solid. For simplicity we will begin by considering elastic scattering in which $H = H_0 + V$ and V depends only on the particle variables. We will return to the problem of the full many-body problem with inelastic scattering in Sect. 8.5. The Schrödinger equation for the elastic wavefunction Ψ_i is

$$(H_0 + V)\Psi_i = E_i\Psi_i \quad , \tag{8.2}$$

where the subscript i corresponds to the plane wave solution of the unperturbed Hamiltonian Φ_i obeying

$$H_0\Phi_i = E_i\Phi_i \quad . \tag{8.3}$$

Far from the surface Ψ_i consists of the plane wave Φ_i incident on the surface and a sum of outgoing scattered plane waves. The wavefunction Ψ_i is thus the solution to (8.2) obeying outgoing wave boundary conditions, and we will show that the coefficients of the outgoing scattered plane waves are proportional to the transition matrix elements evaluated under energy conservation (i. e., evaluated on the energy shell). Explicitly, the transition matrix is given by

$$T_{fi} = (\Phi_f \mid V \mid \Psi_i) = N \int d\mathbf{r}\, \Phi_f(\mathbf{r}) V(\mathbf{r}) \Psi_i(\mathbf{r}) \quad , \tag{8.4}$$

174

where N is a normalization factor.

Some approaches to atom–surface scattering solve directly for the outgoing wavefunctions Ψ_i, but in many cases it is convenient to separate V into a strong part U, which backreflects all incoming particles, together with a remainder v

$$V = U + v \quad . \tag{8.5}$$

This form is ideally suited to the two-potential scattering formalism of Gellmann and Goldberger. We denote the eigenfunctions of U by χ_i and they satisfy the Schrödinger equation

$$(H_0 + U)\chi_i^{\pm} = E_i\chi_i^{\pm} \quad . \tag{8.6}$$

The Green's function form of this equation is the Lippmann–Schwinger equation

$$\chi_i^{\pm} = \Phi_i + \frac{1}{E_i - H_0 \pm i\epsilon}\, U\chi_i^{\pm} \quad . \tag{8.7}$$

The superscripts $(\pm)$ on $\chi_i^{\pm}$ denote incoming $(-)$ or outgoing $(+)$ wave solutions. The incoming wave solutions are useful for some problems, but if the distorting potential U is one dimensional, depending only on the displacement normal to the surface, then χ_i^{+} and χ_i^{-} differ only by a trivial phase. In all that follows we will be concerned with only outgoing wave solutions and hence will drop the $(\pm)$ notation.

Having established the set of distorted wavefunctions χ_j, we can write the Green's function form of the full wavefunction Ψ_i in terms of them. This is the Gellmann–Goldberger equation [8.2].

$$\Psi_i = \chi_i + \frac{1}{E_i - H_0 - U + i\epsilon}\, v\Psi_i \quad . \tag{8.8}$$

All information that can be obtained in a scattering experiment is contained in the asymptotic form of the wavefunction in the region $z \to \infty$ where z is the direction normal to the surface. In order to examine the asymptotic region we begin by introducing the projection operator for the complete set of distorted states,

$$1 = \sum_j |\,\chi_j)(\chi_j\,| \quad ,$$

into the second term of (8.8).

$$\Psi_i = \chi_i + \sum_j \chi_j \, \frac{1}{E_i - E_j + i\epsilon}(\chi_j\,|\,v\,|\,\Psi_i) \quad . \tag{8.9}$$

The matrix element now appearing in (8.9) is the reduced transition matrix element:

$$t_{fi} = (\chi_f\,|\,v\,|\,\Psi_i) = \frac{1}{L^2} \int d\boldsymbol{r}\,\chi_f^{*}(\boldsymbol{r})v(\boldsymbol{r})\Psi_i(\boldsymbol{r}) \quad . \tag{8.10}$$

175

The normalization factor $N = 1/L^2$, where L^2 is the area of the surface, arises because it is convenient to use box normalization in directions parallel to the surface but not in the normal direction. With this normalization all matrix elements have dimensions of energy multiplied by length. The wave equation can be converted into an equation for t_{fi} by multiplying (8.9) from the left by the product $\chi_f v$ and taking the matrix element:

$$t_{fi} = v_{fi} + \sum_j v_{fj} \frac{1}{E_i - E_j + i\epsilon} t_{ji} \quad , \tag{8.11}$$

where the matrix element of the potential is $v_{fi} = (\chi_f \mid v \mid \chi_i)$. The wavefunction (8.8) and the transition matrix equation (8.11) contain identical information, and the choice of which to use depends on the nature of the problem and the methods chosen to solve it. We now make use of the asymptotic behavior of the distorted wavefunctions χ_j. Using a notation in which capital letters denote vector components parallel to the surface, $r = (R, z)$, this is

$$\chi_j(R, z \to \infty) = \exp(i K_j \cdot R)[\exp(-i k_{jz} z) + \exp(i k_{jz} z + i \delta_j)] \quad , \tag{8.12}$$

where the momentum of a plane wave in state j is $k_j = (K_j, k_{jz})$, and δ_j is the phase gained upon scattering of the plane wave by the flat one-dimensional potential U. The asymptotic form of Ψ_i is now obtained upon inserting (8.12) into (8.9) and making use of (8.10),

$$\Psi_i(R, z \to \infty) = \exp(i K_i \cdot R) \ [\exp(-i k_{iz} z) + \exp(i k_{iz} z + i \delta_i)]$$
$$+ \frac{1}{2\pi} \sum_{K_j} \int_0^\infty dq_{jz} \ \frac{\exp(i K_j \cdot R) \ [\exp(-i q_{jz} z) + \exp(i q_{jz} z + i \delta_j)]}{E_i - E_j + i\epsilon} t_{ji} \quad . \tag{8.13}$$

Recognizing that the energy E_j of the distorted state is given by

$$E_j = \frac{\hbar^2}{2m}(K_j^2 + q_{jz}^2) \quad , \tag{8.14}$$

where m is the particle mass, we can cast the above asymptotic expression into a form which is readily evaluated by contour integration

$$\Psi_i(R, z \to \infty) == \exp(i K_i \cdot R) \ [\exp(-i k_{iz} z) + \exp(i k_{iz} z + i \delta_i)]$$
$$- \frac{m}{\pi \hbar^2} \sum_{K_j} \int_0^\infty dq \ \frac{\exp(i K_j \cdot R) \ [\exp(-i q z) + \exp(i q z + i \delta_j)]}{(q - k_{jz} - i\gamma)(q + k_{jz} + i\gamma)} t_{ij} \quad , \tag{8.15}$$

where γ is taken to be small and

176

$$k_{jz} = \sqrt{K_i^2 + k_{iz}^2 - K_j^2} \qquad (8.16)$$

is the perpendicular wave vector given by energy conservation. The contour for carrying out the integration in the complex q-plane is an infinitesimally thin strip extending along the positive real axis. For the integral involving e^{iqz} the return path is parallel to the positive real axis with a small displacement into the upper half plane. The only pole enclosed is that of the energy denominator at $q = k_{jz} + i\gamma$, since all other possible singularities in the upper half plane come from t_{ji} and they are not close to the real axis. A similar argument for the integral involving e^{-iqz} leads to a vanishing contribution since the return path must be in the lower half plane and encloses no poles. The result is an asymptotic wavefunction consisting of a single incoming plane wave and a sum of outgoing scattered plane waves

$$\Psi_i(\boldsymbol{R}, z \to \infty) = \exp(i\boldsymbol{K}_i \cdot \boldsymbol{R})[\exp(-ik_{iz}z) + \exp(ik_{iz}z + i\delta_i)]$$

$$- \frac{im}{\hbar^2} \sum_{\boldsymbol{K}_j} \frac{t_{ji}}{k_{jz}} \exp(i\boldsymbol{K}_j \cdot \boldsymbol{R} + ik_{jz}z + i\delta_j) \quad ,$$

$$(8.17)$$

where the sum over $\boldsymbol{K}_j$ is restricted to values such that $k_{jz}^2 > 0$. (Waves with $k_{jz}^2 < 0$ are evanescent. They decay exponentially away from the surface and do not appear in the asymptotic region.)

Equation (8.17) makes it clear that the asymptotic outgoing plane wave components give directly the transition matrix t_{fi} on the energy conservation shell. The coefficients of the outgoing waves are usually written as a function of parallel wave vector transfer $\boldsymbol{K} = \boldsymbol{K}_j - \boldsymbol{K}_i$ and are given by

$$C(\boldsymbol{K}) = C(\boldsymbol{K}_j - \boldsymbol{K}_i) = \exp(i\delta_j)[\delta_{\boldsymbol{K},0} - \frac{im}{\hbar^2 k_{jz}} t_{ji}] \quad . \qquad (8.18)$$

This also gives the relation between T_{fi} and t_{fi} on the energy shell. Within a trivial multiplicative phase the relation is

$$T_{fi} = \frac{i\hbar^2 k_{iz}}{m} \delta_{\boldsymbol{K},0} + t_{fi} \quad . \qquad (8.19)$$

The measurable scattered intensity is proportional to $|C(\boldsymbol{K})|^2$. What is usually measured is the ratio of incident to final flux crossing a plane parallel to the surface. Since the flux is proportional to the normal wave vector, this gives for $R(\boldsymbol{K})$, the reflected intensity,

$$R(\boldsymbol{K}) = \frac{k_{fz}}{k_{iz}} |C(\boldsymbol{K})|^2 \quad . \qquad (8.20)$$

The reflection intensity of (8.20) is useful for comparing calculated values with experimentally measured sharp diffraction peaks scattered from well-ordered surfaces. For situations in which the scattered intensity is diffuse it is useful to have a differential reflection coefficient related to the intensity entering a detector which subtends solid angle $d\Omega_f$. This can be obtained by recognizing that box normalization in directions parallel to the surface implies $d\mathbf{K} = (2\pi/L)^2$, and the conversion to final solid angle is given by $d\mathbf{K} = k_i^2 \cos\theta_f \, d\Omega_f$

$$\frac{dR}{d\Omega_f} = \frac{L^2}{(2\pi)^2} \, k_i^2 \, \frac{\cos^2\theta_f}{\cos\theta_i} \, |\, C(\mathbf{K}) \,|^2 \quad , \tag{8.21}$$

recalling that k_i is the magnitude of the incident wave vector, and θ_i and θ_f are the initial and final polar angles, respectively.

The form chosen for the asymptotic wavefunction expressed in Eq.(8.17) is especially suited to scattering from the surface, but often in general treatments of scattering theory one expresses the asymptotic wavefunction in elastic scattering with a radial outgoing wavefunction multiplied by a scattering amplitude $f(\theta, \phi) = f(\Omega)$:

$$\Psi_i(r \to \infty) = \exp(i\mathbf{K}_i \cdot \mathbf{R}) \exp(-ik_{iz}z) + \frac{f(\Omega)}{r}\exp(ik_i r) \tag{8.22}$$

with the differential cross section of the scattering center given by $d\sigma(\Omega)/d\Omega = |f(\Omega)|^2$. Such a form can readily be obtained from the asymptotic wavefunction (8.17) upon converting the sum over parallel wavevector to an integral and then evaluating in the region $r \to \infty$ using stationary phase methods, or alternatively by integrating (8.15) directly after converting the integrals to spherical coordinates. The result gives the scattering amplitude in terms of the transition matrix

$$f(\Omega) = -\frac{L^2 m}{2\pi\hbar^2} t_{fi} \quad . \tag{8.23}$$

The relation between the differential reflection coefficient and the differential cross section is

$$\frac{dR}{d\Omega_f} = \frac{1}{L^2\cos\theta_i} \, \frac{d\sigma(\Omega)}{d\Omega} \quad . \tag{8.24}$$

For the study of the elastic diffraction from periodic surfaces, the reflection coefficient (8.20) and the differential reflection coefficient (8.21) are most convenient, but for the study of defects and adsorbates on surfaces the cross section is usually used. This is because $d\sigma(\Omega)/d\Omega$ gives a direct measure of the differential cross section of all the defects on the surface (after subtracting off any intensity that might appear in diffraction peaks arising from periodicity of the underlying substrate).

We note at this point that the geometrical factor $1/\cos\theta_i$ appearing on the right-hand side of (8.24) may be replaced by a different function depending on the configuration of the experiment being considered. For example, in some time-of-flight measurements the detector angle subtends only a small portion of a fully illuminated target surface [8.3]. In this case the factor $1/\cos\theta_i$ is replaced by $1/\cos\theta_f$.

8.2 Exact Solutions for Rigid Periodic Potentials

A large number of interesting theoretical applications involve determining the
elastic diffraction intensities for scattering from rigid periodic potentials. A
number of successful methods have been proposed and we discuss briefly here
three of the most important, the close coupling formalism, summation of the
perturbation series, and numerical propagation of wave packets.

8.2.1 Close Coupling Formalism

The periodic surface potential can be expanded in directions parallel to the
surface in a Fourier series in the surface reciprocal lattice vectors

$$V(\boldsymbol{r}) = V(\boldsymbol{R}, z) = \sum_{\boldsymbol{G}} \exp(i\boldsymbol{G} \cdot \boldsymbol{R}) V_{\boldsymbol{G}}(z) \quad , \tag{8.25}$$

where the Fourier transform $V_{\boldsymbol{G}}(z)$ is

$$V_{\boldsymbol{G}}(z) = \frac{1}{S} \int_{\mathrm{uc}} d\boldsymbol{R} \, \exp(-i\boldsymbol{G} \cdot \boldsymbol{R}) V(\boldsymbol{R}, z) \tag{8.26}$$

and the integral is over a single unit cell (uc) of area S. Because of the two
dimensional periodicity of the surface, the wavefunctions will be of the form of
Bloch waves parallel to the surface

$$\Psi_i(\boldsymbol{R}, z) = \exp(i\boldsymbol{K}_i \cdot \boldsymbol{R}) \sum_{\boldsymbol{G}} \exp(i\boldsymbol{G} \cdot \boldsymbol{R}) \boldsymbol{\Psi}_{\boldsymbol{G}}(z) \quad . \tag{8.27}$$

The perpendicular momentum $k_{\boldsymbol{G}z}$ of the normal wavefunctions $\boldsymbol{\Psi}_{\boldsymbol{G}}(z)$ is given
by the conservation of energy as in (8.16), together with the condition of parallel
momentum conservation as expressed by $\boldsymbol{K}_f = \boldsymbol{K}_i + \boldsymbol{G}$:

$$k_{\boldsymbol{G}z} = \sqrt{\boldsymbol{K}_i^2 + k_{iz}^2 - (\boldsymbol{K}_i + \boldsymbol{G})^2} \quad . \tag{8.28}$$

As in (8.16), imaginary values of $k_{\boldsymbol{G}z}$ are associated with evanescent wave
components which decay exponentially with distance away from the surface in
the asymptotic region.

When the potential of (8.25) and the wavefunction of (8.27) are inserted in
the Schrödinger equation:

$$-\frac{\hbar^2}{2m} \nabla^2 \Psi_i(\boldsymbol{R}, z) + V(\boldsymbol{R}, z) \Psi_i(\boldsymbol{R}, z) = E_i \Psi_i(\boldsymbol{R}, z) \tag{8.29}$$

the result is a set of coupled equations for the $\boldsymbol{\Psi}_{\boldsymbol{G}}(z)$

$$-\frac{d^2}{dz^2} \boldsymbol{\Psi}_{\boldsymbol{G}}(z) + \frac{2m}{\hbar^2} \sum_{\boldsymbol{M}} V_{\boldsymbol{G}-\boldsymbol{M}}(z) \boldsymbol{\Psi}_{\boldsymbol{M}}(z) = k_{\boldsymbol{G}z}^2 \boldsymbol{\Psi}_{\boldsymbol{G}}(z) \quad . \tag{8.30}$$

This is the set of coupled channel equations for the Bloch waves. Early approaches to the solution of this set of equations were developed by *Cabrera* et

al. [8.4,5] and by *Tsuchida* [8.6]. *Wolken* first applied the close coupling formalism [8.7] which was developed for atomic and molecular scattering in the gas phase [8.8]. Since then a variety of different approaches have been successfully used [8.9].

A basic description of all these methods begins by recognizing that the components $\Psi_G(z)$ obey the boundary conditions of an outgoing scattered wave far from the crystal

$$\Psi_G(z \to \infty) = \exp(-ik_{iz}z)\,\delta_{G,0} + C(G)\,\exp(ik_{Gz}z) \tag{8.31}$$

and become vanishingly small deep inside the crystal where the potential is large

$$\Psi_G(z \to \infty) = 0 \quad . \tag{8.32}$$

The infinite set of G vectors is truncated to a set of size N large enough to produce the desired numerical accuracy. The coupled equations for the set of N functions $\Psi_G(z)$ can be integrated by any of a number of standard algorithms.

Typically, as an approximation to the boundary condition (8.32) one chooses initial conditions $\Psi_G(z_0) = 0$, where z_0 is a point far under the surface for which $V(R, z_0) > E_i$. Then, depending upon the type of algorithm chosen, one supposes an arbitrary value for either $d\Psi_G(z_0)/dz$ or for $\Psi_G(z_0 + \delta)$ where δ is the next integration step. With these boundary conditions the set of N functions $\Psi_G(z)$ can be numerically integrated sufficiently far in front of the surface to where the potential becomes negligibly small. The problem arises that in general the functions generated in this manner will not necessarily obey the asymptotic boundary conditions (8.31). This can be resolved by constructing for each G value N distinct solutions which we can denote by $\Psi_{G,M}(z)$, where M is a reciprocal lattice vector from the same set as G. Each of the $\Psi_{G,M}(z)$ is calculated from a different set of initial conditions. Examples of such initial conditions are

$$\Psi_{G,M}(z_0) = 0 \tag{8.33}$$

together with, depending on the algorithm, either

$$\frac{d}{dz}\Psi_{G,M}(z_0) = \delta_{G,M} \tag{8.34}$$

or

$$\Psi_{G,M}(z_0 + \delta) = \gamma\,\delta_{G,M} \quad , \tag{8.35}$$

where γ is a small but arbitrary real number. From the N solutions $\Psi_{G,M}(z)$ one can form just one linear combination which satisfies the boundary condition (8.31), and we write this linear combination as

$$\Psi_G(z) = \sum_M A_M\,\Psi_{G,M}(z) \quad . \tag{8.36}$$

180

In order to determine the desired scattering amplitudes $C(G)$ we match $\Psi_G(z)$ and its derivative to the asymptotic form (8.30) at a point z_1 sufficiently far from the surface:

$$\sum_M A_M \, \Psi_{G,M}(z_1) = \exp(-\mathrm{i}k_{iz}z_1)\delta_{G,0} + C(G)\,\exp(\mathrm{i}k_{Gz}z_1)$$

(8.37)

$$\sum_M A_M \, \frac{d}{dz}\Psi_{G,M}(z_1) = -\mathrm{i}k_{Gz}\,[\exp(-\mathrm{i}k_{iz}z_1)\,\delta_{G,0} - C(G)\exp(\mathrm{i}k_{Gz}z_1)] \quad .$$

(8.38)

These are two sets of N linear algebraic equations from which the $C(G)$ can be evaluated upon elimination of the A_G. In usual practice, the upper equation is multiplied by $-ik_{Gz}$ and then the two are added to produce one set of N equations for the A_G. Once this is solved by matrix inversion, the $C(G)$ are determined by substituting back into either (8.37) or (8.38). The practical limit to the use of the close coupling method is the number of G vectors that must be retained in order to obtain well convergent intensities. For small surface unit cells this method has proven to be relatively rapid and is usually the method of choice, while for larger unit cells it becomes very slow. The problem that arises in cases where the truncated set of G vectors must be large is that it is necessary to calculate N^2 functions $\Psi_{G,M}(z)$ rather than obtaining directly the N Bloch functions $\Psi_G(z)$ which are needed for the given initial conditions. This is because the coupled channel method is a complete solution to the boundary value problem, and such a solution usually contains far more information than is necessary; i. e., once the N^2 functions $\Psi_{G,M}(z)$ are determined, solutions can be obtained for any specified initial conditions. When the coupled channel calculations become too consuming of computer time, and it is still necessary to have numerically exact results, one can choose one of the methods which solves the Schrödinger equation as an initial value problem such as the summation of the perturbation series or the wave packet method discussed in the following.

8.2.2 Inversion of the Transition Matrix Equation

Several approaches to the solution for rigid periodic potentials are based on the transition matrix equation developed above in (8.11)

$$t_{fi} = v_{fi} + \sum_j v_{fj} \frac{1}{E_i - E_j + \mathrm{i}\epsilon} t_{ji} \quad .$$

(8.11)

As shown in (8.20) or (8.21) the scattered intensities are proportional to the square modulus of t_{fi} evaluated on the energy shell. Conceptually, the simplest approach is to directly invert the t-matrix equation. Although this can lead to rather large matrix inversion problems, it has been successfully carried out [8.10]. The procedure is facilitated by the fact that the summation over intermediate states in (8.11) is, in part, a discrete sum in terms of the reciprocal lattice vectors due to the periodicity of the surface. Explicitly, the sum over intermediate quantum states can be expressed as

$$\sum_j \rightarrow \sum_{G_j} \frac{1}{2\pi} \int_0^\infty dk_{jz} + \sum_{G_j} \sum_{b_j} \quad , \tag{8.39}$$

where the G_j are reciprocal lattice vectors, k_{jz} is the perpendicular wave vector, and the b_j denote the discrete set of bound states of the distorted waves. In practice, the set of G_j vectors is truncated to a finite set sufficiently large to produce the desired numerical accuracy, and the integral over k_{jz} is discretized. For all G_j corresponding to real diffracted beams the energy denominator of (8.11) will have a pole for $k_{jz} = k_{G_j z}$ corresponding to energy conservation, with $k_{G_j z}$ given by (8.28). In this case, when the integral over k_{jz} is discretized, care must be taken to separate out the pole contribution and the principal part integral. The resulting discrete set of linear equations is inverted to obtain the t_{fi}.

The major limitation of this method of direct inversion is the size of matrix that must be inverted. It has been applied to one dimensional corrugations, and with large computers it is becoming quite practical for full three dimensional calculations [8.11].

8.2.3 Summation of the Perturbation Series

A second method based on the transition matrix is to iterate the equation in a perturbation series, keeping terms of sufficiently high order to produce the desired numerical precision [8.12,13]. The starting point is the transition matrix equation (8.11), whose perturbation series is

$$t_{fi} = v_{fi} + \sum_j v_{fj} \frac{1}{E_i - E_j + i\epsilon} v_{ji}$$

$$+ \sum_j \sum_q v_{fj} \frac{1}{E_i - E_j + i\epsilon} v_{jq} \frac{1}{E_i - E_q + i\epsilon} v_{qi} + \cdots$$

$$\tag{8.40}$$

Each term in this series is an iteration of order n with the first order given by $t_{fi}^1 = v_{fi}$ and higher terms represented by

$$t_{fi}^{n+1} = v_{fi} + \sum_j v_{fj} \frac{1}{E_i - E_j + i\epsilon} t_{ji}^n \quad . \tag{8.41}$$

The numerical procedure is to truncate and discretize the sums over intermediate states in the same manner as described in (8.39) and the discussion following. Then (8.40) becomes a finite set of coupled linear equations which is iterated to a sufficiently large order n to obtain convergence.

The summation of perturbation series method has been used for a variety of potentials, particularly for problems involving stepped metal surfaces

[8.14]. Methods based on perturbation theory are always subject to questions of convergence, and this is no exception. The radius of convergence of the three dimensional perturbation series as a function of a given parameter remains an unsolved problem in general. For the systems studied with the present method, practical experience indicates that the convergence is numerically stable if the corrugation height of the surface equipotentials is not too large, generally limited to 10–20% of the width of the unit cell for a one dimensional corrugation. The convergence properties are also strongly affected by the choice of separation of the total potential V into distorting potential U and the perturbation v [8.15].

8.2.4 Wave Packet Methods

Up to this point we have discussed methods for solving the Schrödinger equation which are time independent period. One can, alternatively, exploit the time dependence of the Schrödinger equation in order to treat the scattering problem as a wave packet which propagates in time toward the surface, interacts with it, and then scatters away. A highly sophisticated and efficient numerical method for accomplishing this, which is ideally suited to the surface scattering problem, has been developed by *Kosloff* and coworkers [8.16].

The basic approach is quite simple, and conceptually very close to the way in which an actual experiment is performed. The initial state of the incident atomic beam far from the surface is written as a wave packet $\Psi_i(\boldsymbol{r},t)$ with energy centered about the incident energy and directed toward the surface at a given incident angle. Typically, the initial wave packet is chosen to be Gaussian in form, and its spatial extent is sufficiently large that the spread in energies is relatively narrow. The initial wave packet then propagates and interacts with the surface according to the time dependent Schrödinger equation

$$ -\frac{\hbar^2}{2m}\nabla^2\,\Psi_i(\boldsymbol{r},t) + V(\boldsymbol{r})\Psi_i(\boldsymbol{r},t) = \mathrm{i}\hbar\frac{\partial\Psi_i(\boldsymbol{r},t)}{\partial t}\quad . \tag{8.42}$$

For numerical solution the problem is discretized on a grid in both space and time. The spatial grid is truncated at a size which is large enough to contain the wave packet both before and after interaction with the potential. This involves imposing a boundary condition at the truncated edges of the grid. The potential energy term is the simplest to handle since it is a local operation, a simple multiplication at each grid point. The Laplacian operator poses the greatest difficulty, since differentiation on a grid of non-zero step size becomes a nonlocal operation. Standard finite difference methods for obtaining the Laplacian in more than one dimension present several problems, both in error generation and in computation time [8.17]. The artificial nonlocality can be avoided by first making a Fourier transform of the wave packet in the spatial dimensions

$$ \Psi_i(\boldsymbol{r},t) = \sum_{\boldsymbol{k}} \mathrm{e}^{\mathrm{i}\boldsymbol{k}\cdot\boldsymbol{r}}\Psi_i(\boldsymbol{k},t)\quad . \tag{8.43}$$

The Laplacian operation $\nabla^2 \Psi_i(\boldsymbol{r}, t)$ is then equivalent to multiplying the Fourier transform $\Psi_i(\boldsymbol{k}, t)$ by the factor $-k^2$. The numerical procedure for treating the Laplacian term in the Schrödinger equation at each spatial grid point is as follows: (1) the wavefunction $\Psi_i(\boldsymbol{r}, t)$ is transformed to $\boldsymbol{k}$-space using the fast Fourier transform algorithm, (2) the transform $\Psi_i(\boldsymbol{k}, t)$ is multiplied by $\hbar^2 k^2 / 2m$, and (3) the result is transformed back to coordinate space by applying the inverse Fourier transform. Because of the numerical efficiency of the fast Fourier transform algorithm, this procedure can present enormous computing time advantages when compared to finite difference methods [8.17]. Although the Fourier transformation process imposes periodic boundary conditions at the grid edges, this presents no major difficulties since the natural coordinate system imposed by the surface problem is cartesian.

The remaining part of the problem is the propagation of the wavefunction in time. The simplest procedure is the second order differencing form [8.18], in which the time differential is approximated as

$$\frac{\partial \Psi_i(t)}{\partial t} \simeq \frac{\Psi_i(t + \Delta t) - \Psi_i(t - \Delta t)}{2\Delta t} \quad . \tag{8.44}$$

Higher order differencing forms or other schemes have been developed to reduce error [8.19].

The final operation is the determination of the coefficients $C(\boldsymbol{K})$ of the asymptotic wavefunction, which is accomplished by projection of the scattered wave packet onto a plane wave. The initial wave packet $\Psi_i(\boldsymbol{r}, t \to -\infty)$ evolves, after colliding with the surface, into the wave packet $\Psi_i(\boldsymbol{r}, t \to +\infty)$, where $t \to +\infty$ means at a time large enough so that the interaction is negligible and asymptotic behavior has been attained. The projection operation is then

$$C(\boldsymbol{K}) = \int d\boldsymbol{R} dz \, \exp[-\mathrm{i}(\boldsymbol{K}_i + \boldsymbol{G}) \cdot \boldsymbol{R} - \mathrm{i}k_{Gz}z] \, \Psi_i(\boldsymbol{r}, t \to +\infty) \quad . \tag{8.45}$$

In practice this final operation may not be necessary if the initial wave packet is sufficiently extended in space that it appears as a delocalized wave.

A detailed review of both exact and approximate wave packet methods has been presented by *Gerber* et al. [8.17]. In addition to the advantages discussed above, we note that at no point was it necessary to invoke the translational periodicity of the surface. Consequently, the method is well suited to treat surface systems which lack periodicity, and has been extensively applied to problems of rough surfaces and surfaces with adsorbates [8.20].

8.2.5 Resonances in Elastic Scattering

There is one feature of elastic diffraction from ordered periodic surfaces that merits special attention, and this is resonant scattering with the bound states of the potential. This occurs when the incident conditions are such that the particle can diffract into a bound state of the adsorption well while still conserving energy. Denoting the energy of a bound state of the distorting potential U(z) by $- \mid \epsilon_b \mid$ this condition is

$$0 = E_i - E_b = \frac{\hbar^2}{2m} \left[\boldsymbol{K}_i^2 + k_{iz}^2 - (\boldsymbol{K}_i + \boldsymbol{B})^2 \right] + |\epsilon_b| \quad , \tag{8.46}$$

where $\boldsymbol{B}$ is the reciprocal lattice vector involved in the diffraction. According to (8.28) this is the same as

$$k_{\boldsymbol{B}z}^2 + \frac{2m}{\hbar^2} \mid \epsilon_b \mid = 0 \tag{8.47}$$

and for this to occur $k_{\boldsymbol{B}z}^2$ must be negative. Equation (8.46) shows that resonances occur when the bound state is degenerate in energy with the other diffraction channels. The sudden appearance of an additional scattering channel can cause a rearrangement of intensities in the diffraction peaks near resonant conditions.

Several exact and approximate formalisms have been developed for calculating diffraction intensities in the presence of bound state resonances. In general, the close coupling and wave packet methods calculate the resonances without difficulty [8.21,22], but because those calculations are completely numerical often the underlying physics is obscured. Also these methods sometimes take excessive computing time in order to map out the details of the resonant features. We adopt here an approach based on the transition matrix [8.23,24]. We rewrite the transition matrix equation (8.11) with a small subset of resonant bound states separated out of the sum over intermediate states:

$$t_{fi} = v_{fi} + \sum_{j}{}' v_{fj} \frac{1}{E_i - E_j + i\epsilon} t_{ji} + \sum_{b}{}' v_{fb} \frac{1}{E_i - E_b} t_{bi} \quad .$$

$$\tag{8.48}$$

The notation $\sum'$ indicates that those bound state terms which do not appear in the first summation of the right hand side of (8.48) are those present in the second summation term. Note that an intermediate bound state is labeled by two discrete variables, the reciprocal lattice vector and the bound state energy, and there is no need to include a small imaginary term in the energy denominator. It is clear that an attempt to calculate resonant behavior directly using (8.48) will fail because of divergences in the summation when the resonant denominators vanish. The problem is avoided through a projection technique originally developed by *Feshbach* [8.25,26]. A straightforward manipulation converts (8.48) into a set of two coupled equations:

$$t_{fi} = h_{fi} + \sum_{b}{}' h_{fb} \frac{1}{E_i - E_b} t_{bi} \tag{8.49}$$

$$h_{fi} = v_{fi} + \sum_{j}{}' v_{fj} \frac{1}{E_i - E_j + i\epsilon} h_{ji} \quad . \tag{8.50}$$

As a first example, let us consider the most frequently encountered situation, that of a single isolated resonance. Then the sum in (8.48) contains only a single term

$$t_{fi} = h_{fi} + h_{fb} \frac{1}{E_i - E_b} t_{bi} \tag{8.51}$$

to which we add the same equation for t_{bi}

$$t_{bi} = h_{bi} + h_{bb} \frac{1}{E_i - E_b} t_{bi} \tag{8.52}$$

and solving the pair for t_{fi} gives

$$t_{fi} = h_{fi} + \frac{h_{fb} h_{bi}}{E_i - E_b - h_{bb}} \quad . \tag{8.53}$$

We assume that the h_{fi} of (8.50) can be calculated, then (8.53) shows that the amplitude for scattering into the diffraction state f consists of two terms, a direct term h_{fi} and the resonant term. In the resonant term, the unperturbed energy denominator is modified by the self-energy h_{bb} which is complex. The real part of h_{bb} gives the shift of the resonant energy from the unperturbed energy; the imaginary part of h_{bb} gives the energy width of the resonance, or alternatively $\hbar/\mathrm{Im}\{h_{bb}\}$ gives the lifetime in the bound state.

Equation (8.53) is of the form of a Feshbach amplitude, and the intensity of the diffraction peak, which according to (8.17) and (8.20) is proportional to $|\,t_{fi}\,|^2$, will have the form of a Feshbach resonance. Depending on the relative importance of the real and imaginary parts of the h-matrix elements, this form will be a single peak, a single dip, or a combination of the two. Since in most cases the calculated energy shifts $\mathrm{Re}\{h_{bb}\}$ are small, a simple observation of the position of the resonance gives a good estimate of the bound state energies of the surface averaged distorting potential $U(z)$. Additional physical information can be obtained from observations of the shape of the resonance feature, and elaborate sets of rules have been developed for predicting whether the resonance form will be a peak, dip, or mixture of the two [8.27,28].

As a second example we consider the situation in which two unperturbed bound states, denoted by b and n, satisfy the resonant condition simultaneously. The sum over intermediate states in (8.48) now contains two terms. The result of solving for t_{fi} in terms of the h-matrix is similar in form to (8.52); the direct term is unchanged while the resonant term has a more complicated numerator and its denominator is replaced by the determinantal relation

$$(E_i - E_b - h_{bb})(E_i - E_n - h_{nn}) - h_{bn} h_{nb} \quad . \tag{8.54}$$

Equating this denominator to zero determines the complex energies of the two resonances, and is an example of the classic case of an avoided crossing between two degenerate unperturbed levels. For weakly diffractive systems, h_{bb} and h_{nn} are usually small, and $h_{bn} \cong h_{nb}^{*} \cong v_{bn}$. Setting (8.54) equal to zero in this case leads to

$$E_b \cong E_i + |\,v_{bn}\,| \tag{8.55}$$

$$E_n \cong E_i - |\,v_{bn}\,| \tag{8.56}$$

which is the well known result that the two bound states are split by the energy $2\,|v_{bn}|$. Thus, in the weak diffraction case, measurement of bound state splittings give directly the values of the matrix elements which can be used to map out the entire bound state band structure and to determine the form of the potential in the region of the adsorption well. This has been carried out on the He/graphite system [8.29], and later calculations have obtained the corrections coming from higher order terms in the evaluation of the h-matrices [8.30].

Purely elastic resonance theories of the type discussed here work quite well for helium scattering from rigid materials with high Debye temperatures such as LiF. In other cases, particularly scattering from metals, the linewidths as predicted by an elastic theory may be too narrow or may even have the wrong signature [8.31,32]. The problems are usually ascribed to inelastic scattering or scattering by defects or other surface disorder [8.33,34].

8.3 Approximate Methods

Thus far the methods discussed for solving elastic problems have been exact, in the sense that for a given model potential, numerical results for the diffraction intensities can be obtained to any desired precision. Now we consider a variety of approximations. These solutions can be extremely useful as often they accurately describe the underlying physics of the scattering process while giving enormous savings in computational effort.

8.3.1 The Hard Corrugated Wall Model

We begin by considering an approximate model for the surface potential, for which exact numerical solutions are readily obtained. The hard corrugated wall potential is represented by an infinitely repulsive barrier with periodic corrugations representing the underlying crystalline structure. Explicitly

$$V(\boldsymbol{R}, z) = \begin{cases} 0; z > \xi(\boldsymbol{R}) \\ \infty; z \le \xi(\boldsymbol{R}) \end{cases} , \tag{8.57}$$

where $\xi(\boldsymbol{R})$ is the periodic corrugation function. A straightforward way to obtain the Schrödinger wavefunction $\Psi_i(\boldsymbol{r})$ is to begin with the Lippmann–Schwinger equation which is the analog of (8.7) for the potential $V(\boldsymbol{r})$

$$\Psi_i = \Phi_i + \frac{1}{E_i - H_0 \pm i\epsilon} V \Psi_i , \tag{8.58}$$

and introducing the projection operator for the complete set of plane waves Φ_i in the second term on the right hand side leads to

$$\Psi_i(\boldsymbol{r}) = \Phi_i(\boldsymbol{r})$$
$$+ \sum \boldsymbol{k}_j \, \Phi_j(\boldsymbol{r}) \frac{1}{E_i - E_j \pm i\epsilon} (\Phi_j(\boldsymbol{r}) \mid V(\boldsymbol{r}) \mid \Psi_i(\boldsymbol{r})) . \tag{8.59}$$

Appearing on the right hand side of (8.59) is the transition matrix of (8.4), $T_{ji} = (\Phi_j \mid V \mid \Psi_i)$. The simplification of the problem comes in the singular nature of the product $V(\boldsymbol{r})\Psi_i(\boldsymbol{r})$; it vanishes in front of the surface for $z > \xi(\boldsymbol{R})$ because the potential is zero there, and it must vanish behind the surface for $z < \xi(\boldsymbol{R})$ because the wavefunction must be zero if the potential is infinite. Thus the only way the transition matrix T_{ji} can be nonvanishing is for the product $V(\boldsymbol{r})\Psi_i(\boldsymbol{r})$ to be proportional to a Dirac δ-function whose argument vanishes at the surface:

$$V(\boldsymbol{r})\Psi_i(\boldsymbol{r}) = -\frac{im}{\hbar^2 S} \exp(i\boldsymbol{K}_i \cdot \boldsymbol{R})\, F(\boldsymbol{R})\, \delta(z - \xi(\boldsymbol{R}))\quad , \tag{8.60}$$

where S is the area of a surface unit cell and $F(\boldsymbol{R})$ is called the source function. Because of the periodicity, the sum over parallel wave vector in (8.59) becomes a sum over reciprocal lattice vectors; and the integral over perpendicular wave vectors can be carried out using the energy conservation pole of the denominator. The result is

$$\Psi_i(\boldsymbol{r}) = \exp(i\boldsymbol{K}_i \cdot \boldsymbol{R} - ik_{iz}z) - \sum_{\boldsymbol{G}} \exp[i(\boldsymbol{K}_i + \boldsymbol{G}) \cdot \boldsymbol{R}]$$

$$\times \int_{uc} d\boldsymbol{R}' F(\boldsymbol{R}')\exp(-i\boldsymbol{G} \cdot \boldsymbol{R}')\, \exp(ik_{Gz} \mid z - \xi(\boldsymbol{R}') \mid) \quad .$$

$$\tag{8.61}$$

where the surface integral over $\boldsymbol{R}$ is taken over a unit cell. The amplitude coefficients of the diffracted beams are obtained from the asymptotic region, which in this case is simply $z > \xi(\boldsymbol{R})$.

$$C(\boldsymbol{G}) = \frac{1}{k_G} \int_{uc} d\boldsymbol{R}\, F(\boldsymbol{R}) \exp(-i\boldsymbol{G} \cdot \boldsymbol{R})\, \exp[ik_{Gz}\xi(\boldsymbol{R})] \quad . \tag{8.62}$$

The problem of determining the diffraction intensities now becomes one of determining the source function $F(\boldsymbol{R})$. This is accomplished using the final boundary condition that $\Psi_i(\boldsymbol{r})$ vanish on the surface

$$\Psi_i(\boldsymbol{R}, z = \xi(\boldsymbol{R})) = 0 \quad . \tag{8.63}$$

Applying this condition to the wavefunction of (8.61) gives the defining equation for $F(\boldsymbol{R})$:

$$0 = \exp[-ik_{iz}\xi(\boldsymbol{R})] - \sum_{\boldsymbol{G}} \frac{\exp(i\boldsymbol{G} \cdot \boldsymbol{R})}{k_{Gz}}$$

$$\times \int_{uc} d\boldsymbol{R}' F(\boldsymbol{R}') \exp(-i\boldsymbol{G} \cdot \boldsymbol{R}')\, \exp(ik_{Gz} \mid \xi(\boldsymbol{R}) - \xi(\boldsymbol{R}') \mid) \quad .$$

$$\tag{8.64}$$

An efficient numerical solution for a general corrugation has been developed by *Garcia* and *Cabrera* [8.35]. Their method consists in defining a kernel function $M(\boldsymbol{R}, \boldsymbol{R}')$ according to

188

$$M(\boldsymbol{R}, \boldsymbol{R}') = \sum_{\boldsymbol{G}} \exp[\mathrm{i}\boldsymbol{G} \cdot (\boldsymbol{R} - \boldsymbol{R}')] k_{Gz} \exp(\mathrm{i}k_{Gz} \mid \xi(\boldsymbol{R}) - \xi(\boldsymbol{R}') \mid) \quad .$$

$$(8.65)$$

Then the boundary condition (8.64) becomes

$$0 = \exp[-\mathrm{i}k_{iz}\xi(\boldsymbol{R})] - \int_{\mathrm{uc}} d\boldsymbol{R}' \; F(\boldsymbol{R}') \, M(\boldsymbol{R}, \boldsymbol{R}') \quad . \tag{8.66}$$

This integral equation is converted into a matrix equation by discretizing on a grid in $\boldsymbol{R}$-space and solved by matrix inversion. The sum over reciprocal lattice vectors in (8.65) is handled by truncation at a number sufficiently large to insure numerical accuracy.

The Garcia–Cabrera method has been widely used on a large number of problems. Another approach to the solution of the corrugated hard wall problem is to transform the boundary condition (8.64) to reciprocal space. In this case the numerical solution involves a matrix inversion of dimension equal to the number of reciprocal lattice vectors retained after truncation [8.36]. Similar numerical methods have been developed for solving the hard corrugated wall problem in cases where the corrugation function has discontinuities or is multiple valued [8.37].

8.3.2 Rayleigh Ansatz

The corrugated hard wall model lends itself to a number of approximate solutions. The original approach of Lord Rayleigh to this problem was to assume that the asymptotic form of the wavefunction (8.17), which we write here in terms of the amplitude $C(\boldsymbol{K})$,

$$\Psi_i(\boldsymbol{R}, z) = \exp(\mathrm{i}\boldsymbol{K}_i \cdot \boldsymbol{R})\exp(-\mathrm{i}k_{iz}z)+$$

$$+ \sum_{\boldsymbol{G}} C(\boldsymbol{G})\exp[\mathrm{i}(\boldsymbol{K}_i + \boldsymbol{G}) \cdot \boldsymbol{R} + \mathrm{i}k_{Gz}z] \quad , \tag{8.67}$$

can be extended into the selvege region right up to the surface [8.38]. Applying the hard wall boundary condition (8.63) gives the defining equation for the amplitudes $C(\boldsymbol{G})$:

$$0 = 1 + \sum_{\boldsymbol{G}} C(\boldsymbol{G}) \exp(\mathrm{i}\boldsymbol{G} \cdot \boldsymbol{R}) \exp[\mathrm{i}(k_{Gz} + k_{iz})\xi(\boldsymbol{R})] \quad . \tag{8.68}$$

The summations in (8.67) and (8.68) are over the discrete but infinite set of reciprocal lattice vectors, so all evanescent waves are included. There is certainly reason to suspect that this assumption is an approximation, since it includes only outgoing scattered waves, and in the selvege region one would in general also expect to have incoming scattered waves. However, it can be shown that in at least one special situation the Rayleigh method leads to the exact result, and this is for the case of a one dimensional sinusoidal corrugation of period a, $\xi(\boldsymbol{R}) = ha\ \sin(2\pi x/a)$, subject to the limitation that $h < 0.0713$. For larger values of h the method is divergent [8.39,40]. For a two-dimensional sinusoidal corrugation there is also a strong indication that there is a region of corrugation heights within which the method is convergent [8.41]. Because the Rayleigh method can under certain conditions lead to exact results, it is usually called the Rayleigh ansatz rather than an approximation. However, even under conditions in which the method is known to be divergent, in practice it can give accurate approximate results.

The most straightforward way of solving for the $C(\boldsymbol{G})$ is to Fourier transform the boundary condition equation (8.68) which converts it into a matrix equation in the reciprocal lattice vectors

$$0 = \delta_{\boldsymbol{G},0} + \sum_{\boldsymbol{G}'} C(\boldsymbol{G}')\, Q_{\boldsymbol{G}\boldsymbol{G}'} \quad , \tag{8.69}$$

with

$$Q_{\boldsymbol{G}\boldsymbol{G}'} = \frac{1}{S} \int_{\mathrm{uc}} d\boldsymbol{R}\ \exp[\mathrm{i}(\boldsymbol{G} - \boldsymbol{G}') \cdot \boldsymbol{R})]\ \exp[\mathrm{i}(k_{\boldsymbol{G}z} + k_{iz})\xi(\boldsymbol{R})] \tag{8.70}$$

and the $C(\boldsymbol{G})$ coefficient can be obtained by matrix inversion after truncating the sum to a sufficiently large set of $\boldsymbol{G}$-vectors. A number of other methods adapted to the atom–surface problem have been proposed for numerical solution of (8.68) or (8.69). Particularly useful has been a rapid method developed by *Garcia*, *Cabrera* and coworkers which is based on inversion of the boundary condition (8.68) in real space [8.42,43], similar to that developed for the exact solution [8.35].

The questions of convergence of the Rayleigh ansatz and its usefulness as a numerical approximation under conditions when it is not necessarily convergent have been investigated by several authors [8.41,44]. As a practical technique, the Rayleigh ansatz can give good results for corrugations with amplitudes considerably larger than the proven radius of convergence. However, if the corrugation amplitude becomes too large the solution fails to converge. Outside of its domain of convergence the Rayleigh ansatz behaves much like an asymptotic series as a function of the number N of reciprocal lattice vectors retained in the truncation. If the corrugation is not too large the result will appear to converge closer and closer to the exact solution as N is increased up to a certain value, after which the solution diverges with further increase in N.

8.3.3 The Eikonal Approximation

One of the most useful and certainly the most widespread approximation for interpreting atom–surface scattering phenomena is the eikonal approximation to the hard corrugated wall model. This is obtained most directly from the boundary condition derived for the Rayleigh ansatz, equation (8.68), which we rewrite here as

$$0 = \exp[-\mathrm{i}k_{iz}\xi(\boldsymbol{R})] + \sum_{\boldsymbol{G}} C(\boldsymbol{G})\, \exp(\mathrm{i}\boldsymbol{G} \cdot \boldsymbol{R})\, \exp(\mathrm{i}k_{\boldsymbol{G}z}\xi(\boldsymbol{R})) \quad . \tag{8.71}$$

The eikonal approximation consists of assuming that in the semiclassical limit where $k_i \gg G$, k_{Gz} is a slowly varying function of G and hence $\exp[ik_{Gz}\xi(\mathbf{R})]$ can be taken out of the summation. This enables evaluation of $C(\mathbf{G})$ immediately by Fourier transformation

$$C(\mathbf{G}) = -\frac{1}{S}\int_{uc} d\mathbf{R}\, \exp(-i\mathbf{G}\cdot\mathbf{R})\, \exp[-i(k_{Gz} + k_{iz})\xi(\mathbf{R})] \quad . \tag{8.72}$$

This is an extremely simple form, the Fourier transform of a phase factor whose argument is the product of the normal momentum transfer, $\Delta p = \hbar(k_{Gz} + k_{iz})$, and the surface corrugation function. For simple surface profiles it can be integrated analytically. The eikonal approximation is not necessarily restricted to cases where the diffraction is weak, and it can work very well even under situations where the specular intensity is quite small compared to other diffraction peaks. Rather, the condition that k_{Gz} be a slowly varying function of G implies that all the diffracted peaks are closely grouped around one direction. This will occur, for example, at high incident energies in the neighborhood of the specular beam, or another important case is when all important diffraction beams are near a rainbow direction. The eikonal approximation does not satisfy the unitarity condition, but even under conditions where it gives poor unitarity, the relative strengths of the diffraction intensities can be quite close to that of an exact calculation.

The eikonal approximation was extensively investigated and applied to atom–surface scattering problems by *Garibaldi* et al. [8.45], and they also gave several improvements to the above treatment. A large number of examples and applications are given in the review by *Engel* and *Rieder* [8.1].

The eikonal approximation of (8.72) is essentially equivalent to the Kirchhoff approximation commonly used in optical scattering. A more restrictive approximation, called the primitive semiclassical approximation, is obtained from the eikonal expression by evaluating the coefficients $C(\mathbf{G})$ in (8.72) using the method of stationary phase. The corrugation function $\xi(\mathbf{R})$ is expanded to second order in the variables $\mathbf{R}$, and the point of stationary phase is the position at which the first order terms vanish. This point corresponds to classical mirror reflection in which the angles of reflection and incidence, measured with respect to the local surface normal, are equal [8.46]. The stationary phase integration leads to closed form expressions for $C(\mathbf{G})$. The primitive semiclassical method fails if the stationary phase point is an inflection point of the profile (this corresponds to conditions for classical rainbow scattering) but in this case convergence can often be obtained by carrying the expansion of $\xi(\mathbf{R})$ to higher order.

8.3.4 The Sudden Approximation

The simplicity and usefulness of the eikonal approximation to the hard wall model suggests that it might be fruitful to apply similar approximations to the Schrödinger equation for a more general potential than the corrugated hard wall. Several such semiclassical methods, called "sudden approximations" have been introduced into atom–surface scattering theory by *Gerber* and coworkers [8.47,48]. These are adaptations of the sudden approximations developed for treating rotational and vibrational energy exchange in atomic and molecular collisions [8.49].

The essential assumption of the sudden approximation is that the perpendicular wave vector k_{Gz} is a slowly varying function of G, exactly the same as for the eikonal approximation. The obvious choice in the high energy limit is $k_{Gz} \cong k_{iz}$. We have established in (8.27) that the Schrödinger wavefunction for a periodic surface potential can be written as a Bloch function in terms of the normal wave components $\Psi_G(z)$, which obey the coupled channel equation (8.29)

$$-\frac{d^2}{dz^2}\Psi_G(z) + \frac{2m}{\hbar^2}\sum_M V_{G-M}(z)\,\Psi_M(z) = k_{Gz}^2\,\Psi_G(z) \cong k_{iz}^2\,\Psi_G(z).$$

$$(8.73)$$

This coupled channel equation with the semiclassical approximation represented by the second equality on the right hand side is identical to the following equation for the full wavefunction $\Psi_i(\mathbf{R}, z)$:

$$-\frac{d^2}{dz^2}\Psi_i(\mathbf{R}, z) + \frac{2m}{\hbar^2}V(\mathbf{R}, z)\,\Psi_i(\mathbf{R}, z) = k_{iz}^2\,\Psi_i(\mathbf{R}, z) \quad . \tag{8.74}$$

Equation (8.73) is obtained from (8.74) by Fourier transformation and multiplication by $\exp(\mathrm{i}\mathbf{K}_i\cdot\mathbf{R})$. Equation (8.74) can be regarded as a one-dimensional differential equation in which the solution $\Psi_i(\mathbf{R}, z)$ depends parametrically on $\mathbf{R}$. For each value chosen for $\mathbf{R}$ in the unit cell, the equation can be integrated (generally by numerical methods), and in the asymptotic region the solutions must be of the form

$$\Psi_i(\mathbf{R}, z \to \infty) = \exp(\mathrm{i}\mathbf{K}_i \cdot \mathbf{R})\,[\,\exp(-\mathrm{i}k_{iz}z) + \exp(2\mathrm{i}\eta(\mathbf{R}) + \mathrm{i}k_{iz}z)],$$

$$(8.75)$$

i.e., the differential equation (8.74) simply backreflects the incident wave and multiplies it by a phase $\eta(\mathbf{R})$ which depends parametrically on $\mathbf{R}$. The phase will be periodic since $V(\mathbf{R}, z)$ is periodic in $\mathbf{R}$. The diffraction amplitudes are obtained by comparing (8.75) with the known asymptotic form of the exact wavefunction (8.67)

$$\Psi_i(\mathbf{R}, z \to \infty) = \exp(\mathrm{i}\mathbf{K}_i \cdot \mathbf{R})\,\exp(-\mathrm{i}k_{iz}z)$$

$$+ \sum_G C(G)\,\exp[\mathrm{i}(\mathbf{K}_i + \mathbf{G})\cdot\mathbf{R} + \mathrm{i}k_{Gz}z] \quad . \tag{8.67}$$

After setting $k_{Gz} \cong k_{iz}$ in (8.67) the comparison gives

$$\sum_G C(G)\,\exp(\mathrm{i}\mathbf{G}\cdot\mathbf{R}) = \exp[2\mathrm{i}\eta(\mathbf{R})] \tag{8.76}$$

which leads, after Fourier transformation, to

$$C(\boldsymbol{G}) = \frac{1}{S} \int_{\mathrm{uc}} d\boldsymbol{R} \, \exp(-\mathrm{i}\boldsymbol{G} \cdot \boldsymbol{R}) \, \exp[2\mathrm{i}\eta(\boldsymbol{R})] \quad . \tag{8.77}$$

A further approximation which greatly simplifies the evaluation of the phase $\eta(\boldsymbol{R})$ is to evaluate it in the WKB approximation [8.47]

$$\eta(\boldsymbol{R}) = \lim_{z \to \infty} \int_{z_{\mathrm{R}}}^{z} dz \, [\sqrt{k_{iz}^2 - 2mV(\mathrm{R},z)/\hbar^2} - k_{iz}] + k_{iz}z_{\boldsymbol{R}} \quad , \tag{8.78}$$

where $z_{\boldsymbol{R}}$ is the classical turning point of the potential $V(\boldsymbol{R},z)$ at the energy $\hbar^2 k_{iz}^2/2m$. The connection with the eikonal approximation is now evident, If the potential is represented by a hard repulsive wall with no attractive part, then $V(\boldsymbol{R},z) = 0$ in (8.78) and the phase is determined by the hard wall corrugation function

$$\eta(\boldsymbol{R}) = k_{iz}z_{\boldsymbol{R}} = k_{iz}\xi(\boldsymbol{R}) \quad , \tag{8.79}$$

and the diffraction amplitudes become identical with the eikonal expression (8.72) with $k_{\boldsymbol{G}z} \cong k_{iz}$. The sudden approximation as embodied in (8.77) and (8.78) has been used in discussing a number of problems involving periodic surfaces, adsorbed layers, defects and rough surfaces [8.50]. The method has the advantage of being numerically simple and fast, and it is not limited to hard wall potentials. Thus it is suited to problems where a large number of comparisons with experiment must be made, an example is the determination of the atom–surface potential from the experimental scattering data through the use of inverse scattering theory [8.51]. We have discussed in this section the simplest form of the sudden approximation. Numerous improvements have been proposed to increase its accuracy [8.52,53].

8.3.5 Other Approximate Methods

A number of other approximate solution methods have been applied to the atom–surface scattering problem and we discuss briefly three of them here, the distorted wave Born approximation, semiclassical wave packet methods, and the path integral approach.

One of the most useful and successfully applied approximations in all of quantum physics is the Born approximation, which consists of replacing the transition matrix by its lowest order approximation, a matrix element of the potential. Because of the infinite two-dimensional spatial extent of the potential, and because of its strength, the Born approximation based on plane wave unperturbed states fails in atom–surface scattering. However, the distorted wave Born approximation is useful for weakly diffractive systems and can readily be formulated. The reduced transition matrix of (8.10) or (8.11) is approximated by $t_{fi} \cong v_{fi}$ and the intensity of a diffraction beam is obtained from (8.17) and (8.20):

$$I_G = \frac{k_{Gz}}{k_{iz}} \mid \delta_{G,0} - \frac{im}{\hbar^2 k_{Gz}} \, v_{fi} \mid^2 \, , \tag{8.80}$$

where the subscript f signifies the set of quantum numbers (G, k_{Gz}). As it stands the distorted wave Born approximation of (8.80) is not unitary and the sum of all diffraction intensities will be greater than one. This can be corrected by using a unitarized Born approximation, the simplest version being to divide each non-unitary diffraction intensity by the sum of all such intensities, $I'_G = I_G / \sum_G I_G$. If the distorting potential is one-dimensional, the intensities given by (8.80) will be reasonably accurate when all diffraction intensities are small compared to that of the specular beam. One is not limited to a one-dimensional distorting potential, however, and some work has been done with distorted waves which are exact solutions of a full three dimensional potential [8.54]. As seen in Sect. 8.4 below, the distorted wave Born approximation is most useful in treatments of inelastic scattering.

A semiclassical method, based on numerical propagation of Gaussian wave packets, has been developed by *Drolshagen* and *Heller* [8.55,56]. This is an adaptation of a similar method developed earlier for atomic and molecular scattering in the gas phase [8.57]. A time dependent theory is used in which the initial state of the wavefunction is represented as a superposition of Gaussian wave packets. The individual wave packets are taken to be much more localized than in the exact wave-packet formalism developed above in Sect.8.3.4. These wave packets are propagated individually up to the surface and then back out into the asymptotic region. At every point in the process the interaction potential is approximated by a locally harmonic expansion (the second order expansion in the Taylor series) which retains the Gaussian shape of the wave packets as they propagate along the classical trajectories. This procedure has been applied to a number of problems involving both periodic and non-periodic surfaces [8.58]. It appears to closely approximate exact solutions, except for low energies and glancing angles of incidence. It cannot by itself be used to calculate resonances with the bound states, but modifications which would allow the approximate treatment of such resonances have been proposed [8.59]. Because the method is not inherently dependent on the periodicity of the potential it is ideally suited to treating surface defects, such as adsorbates or roughness [8.60]. It has also been applied to the problem of inelastic scattering [8.61].

The path integral method was first applied to the surface problem by *Doll* [8.62] who adapted the formalism developed by *Miller* [8.63]. The path integral method is a formally exact statement of quantum mechanics [8.64] in which, for a scattering process, the amplitude is given by summing $\exp(-i\hbar \, A)$ over all possible paths between the initial and final conditions. The function A is the classical action, and the paths may be either classical or classically forbidden, depending on whether A is real or complex, respectively. In general the approximations that must be made in order to obtain tractable results involve reducing the number of paths, often to just those which are classically allowed. In the most severe approximation the sum over paths is evaluated by stationary phase methods and the primitive semiclassical result is obtained. Other, less

194

stringent approximations lead to the eikonal result or the sudden approximation. In fact, one of the most useful aspects of the path integral approach is its
ability to lead to approximations at varying levels [8.65].

8.4 Inelastic Scattering

We now take up the question of inelastic scattering in which energy is exchanged in the collision process between particle and surface. This means that
we must deal with the full dynamical Hamiltonian of (8.1). The eigenstates
of the unperturbed crystal $|n_j\rangle$ are solutions of the many body Schrödinger
equation for the crystal Hamiltonian H^c:

$$H^c|n_j\rangle = E_j^c|n_j\rangle \quad . \tag{8.81}$$

It can be shown that the asymptotic wavefunction of the many body Hamiltonian (8.1) can be obtained through manipulations similar to those which
resulted in (8.17) for the elastic problem. The result is

$$\Psi_i(z \to \infty) = \exp(\mathrm{i}\boldsymbol{K}_i \cdot \boldsymbol{R}) \left[\exp(-\mathrm{i}k_{iz}z) + \exp(\mathrm{i}k_{iz}z + \mathrm{i}\delta_i) \right] |n_i\rangle$$
$$- \frac{\mathrm{i}m}{\hbar^2} \sum_{n_f} \sum_{\boldsymbol{K}_f} \frac{t_{fi}}{k_{fz}} \exp(\mathrm{i}\boldsymbol{K}_f \cdot \boldsymbol{R} + \mathrm{i}k_{fz}z) |n_f\rangle \quad ,$$

$$\tag{8.82}$$

where energy conservation for the total system, $E_i + E_i^c = E_f + E_f^c$, fixes the
value of k_{fz} as

$$k_{fz} = \sqrt{\boldsymbol{K}_i^2 + k_{iz}^2 - \boldsymbol{K}_f^2 + 2m(E_i^c - E_f^c)/\hbar^2} \quad , \tag{8.83}$$

where, as in (8.17), $k_{fz}^2 > 0$. The transition matrix t_{fi} appearing in (8.82) is a
matrix element taken with respect to states of both particle and crystal, and
the subscripts f or i refer to the collective set of quantum numbers of both
particle and crystal. For example, the lowest order approximation to t_{fi} is the
matrix element of the potential

$$t_{fi} \cong v_{fi} = \langle n_f | (X_f | v | X_i) | n_i \rangle \quad . \tag{8.84}$$

The transition rate from initial state i to final state f is given by the generalized
golden rule:

$$w_{fi} = \frac{2\pi}{\hbar L^2} | t_{fi} |^2 \, \delta(E_i + E_i^c - E_f - E_f^c) \tag{8.85}$$

and the factor of L^2 appearing in the denominator arises because of the normalization of the matrix elements. Since in an experiment no attempt is made
to measure the final and initial crystal states, we need the transition rate from

initial particle state $\boldsymbol{k}_i$ to final particle state $\boldsymbol{k}_f$. This is obtained from (8.85) by averaging over initial crystal states and summing over final crystal states:

$$w(\boldsymbol{k}_f, \boldsymbol{k}_i) = \left\langle \left\langle \sum_{n_f} w_{fi} \right\rangle \right\rangle \quad . \tag{8.86}$$

The quantity measured in an experiment is the energy resolved differential reflection coefficient which is obtained from (8.86) upon dividing by the incident flux $j_i = \hbar k_{iz}/mL$, and multiplying by the available volume in phase space

$$\frac{dR}{d\Omega_f dE_f} = \frac{L^4}{(2\pi\hbar)^3} \frac{m^2 \, |\boldsymbol{k}_f|}{k_{iz}} w(\boldsymbol{k}_f, \boldsymbol{k}_i) \quad . \tag{8.87}$$

A full and complete treatment of inelastic scattering is quite lengthy, and has been given from a variety of approaches by several authors [8.66]. For the purposes of this review we will discuss in abbreviated terms the salient features of the two most important inelastic effects, the Debye–Waller factor and single phonon inelastic scattering.

8.4.1 The Debye–Waller Factor

The most readily observable effect due to inelastic scattering is a thermal attenuation of the coherent elastic diffraction peaks. The intensities of these peaks decrease strongly with increasing surface temperature. This decrease is qualitatively explained with simple unitarity arguments, at higher temperatures the total inelastic intensity is large and fewer particles can be scattered elastically. The thermal attenuation is well known in neutron and X-ray diffraction from bulk solids, where it is described by the Debye–Waller factor. The intensity $I_k(T)$ of a diffraction peak with momentum transfer $\hbar\boldsymbol{k}$, and for crystal temperature T is given by

$$I_k(T) = I_k(0)\exp[-2W_k(T)] \quad , \tag{8.88}$$

where $I_k(0)$ is the diffraction intensity from the same crystal with all atoms fixed rigidly in their equilibrium positions and $\exp[-2W_k(T)]$ is the Debye–Waller factor, in which

$$W_k(T) = \frac{1}{2}\langle\langle (\boldsymbol{k} \cdot \boldsymbol{u})^2 \rangle\rangle \quad . \tag{8.89}$$

For a simple Bravais lattice $\boldsymbol{u}$ is the vibrational amplitude of an atom in the crystal. Equation (8.89) indicates that the intensity should decrease exponentially with temperature, since the mean square average of the displacement $\boldsymbol{u}$ in a harmonic lattice is linear in temperature for sufficiently large temperatures.

For neutron and X-ray diffraction, the Debye–Waller factor of (8.88) is a rigorous mathematical result following directly from the Born approximation for calculating the scattering intensities. Its success makes it tempting to try to describe thermal attenuation in surface scattering with a similar term. Unfortunately, the theoretical basis for the atom–surface Debye–Waller factor is not so firm. In neutron diffraction single scattering events are dominant, the collisions are of very short duration in time, and the whole calculation can usually be treated in the Born approximation. For atom–surface scattering in the thermal energy range the situation is quite different; the incoming atom may interact with several surface atoms simultaneously, the collision times are long and comparable to phonon vibration periods, and all of the incident atoms are strongly scattered. A strong scattering theory must be used, and thus it is not appropriate to directly map over the results from neutron scattering. In spite of these objections, the ad hoc assumption of (8.88) with $W_k(T)$ given by (8.89) qualitatively describes many experiments rather well. For this reason it is of interest to treat the inelastic scattering in an approximation which produces the standard Debye–Waller factor, and at the same time shows its limitations.

We begin by applying the Van Hove transformation to the transition rate of (8.86) [8.67]. This consists of representing the energy δ-function by a Fourier transform over time τ and defining time dependent operators in the interaction picture according to $t(\tau) = \exp(\mathrm{i}H^c\tau/\hbar)\, t \, \exp(-\mathrm{i}H^c\tau/\hbar)$:

$$
w(\boldsymbol{k}_f, \boldsymbol{k}_i) = \frac{1}{\hbar^2} \int_{-\infty}^{\infty} d\tau \; \exp[\mathrm{i}(E_f - E_i)\tau/\hbar]\langle\!\langle t_{if}(0)\, t_{fi}(\tau)\rangle\!\rangle \quad .
$$

$$(8.90)$$

Now, examination of the asymptotic form of the wavefunction (8.17) or (8.82) shows that if the crystal is rigidly translated through a displacement $\boldsymbol{U}$ parallel to the surface and u_z perpendicular, $\boldsymbol{u} = (\boldsymbol{U}, u_z)$, the effect on the transition matrix of (8.19) is

$$
T_{fi} \rightarrow T_{fi} \exp[\, \mathrm{i}(\boldsymbol{K}_f - \boldsymbol{K}_i)\cdot\boldsymbol{U} + \mathrm{i}(k_{fz} + k_{iz}]u_z \,] = T_{fi}\exp(\, \mathrm{i}\boldsymbol{k}\cdot\boldsymbol{u}\,) \quad ,
$$

$$(8.91)$$

where $\boldsymbol{k} = \boldsymbol{k}_f - \boldsymbol{k}_i$. At this point we make the assumption that the displacement $\boldsymbol{u}$ is the vibrational displacement of the surface. Clearly this is an approximation since a vibration appears like a rigid displacement only for very long wavelength phonons. However, it is well known that the major part of the thermal attenuation is due to long wavelength and low energy phonons. Furthermore, we will assume that the only important time dependence of the transition operator is in the displacement appearing in (8.91). This is equivalent to stating that a given unit cell and its environment are not distorted by the vibration, and again this is a reasonable assumption for long wavelength phonons only. With these further approximations (8.90) becomes

$$w(\boldsymbol{k}_f, \boldsymbol{k}_i) = \frac{1}{\hbar^2} \mid t_{fi} \mid^2$$

$$\times \int_{-\infty}^{\infty} d\tau \, \exp[\mathrm{i}(E_f - E_i)\tau/\hbar] \, \langle\langle \, e^{-\mathrm{i}\boldsymbol{k}\cdot\boldsymbol{u}(0)} \, e^{-\mathrm{i}\boldsymbol{k}\cdot\boldsymbol{u}(\tau)} \, \rangle\rangle \quad .$$

(8.92)

For surface vibrational displacements within the harmonic approximation, the average over crystal states can be carried out to give [8.68]

$$w(\boldsymbol{k}_f, \boldsymbol{k}_i) = \frac{1}{\hbar^2} \mid t_{fi} \mid^2 \, e^{-2W}$$

$$\times \int_{-\infty}^{\infty} d\tau \, \exp[\mathrm{i}(E_f - E_i)\tau/\hbar] \, e^{\langle\langle \, \boldsymbol{k}\cdot\boldsymbol{u}(0) \, \boldsymbol{k}\cdot\boldsymbol{u}(\tau) \, \rangle\rangle} \quad ,$$

(8.93)

where e^{-2W} is the standard Debye–Waller factor with W given by (8.88). The derivation leading to (8.93) shows the extent of the approximations necessary to obtain the Debye–Waller factor. It also shows that the exchange of real phonons is associated with the displacement correlation function $\langle\langle \boldsymbol{k} \cdot \boldsymbol{u}(0) \, \boldsymbol{k} \cdot \boldsymbol{u}(t) \rangle\rangle$ appearing under the integral. For example, if $\exp(\langle\langle \, \boldsymbol{k}\cdot\boldsymbol{u}(0) \, \boldsymbol{k}\cdot\boldsymbol{u}(t) \, \rangle\rangle)$ is expanded in terms of powers of its argument, the zeroth order term describes elastic scattering while the first order term will describe single phonon inelastic processes. The zeroth order term gives

$$w(\boldsymbol{k}_f, \boldsymbol{k}_i) = \frac{2\pi}{\hbar} \mid t_{fi} \mid^2 \, e^{-2W} \, \delta(E_f - E_i) \quad . \tag{8.94}$$

Inserting (8.94) in the differential reflection coefficient of (8.87) and integrating leads directly to the elastic reflection coefficient of (8.20), except that now it will be multiplied by the Debye–Waller factor. The question of inelastic exchange of a single phonon is taken up in Sect. 8.4.2 below.

In interpreting experiments, several further approximations to the Debye–Waller factor have proven useful. Usually, in the collision process the atomic projectiles are sensitive only to the normal vibrations of the surface. If $W_k(T)$ of (8.89) is evaluated for a one dimensional vibration normal to the surface and for a bulk Debye phonon distribution, the following simple form is obtained:

$$W(T) = \frac{3\hbar^2 (k_{iz} + k_{fz})^2 T}{2Mk_{\mathrm{B}}\Theta_{\mathrm{D}}^2} \quad , \tag{8.95}$$

where Θ_{D} is the Debye temperature, M is the mass of a surface atom, k_{B} is the Boltzmann constant, and the evaluation is for high temperature, $T > \Theta_{\mathrm{D}}$. The attractive well near the surface can be taken into account by assuming that its only effect is to accelerate the particle before the collision. This is done by replacing k_{ji} by $k_{ji}^* = \sqrt{k_{ji}^2 + 2mD/\hbar^2}$, where D is the well depth.

We have shown here one very restrictive model which can give rise to a Debye–Waller factor in atom–surface scattering, and which shows the nature of the approximations that must be made. Similar results can also be obtained from models in which the collision with the surface is instantaneous, such as the hard corrugated wall [8.69] or the sudden approximation [8.70]. In general, a Debye–Waller model with (8.95) for the exponent does not always agree well with experiment, neither in the momentum dependence nor in the temperature dependence. A full treatment of thermal attenuation for realistic potentials requires a detailed investigation of multiple phonon exchange, and such calculations have been attempted by several groups [8.71].

8.4.2 Single Phonon Exchange

Many different approaches to the single phonon inelastic scattering problem have been proposed [8.72], and in view of the fact that inelastic scattering is much more complicated than the elastic diffraction problem, separate approaches may be appropriate for each specific system to be examined. The basic kinematical features of single phonon exchange are the conservation of momentum and conservation of energy as expressed by

$$\boldsymbol{K}_f - \boldsymbol{K}_i = \boldsymbol{Q} + \boldsymbol{G} \tag{8.96}$$

and

$$E_f - E_i = \pm \hbar \omega \quad , \tag{8.97}$$

where $\boldsymbol{Q}$ is the parallel momentum of the exchanged phonon and ω is its frequency. For a surface phonon such as a Rayleigh mode the frequency depends only on parallel momentum, $\omega = \omega(\boldsymbol{Q})$, and the kinematical relations completely specify the relation between the final and initial particle momentum states. This means that such modes are ideally observed as sharp peaks in the differential reflection coefficient. Scattering from the bulk modes generally contributes a diffuse background, except in situations where there are strong singularities in the bulk spectral density.

The inelastic intensities depend on the dynamics of the collision process. In general the intensity is proportional to a convolution of the phonon spectral density at the surface and a form factor (usually a square matrix element of the interaction potential). Most calculations use the distorted wave Born approximation, with either a model potential function or with a potential which is a summation of pairwise interactions between the projectile and the atoms of the surface. The procedure of averaging the transition rate over crystal states follows similarly to that which led to (8.93) above. When the single phonon contribution is extracted, the dominant contribution to the intensity is due to vertical surface displacement, and the differential coefficient has the following appearance:

$$\frac{dR}{d\Omega_f dE_f} = \frac{m^2}{2\hbar M k_{iz}} \sum_{G} \sum_{Q} \int d\omega \frac{|\,\boldsymbol{k}_f\,|}{\omega} e^{-2W}$$

$$\times \left\{ \rho_{zz}(\boldsymbol{Q},\omega) \; |\,(f\,|\,\frac{d}{dz}U_{-\boldsymbol{Q}-\boldsymbol{G}}(z)\,|\,i)\,|^2 \; [n(\omega)+1] \right.$$

$$\times \delta\,(\boldsymbol{K}_f - \boldsymbol{K}_i + \boldsymbol{Q} + \boldsymbol{G})\delta(E_f - E_i + \hbar\omega)$$

$$+ \rho_{zz}(\boldsymbol{Q},\omega) \; |\,(f\,|\,\frac{d}{dz}U_{\boldsymbol{Q}-\boldsymbol{G}}(z)\,|\,i)\,|^2 \; n(\omega)$$

$$\times \; \delta(\boldsymbol{K}_f - \boldsymbol{K}_i - \boldsymbol{Q} + \boldsymbol{G})\,\delta(E_f - E_i - \hbar\omega) \quad ,$$

$$(8.98)$$

where $\rho_{zz}(\boldsymbol{Q},\omega)$ is the phonon spectral density for displacement normal to the surface, $n(\omega)$ is the Bose–Einstein function, and $(f\,|\,\frac{d}{dz}U_{-\boldsymbol{Q}-\boldsymbol{G}}(z)\,|\,i)$ is a matrix element of the normal gradient of the thermally averaged interaction potential. The appearance of the normal derivative indicates that the effective potential for single phonon scattering can be expressed as the first order term in a Taylor expansion in powers of the vibrational displacement, suitably averaged over initial crystal states. The kinematical conditions (8.96) and (8.97) are contained in the δ-functions. The term in $n(\omega)$ is for phonon annihilation (in which case the incoming particle gains energy from the surface) and the term in $n(\omega)+1$ is for phonon creation.

Equation (8.98) shows that the single phonon differential reflection coefficient is directly proportional to the squared matrix element (or in more general cases, a form factor) and to the phonon spectral density in the case of surface phonons for which $\omega = \omega(\boldsymbol{Q})$, i. e., for surface phonons in which ω depends only on parallel momentum. In this case, if the potential $U_K(z)$ is reasonably well known from other experiments, for example elastic diffraction, then measurements of the one phonon intensity are directly proportional to the surface phonon spectral density $\rho_{zz}(\boldsymbol{Q},\omega)$.

We have outlined in this section an approach to calculating single phonon inelastic intensities which illustrates the global features of most treatments. Ramifications and improvements usually involve modifications of the approximation of a pairwise sum of atomic potentials or changes in the distorting potential. A pairwise summation of potentials is particularly questionable in the case of metal surfaces, and often will not give rise to the correct attractive part. Thus one approximation is to add a static attractive term into the distorting potential [8.73]. An important improvement is to use the thermally averaged total potential as the static distorting potential, instead of a one dimensional distorting potential. This has been highly successful in explaining inelastic scattering from highly corrugated surfaces where there is also strong diffraction [8.74].

8.5 Scattering from Disordered Surfaces

The extreme sensitivity of atomic scattering to surface defects makes it an ideal tool for characterizing the nature of surface disorder. Adsorbates, steps, vacancies and defects, as well as islands and domains have been studied, and each has its characteristic signature in the observed scattered intensity distributions. Often the characteristics of a particular defect can be used to follow the dynamics of surface processes, such as the diffusion of adsorbates or monitoring phase transitions. There are a large range of theoretical techniques that have been developed to treat various specialized problems of surface disorder [8.75]. We treat here the most frequently encountered situation, the elastic and inelastic diffuse intensity that arises from scattering by a dilute distribution of defects such as adsorbates or steps.

We examine the situation in which the defects are sufficiently separated so that multiple scattering between different defects is negligible. Then the transition matrix can be written as a sum of atomic-like scattering amplitudes, each multiplied by a phase factor

$$t_{fi} = \tau_{fi} \sum_n \exp[-i\boldsymbol{k} \cdot (\boldsymbol{r}_n + \boldsymbol{u}_n)] \quad , \tag{8.99}$$

where as before $\boldsymbol{k}$ is the total exchanged wavevector, $\boldsymbol{r}_n$ is the surface defect position, and $\boldsymbol{u}_n$ is the displacement of the defect from its equilibrium position due to thermal motion. All of the scattering centers are assumed identical and located at identical sites. Note that t_{fi} is not simply an atomic scattering amplitude for the defect; it must also include multiple scattering with the surrounding substrate surface. Further, the phase factor in (8.99) includes the possibility that the defects could be at different heights on the surface, as in the case of steps which generate different terrace heights. Inserting the transition matrix of (8.99) into the transition rate of (8.85) and carrying out the averaging process of (8.86) just as was done to produce (8.92), gives the following:

$$w(\boldsymbol{k}_f, \boldsymbol{k}_i) = \frac{1}{\hbar^2} \mid \tau_{fi} \mid^2 \int_{-\infty}^{\infty} d\tau \, \exp[i(E_f - E_i)\tau / \hbar]$$

$$\times \sum_{n;j} \exp[i\boldsymbol{k} \cdot (\boldsymbol{r}_n - \boldsymbol{r}_j)] \, \langle\langle \exp[i\boldsymbol{k} \cdot \boldsymbol{u}_n(0)] \exp[-i\boldsymbol{k} \cdot \boldsymbol{u}_j(\tau)]\rangle\rangle \quad .$$

$$\tag{8.100}$$

Equation (8.100) has the classic form of the fundamental equation governing a large class of scattering treatments involving many-body targets:

$$w(\boldsymbol{k}_f, \boldsymbol{k}_i) = \frac{2\pi}{\hbar} \mid \tau_{fi} \mid^2 S(\boldsymbol{k}, \omega) \quad , \tag{8.101}$$

where $\omega = (E_f - E_i)/\hbar$. This is a form factor $|\tau_{fi}|^2$ for the scattering centers, multiplying the Fourier transform of a time dependent dynamical structure factor $S(\boldsymbol{k}, \tau)$, which depends on the average over positions of the scattering centers. Identifying $S(\boldsymbol{k},\tau)$ from the integrand in (8.100), and carrying out the averaging under the assumption of harmonic vibrational motion we have

$$S(\boldsymbol{k},\tau) = \mathrm{e}^{-2W(\boldsymbol{k})}$$

$$\times \sum_{n;j} \exp[\mathrm{i}\boldsymbol{k}\cdot(\boldsymbol{r}_n - \boldsymbol{r}_j)]\exp(\langle\langle\boldsymbol{k}\cdot\boldsymbol{u}_n(0)\cdot\boldsymbol{k}\cdot\boldsymbol{u}_j(\tau)\,\rangle\rangle) \quad,$$

$$(8.102)$$

where $\exp(-2W)$ is the Debye–Waller factor of (8.89) and the vibrational displacement correlation function is similar to that appearing in (8.93).

Equation (8.100) with (8.101) and (8.102) gives a complete description of the elastic and inelastic diffuse scattering for a small coverage of defects on the surface. Correlations in the scattering between different defects are contained in (8.102) in the double sum over sites, and inelastic exchange is given by the exponentiated vibrational displacement correlation function.

A useful example is to consider the elastic case, where setting $\boldsymbol{u}_n(0)$ and $\boldsymbol{u}_j(\tau)$ equal to zero in (8.100–102) leads to

$$\frac{dR}{d\Omega_f} = \frac{m^2\,|\,\boldsymbol{k}_f\,|}{4\hbar^4 k_{iz}\pi^2}\;|\,\tau(\boldsymbol{k}_f,\boldsymbol{k}_i)\,|^2\;\mathrm{e}^{-2W(k)} \sum_{n;j} \exp[\mathrm{i}\boldsymbol{k}\cdot(\boldsymbol{r}_n - \boldsymbol{r}_j)] \quad.$$

$$(8.103)$$

We note that the diffuse defect scattering intensity is multiplied by a Debye–Waller factor similar to that multiplying the coherent diffraction peaks. In the simplest case of randomly positioned defects, the structure factor in (8.103) is just a constant equal to the total number of defects. Then the diffuse intensity is given by the form factor $|\,\tau(\boldsymbol{k}_f,\boldsymbol{k}_i)\,|^2$. The multiple scattering between the defect and substrate, as well as rainbow scattering give rise to characteristic oscillations in the intensity as a function of the detector position. These characteristic oscillations have been observed in the cases of He scattered both by steps [8.76] and by adsorbates [8.77].

Usually, defects will appear at well defined lattice sites on the surface. This will give rise to correlations in the observed intensity which are accounted for in the structure factor of (8.103). In the case of steps on metal surfaces, fine structure in the scattered intensity clustered around diffraction peak positions has been attributed to the lattice gas behavior of the distribution of the steps and associated terraces. Equation (8.103) with a lattice gas structure factor has been used to verify this interpretation [8.78].

For the treatment of diffuse inelastic scattering the vibrational displacement correlation function in (8.102) must be considered. For example, an ordered series in number of phonons exchanged is obtained by expanding the exponential $\exp(\langle\langle\boldsymbol{k}\cdot\boldsymbol{u}_n(0)\cdot\boldsymbol{k}\cdot\boldsymbol{u}_j(\tau)\rangle\rangle)$ in (8.102) in a power series in its argument. The first order, or single phonon, term is given by replacing the exponential by its argument, which shows again that single quantum scattering is given directly by the displacement correlation function. When this process is carried out the resulting differential reflection coefficient is useful for describing the scattering by defect centers which exhibit Einstein-like vibrational modes. Experimental examples of such systems are CO adsorbed on smooth metal surfaces [8.79], or physisorbed layers of rare gases [8.80,81].

Acknowledgment

The author would like to thank the Alexander von Humboldt Foundation for providing support during the course of this work.

References

8.1   Recent reviews include the following: V. Bortolani, A. C. Levi: Rivista del Nuovo Cimento **9**, 1 (1986); J. A. Barker, D. J. Auerbach: Surf. Sci. Reports **4**, 1 (1984); G. Boato, P. Cantini: Adv. Electron. Electron Phys. **60**, 95 (1983); T. Engel, K. H. Rieder: Springer Tracts Mod. Phys. **19**, 1 (1980); H. Hoinkes: Rev. Mod. Phys. **52**, 933 (1980); R. B. Gerber: Chem. Rev. **87**, 29 (1987); J. P. Toennies: Physica Scripta **T19**, 39 (1987); V. Celli: In *Many Body Phenomena at Surfaces*, ed. by D. Langreth, H. Suhl (Academic, New York 1984) p. 315

8.2   M. Gell-Mann, M. L. Goldberger: Phys. Rev. **91**, 398 (1953)

8.3   G. Lilienkamp, J. P. Toennies: J. Chem. Phys **78**, 5210 (1983)

8.4   N. Cabrera, V. Celli, R. Manson: Phys. Rev. Lett. **22**, 396 (1969)

8.5   N. Cabrera, V. Celli, F. O. Goodman, R. Manson: Surf. Sci. **19**, 67 (1970)

8.6   A. Tsuchida: Surf. Sci. **14**, 375 (1969)

8.7   G. Wolken, Jr.: J. Chem. Phys. **58**, 3047 (1973); G. Wolken, Jr.: J. Chem. Phys. **59**, 1159 (1973)

8.8   R. G. Gordon: J. Chem. Phys. **51**, 14 (1969); R. G. Gordon: Methods Comput. Phys. **10**, 81 (1971)

8.9   A. Liebsch, J. Harris: Surf. Sci. **123**, 355 (1982); R. Laughlin: Phys. Rev. B **25**, 2222 (1982); G. Drolshagen, A. Kaufhold, J. P. Toennies: Isr. J. Chem. **22**, 283 (1982)

8.10   G. Armand, J. R. Manson: Phys. Rev. Lett. **43**, 1839 (1979)

8.11   G. Armand: Private communication

8.12   G. Armand: J. de Phys. **41**, 1475 (1980)

8.13   G. Armand, J. R. Manson: J. de Phys. **44**, 473 (1983); G. Armand, J. R. Manson: Surf. Sci. **119**, L299 (1982)

8.14   B. Salanon, G. Armand, J. Perreau, J. Lapujoulade: Surf. Sci. **127**, 135 (1983); B. Salanon, J. Lapujoulade: Phys. Rev. B **36**, 4281 (1987)

8.15   G. Armand, J. R. Manson: Phys. Rev. B **25**, 6195 (1982)

8.16   D. Kosloff, R. Kosloff: J. Comp. Phys. **52**, 35 (1983); D. Kosloff, R. Kosloff: J. Chem. Phys. **79**, 1823 (1983)

8.17   R. B. Gerber, R. Kosloff, M. Berman: Computer Phys. Reports **5**, 59 (1986)

8.18   A. T. Yinnon, R. Kosloff: Chem. Phys. Lett. **102**, 216 (1983)

8.19   R. Kosloff, H. Tal-Ezer: Chem. Phys. Lett. **127**, 223 (1986)

8.20   A. T. Yinnon, R. Kosloff, R. B. Gerber: Surf. Sci. **148**, 148 (1984); A. T. Yinnon, R. Kosloff, R. B. Gerber: J. Chem. Phys. **88**, 7209 (1988); A. T. Yinnon, R. B. Gerber, D. K. Dacol, H. Rabitz: J. Chem. Phys. **84**, 5955 (1986)

8.21   H. Chow, E. D. Thompson: Surf. Sci. **59**, 225 (1976); H. Chow: Surf. Sci. **62**, 487 (1977)

8.22 J. M. Hutson, C. Schwartz: J. Chem. Phys. **79**, 5179 (1983)

8.23 V. Celli: In *Rarefied Gas Dynamics*, ed. by S. Fisher, (AAIA, New York 1981) Part I, p. 50

8.24 J. Mantovani, J. R. Manson, G. Armand: Surf. Sci. **143**, 536 (1984)

8.25 H. Feshbach: Ann. Phys. (New York) **5**, 357 (1958); **19**, 287 (1962)

8.26 C. E. Harvie, J. H. Weare: Phys. Rev. Lett. **40**, 187 (1978)

8.27 K. L. Wolfe, J. H. Weare: Phys. Rev. Lett. **41**, 1663 (1978); K. L. Wolfe, J. H. Weare: Surf. Sci. **84**, 581 (1980)

8.28 V. Celli, N. Garcia, J. Hutchison: Surf. Sci. **87**, 112 (1979)

8.29 W. E. Carlos, M. W. Cole: Surf. Sci. **77**, L173 (1978)

8.30 W. E. Carlos, M. W. Cole: Phys. Rev. Lett. **43**, 697 (1979)

8.31 J. Hutchison: Phys. Rev. B **22**, 5671 (1980)

8.32 P. Cantini, S. Terrini, C. Salvo: Surf. Sci. **119**, L299 (1982)

8.33 J. Lapujoulade, G. Armand, J. R. Manson: Phys. Rev. B **39**, 10514 (1989)

8.34 V. Celli, D. Evans: Israel J. of Chem. **22**, 289 (1982)

8.35 N. Garcia, N. Cabrera: Phys. Rev. B **18**, 576 (1978)

8.36 G. Armand, J. R. Manson: Phys. Rev. B **18**, 6510 (1978)

8.37 G. Armand, J. Lapujoulade, J. R. Manson: Surf. Sci. **82**, L265 (1979)

8.38 J. W. Strutt (Baron Rayleigh): Proc. Roy. Soc. London, Ser. A **79**, 388 (1907); *The Theory of Sound* (Dover, New York 1945), Vol. II

8.39 R. Petit, M. Cadilhac: C. R. Acad. Sci. B **262**, 468 (1966)

8.40 R. F. Millar: Proc. Camb. Philos. Soc. **69**, 217 (1971)

8.41 N. R. Hill, V. Celli: Phys. Rev. B **17**, 17 (1978)

8.42 N. Garcia, J. Ibanez, J. Solana, N. Cabrera: Solid State Commun. **20**, 1159 (1976); N. Garcia, J. Ibanez, J. Solana, N. Cabrera: Surf. Sci. **60**, 385 (1976)

8.43 N. Garcia: J. Chem. Phys. **67**, 897 (1977)

8.44 G. Armand, J. R. Manson: Phys. Rev. B **19**, 4091 (1979)

8.45 U. Garibaldi, A. C. Levi, R. Spadacini, G. E. Tommei: Surf. Sci. **48**, 649 (1975); Surf. Sci. **55**, 40 (1976)

8.46 P. M. Morse, H. Feshbach: *Methods of Theoretical Physics* (McGraw-Hill, New York 1953), Vol. 2, p. 1554

8.47 R. B. Gerber, A. T. Yinnon, J. N. Murrell: Chem. Phys. **31**, 1 (1978)

8.48 A. T. Yinnon, S. Bosanac, R. B. Gerber, J. N. Murrell: Chem. Phys. Lett. **58**, 364 (1978)

8.49 K. Takayanagi: Prog. Theoret. Phys. (Kyoto), Suppl. **25**, 40 (1963); T. P. Tsien, R. T. Pack: Chem.Phys. Lett. **6**, 54 (1970)

8.50 R. B. Gerber, A. T. Yinnon, Y. Shimoni, D. J. Kouri: J. Chem. Phys. **73**, 4397 (1980); A. T. Yinnon, R. B. Gerber, D. K. Dacol, H. Rabitz: J. Chem. Phys. **84**, 5955 (1986); J. I. Gerstein, R. B. Gerber, D. K. Dacal, H. Rabitz: J. Chem. Phys. **76**, 4277 (1983); D. K. Dacol, H. Rabitz, R. B. Gerber: J. Chem. Phys. **86**, 1616 (1987)

8.51 R. B. Gerber, A. T. Yinnon: J. Chem. Phys. **73**, 3232 (1980)

8.52 T. P. Tsien, G. A. Parker, R. T. Pack: J. Chem. Phys. **59**, 5373 (1973); D. Secrest: J. Chem. Phys. **62**, 710 (1975)

8.53 B. J. Hinch: Surf. Sci., to be published

8.54 D. Eichenhauer, U. Harten, J. P. Toennies, V. Celli: J. Chem. Phys. **86**, 3693 (1987)

8.55 G. Drolshagen, E. J. Heller: J. Chem. Phys. **79**, 2072 (1983)

8.56 G. Drolshagen, E. J. Heller: Surf. Sci. **139**, 260 (1984)

8.57 E. J. Heller: J. Chem. Phys. **62**, 1544 (1975); E. J. Heller: J. Chem. Phys. **65**, 4979 (1976)

8.58 R. Heather, H. Metiu: J. Chem. Phys. **84**, 3250 (1986)

8.59 J. R. Manson, G. Drolshagen: Surf. Sci. **184**, L383 (1987)

8.60 G. Drolshagen, R. Vollmer: J. Chem. Phys. **87**, 4948 (1987); D. Huber, E. J. Heller, R. G. Littlejohn: J. Chem. Phys. **89**, 2003 (1988); H. Xu, D. Huber, E. J. Heller: J. Chem. Phys. **89**, 2550 (1988)

8.61 B. Jackson, H. Metiu: J. Chem. Phys. **84**, 3535 (1986)

8.62   J. D. Doll: Chem. Phys. **3**, 257 (1974); J. D. Doll: J. Chem. Phys. **61**, 954 (1974)

8.63   W. H. Miller: Adv. Chem. Phys. **25**, 69 (1974)

8.64   R. P. Feynman, A. R. Hibbs: *Quantum Mechanics and Path Integrals* (McGraw-Hill, New York, 1965)

8.65   K. J. McCann, V. Celli: Surf. Sci. **61**, 10 (1976)

8.66   A. C. Levi, H. Suhl: Surf. Sci. **88**, 221 (1979); R. Manson, V. Celli: Surf. Sci. **24**, 495 (1971); V. Bortolani, A. Francini, F. Nizzoli, G. Santoro: Surf. Sci. **152**, 811 (1985)

8.67   L. Van Hove: Phys. Rev. **95**, 249 (1954)

8.68   A. A. Maradudin, E. W. Montroll, G. H. Weiss: *Theory of Lattice Dynamics in the Harmonic Approximation*, Solid State Physics, ed. by F. Seitz, D. Turnbull, Suppl. 3 (Academic, New York 1963)

8.69   G. Benedek, N. Garcia: Surf. Sci. **80**, 543 (1980); G. Armand, J. R. Manson: Surf. Sci. **80**, 532 (1979)

8.70   H.-D. Meyer: Surf. Sci. **104**, 117 (1981); R. Schinke: Surf. Sci, **127**, 283 (1983)

8.71   V. Celli, A. A. Maradudin: Phys. Rev. B **31**, 825 (1985); J. Idiodi, V. Bortolani, A. Franchini, G. Santoro, V. Celli: Phys. Rev. B **35**, 6029 (1987); G. Armand, J. R. Manson: Phys. Rev. B **37**, 4363 (1988)

8.72   M. Lagos, L. Bernstein: Surf. Sci. **52**, 391 (1975); M. Lagos: Surf. Sci. **65**, 124 (1977); A. M. Marvin, F. Toigo: Nuovo Cimento B **53**, 25 (1979); A. C. Levi: Nuovo Cimento B **54**, 357 (1979); W. Brenig: Z. Phys. B **36**, 81 (1979)

8.73   V. Bortolani, A. Franchini, G. Santoro: In *Dynamical Phenomena at Surfaces, Interfaces, and Superlattices*, ed. by F. Nizzoli, K. H. Rieder, R. F. Willis, Springer Ser. Surf. Sci. Vol. 3 (Springer, Berlin, Heidelberg 1985) p. 92

8.74   D. Eichenauer, J. P. Toennies: J. Chem. Phys. **85**, 532 (1987)

8.75   For a recent review of scattering from disordered surfaces see: B.Poelsema, G. Comsa: Springer Tracts Mod. Phys. **115**, ed. by G. Höhler (Springer, Berlin, Heidelberg 1989)

8.76   A. Lahee, J. R. Manson, J. P. Toennies, Ch. Wöll: Phys. Rev. Lett. **57**, 471 (1986)

8.77   A. Lahee, J. R. Manson, J. P. Toennies, Ch. Wöll: J. Chem. Phys. **86**, 7194 (1987)

8.78   B. J. Hinch: Phys. Rev. B, to be published; C. W. Skorupka, J. R. Manson: Phys. Rev. B **41**, 9183 (1990)

8.79   A. Lahee, J. P. Toennies, Ch. Wöll: Surf. Sci. **177**, 371 (1986)

8.80   K. D. Gibson, S. J. Sibener: Phys. Rev. Lett. **55**, 1514 (1985); S. J. Sibener, D. L. Mills, B. M. Hall, J. E. Black: J. Chem. Phys. **83**, 4256 (1985)

8.81   K. Kern, P. Zeppenfeld, R. David, G. Comsa: Phys. Rev. B **35**, 886 (1987)

9. Surface Phonons: From Theory to Spectroscopy and Back

G. Benedek, L. Miglio and G. Seriani

9.1 Early Theoretical Predictions

The mathematical demonstration given by Lord Rayleigh about one century ago [9.1] that the long-wave (L) component of earthquakes is due to sagittal surface elastic waves is commonly considered the starting point of surface physics. The theory of Rayleigh waves and other types of surface waves in semiinfinite elastic continua very soon left the realm of geophysics and attracted the interest of solid state physicists and materials scientists due to the potential of surface acoustic waves (SAW) for applications in delay lines, optoacoustic systems and signal processing devices [9.2].

The main lines of theoretical research initiated by the fundamental work of Rayleigh were

— the extension to different geometrical configurations,
— the extension to anisotropic crystals, piezoelectric materials, etc.,
— the determination of the dispersion in the high-frequency (short-wave) limit where the lattice structure of solids plays a crucial role.

Surface lattice dynamics was pioneered in the USSR in the 1940s by the theoretical work of *Lifshitz* and *Rozenzweig* [9.3], based on the Green's function (GF) method. The basic idea of that work was to consider the free surface as a perturbation of a lattice with cyclic boundary conditions and therefore to derive the surface lattice vibrations from the spectrum of bulk vibrations. This method has received various elegant formulations and extensions by *Maradudin* et al., *Wallis* and *Dobrzynski* [9.4, 5], *Garcia–Moliner* (GF matching [9.6]), *Armand* (GF generating coefficients [9.7]), *Allen* (continued fraction method [9.8]), and an appreciable number of applications to systems with short-range interactions [9.9–13].

The GF method has the advantage of reducing the size of the dynamical problem from that of the whole crystal to that of the perturbation space and is therefore suitable for materials with short-range forces. However we have devised an extension of the GF method to ionic crystals with long-range coulomb

Springer Series in Surface Sciences, Vol. 27   **Helium Atom Scattering from Surfaces**
Editor: E. Hulpke   © Springer-Verlag Berlin, Heidelberg 1992

interactions, such as alkali halides [9.14–19], and to crystals with complex many-body interactions such as refractory superconductors [9.20]. This is based on an effective short-range perturbation self-consistently defined through the translational (TI) and rotational (RI) invariance conditions. The advantage of dimensional reduction in GF methods is restrained by computational difficulties due to the singular nature of GFs.

Another method for surface lattice dynamics, conceptually similar to Lord Rayleigh's approach for continua, is based on trial surface-localized waves with a number of components comparable to the largest order of interacting neighbors. An extension to long-range forces has been given by *Feuchtwang* [9.21] and occasionally applied by the Modena group for transition metals [9.22]. In the late 1950s *Wallis* predicted with this method the existence of optical surface modes with shear vertical (SV) polarization in diatomic crystals [9.23]. Later on optical surface modes polarized in the surface plane were predicted by *Lucas* [9.24] and termed Lucas modes (LM). Wallis and Lucas modes, which also exist in non-polar diatomic crystals like silicon, have microscopic character in that their penetration inside the crystal remains comparable to the lattice distance even at infinite wavelength.

In contrast, the penetration depth of macroscopic modes like Rayleigh waves is proportional to the wavelength in the long wave (continuum) limit. On the other hand macroscopic optical modes do exist in polar crystals as mixed photon–phonon modes (surface polaritons). They were also discovered theoretically, in the late sixties, by *Fuchs* and *Kliever* (FK modes [9.25, 26]; for their derivation by the GF method see [9.19]).

With the advent of fast computers, the calculation of surface vibrations by the direct diagonalization of the dynamical matrix for a sufficiently thick slab has become more expedient and offers a pictorial representation of surface phonon dispersion. The fundamental work with this method is due to *de Wette* and coworkers. Their systematic studies on the low-index surfaces of Lennard-Jones fcc crystals [9.27] and of alkali halides [9.28–30] represent a basic reference point. With the slab method *Alldredge* found another surface acoustic mode with shear-horizontal (SH) polarization, orthogonal to the sagittally polarized Rayleigh waves [9.31].

9.2 Early Experimental Studies

In the early seventies there were quite clear ideas about the nature and dispersion of surface phonons for ideal surfaces in different classes of crystals, but not a single piece of experimental evidence for surface phonons in the dispersive region. Their elusive nature challenged spectroscopists for about ten years. The conventional spectroscopic techniques for bulk phonons, such as neutron, X-ray and Raman scattering, are rather insensitive to surfaces. Surface-sensitive probes like low-energy electrons and atoms were considered in the late sixties as candidates for surface–phonon spectroscopy. Electron energy loss spectroscopy (EELS) experiments performed by *Ibach* in polar crystals actually showed the expected FK modes [9.32], in agreement with earlier attenuated total reflection (ATR) observations [9.33], but even nonpolar crystal surfaces such as Si(111)2 × 1 showed, much to everyone's surprise, a strong, almost isotropic optical surface mode at about 55 meV [9.34]. This classic experiment was soon regarded as evidence for a genuine microscopic surface mode [9.35], but it took about eighteen years before theoreticians could offer an acceptable explanation.

Cabrera, *Celli* and *Manson* (1969–1971) gave a theoretical demonstration that helium atom scattering from single phonons is the dominant process at thermal collision energies and could be exploited for surface phonon detection [9.36, 37]. *Benedek* and *Seriani* [9.38] combined the Celli and Manson theory with their surface-phonon GF calculations and predicted inelastic He-scattering spectra for LiF(001). At that time the venerable technique of atom scattering, used in the thirties by Stern and coworkers at Hamburg to prove the wave nature of matter, had made a breakthrough in resolution, thanks to the development of nozzle beam sources. *Williams* was able to determine the long-wavelength part of the Rayleigh wave dispersion in LiF(001) [9.39] and NaF(001) [9.40] from a difficult analysis of out-of-plane angular distributions of the scattered He atoms.

The study of inelastic processes in angular distributions was one of the hot subjects at the memorable 1973 session of the Enrico Fermi International School at Varenna. There was great interest in the high-resolution angular distributions of helium and neon scattering from LiF(001) (Fig.9.1a) presented by *Boato*, who at that time was assembling an important group in Genoa. One of us (GB) was puzzled by the inextricable wealth of inelastic features observed in the gaps between the sharp diffraction peaks, and tried an analysis in terms of van Hove singularities (alias kinematic focussing (KF), according to a pictorial terminology invented by *Miller* [9.42]). According to his optimistic view the alignments of KF peaks in the $\sin\Theta_i$ versus $\sin\Theta_f$ plot (Fig.9.1b) were indicative of the frequencies of Rayleigh and Lucas modes at the symmetry points [9.42], as displayed in Table 9.1. *Alldredge* and *de Wette* were delighted by the good agreement of their calculations for LiF(001) with GB's analysis [9.29]. GB, however, did not dare to publish these data, because his calculated angular distributions — the result of a remarkable tour-de-force for the computers of that time — bore little resemblance to the experimental data (Fig.9.2).

Cantini et al. [9.43] actually provided a very attractive explanation in terms of inelastic resonances involving Rayleigh waves up to the zone boundary, an effect not included in GB's calculations based on the distorted-wave Born approximation. GB and Boato, in a short review of 1977, came to the dull conclusion that little can be learnt about surface phonons from angular distributions, and that only time-of-flight spectroscopy will enable the observation of surface phonon disperion curves [9.44]; this, however, would require a still better resolution than that available in the Genoa experiments at that time. This problem was long since clear to Peter Toennies (who, by the way, attended the Varenna course as a student!), and led him, along with *Winkelmann* [9.45], to set the

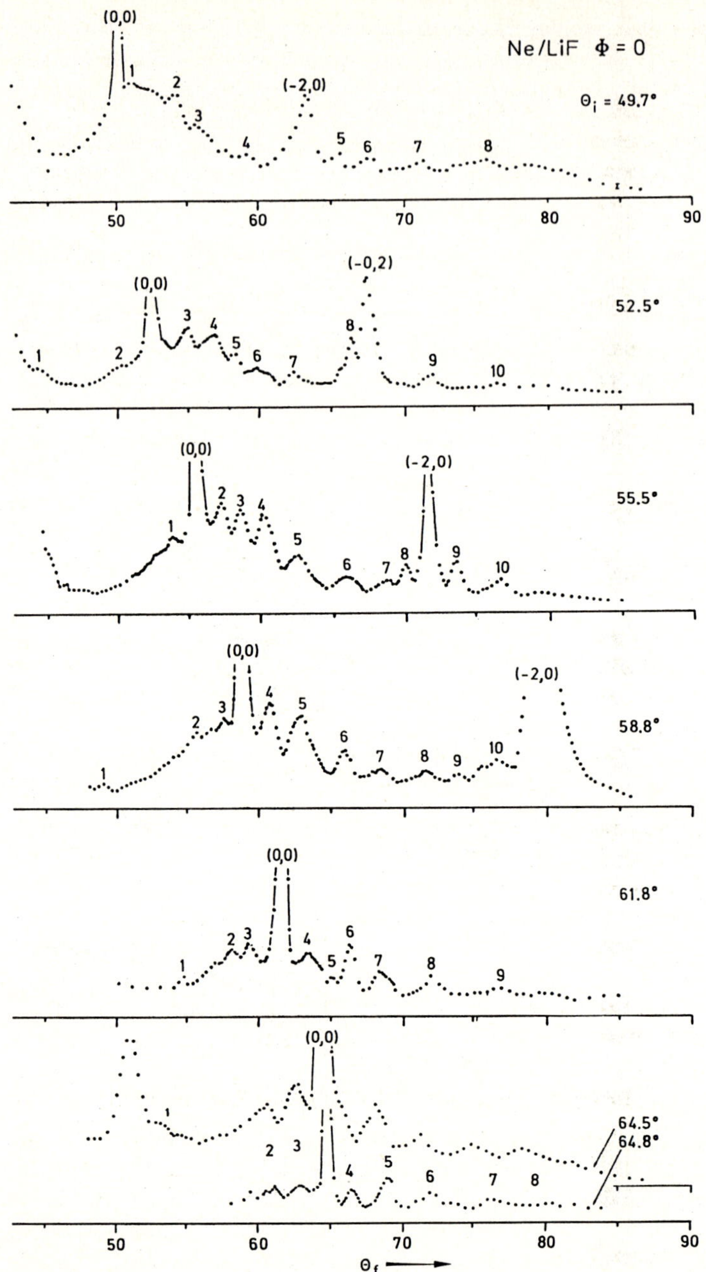

Fig.9.1. Upper panel: (a) Inelastic angular distribution of Ne atoms scattered from Li(001) for different incidence angles (from [9.41] and private communication); the vectors (0,0) and (-2,0) denote the elastic and first diffraction peaks. Lower panel: (b) Plot of $\sin\Theta_i$ versus $\sin\Theta_f$ showing the alignments of peak positions. Points are numbered as the corresponding peaks in (a)

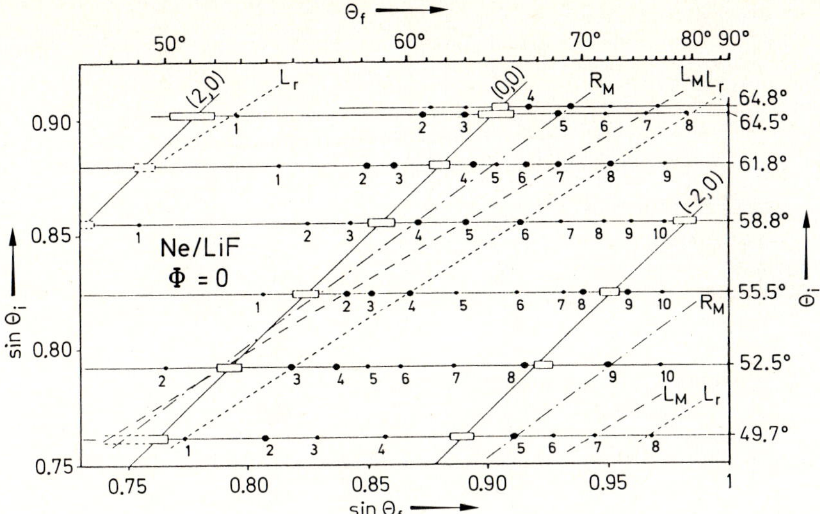

Fig.9.1. (cont.)

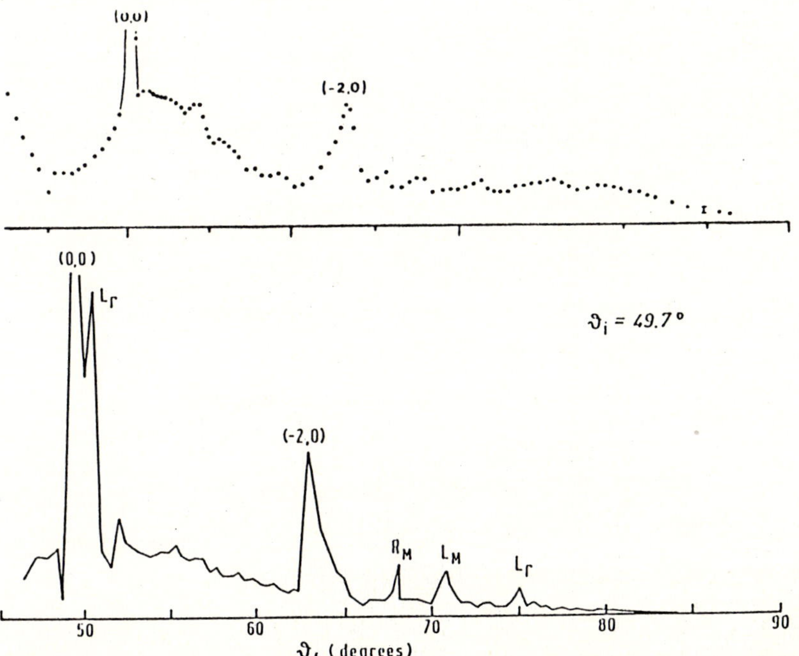

Fig.9.2. Experimental and theoretical inelastic Ne scattering angular distributions from LiF(001) for different incidence angles. Calculations were performed within the distorted wave Born approximation for the Ne–surface Lennard–Jones potential and using surface phonon Green's functions calculated with the breathing shell model [9.38, 42]. Van Hove singularity (kinematic focussing) peaks associated with Rayleigh and Lucas modes at the Γ and M symmetry points are indicated

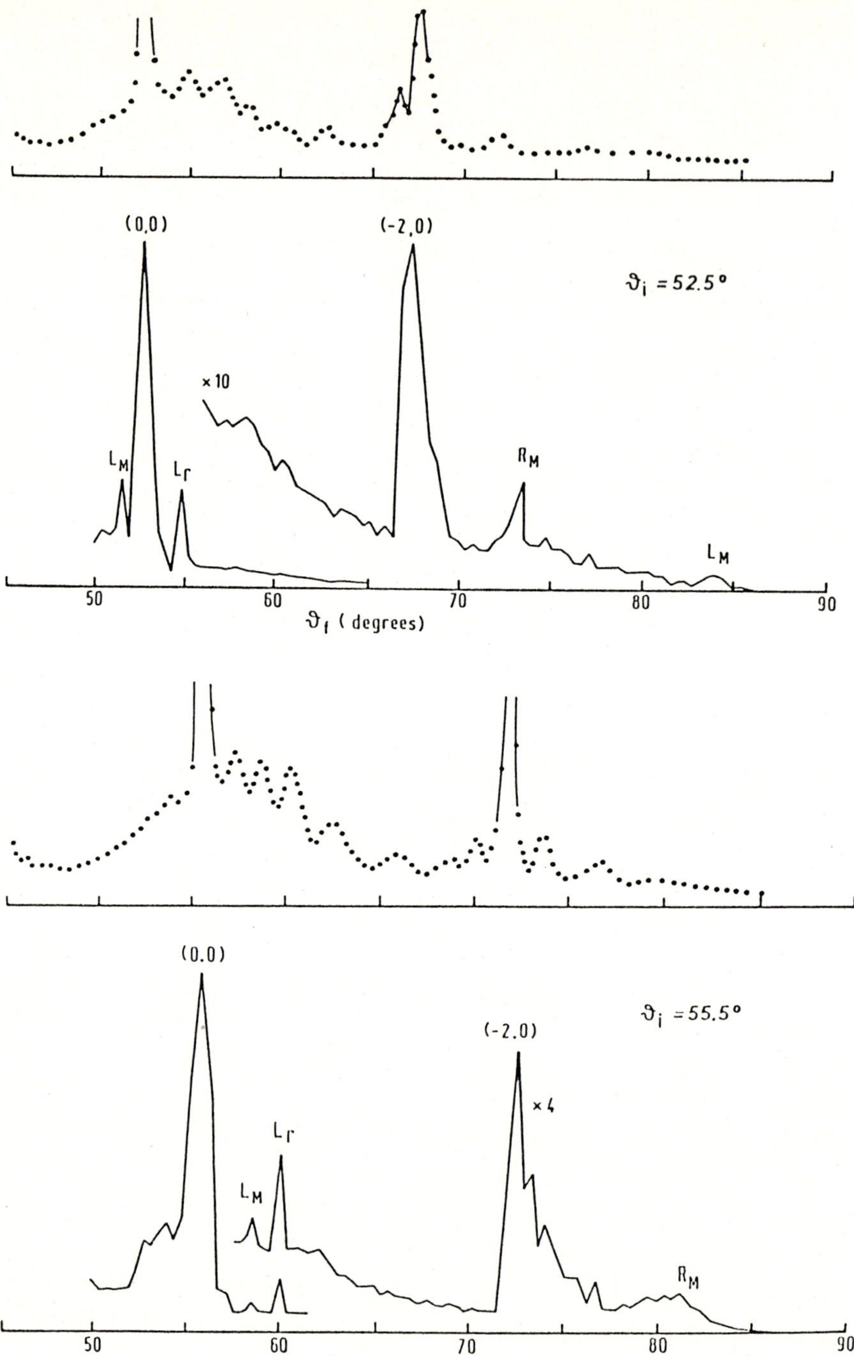

Fig.9.2. (cont.)

212

(0.0)
R_M
L_M
L_r
× 10
(-2.0)
$\vartheta_i = 58.8°$
ϑ_f (degrees)
(2,0)
(0,0)
R_M
L_r
L_M
L_r
$\vartheta_i = 64.5°$
ϑ_f (degrees)
50
60
70
80
9

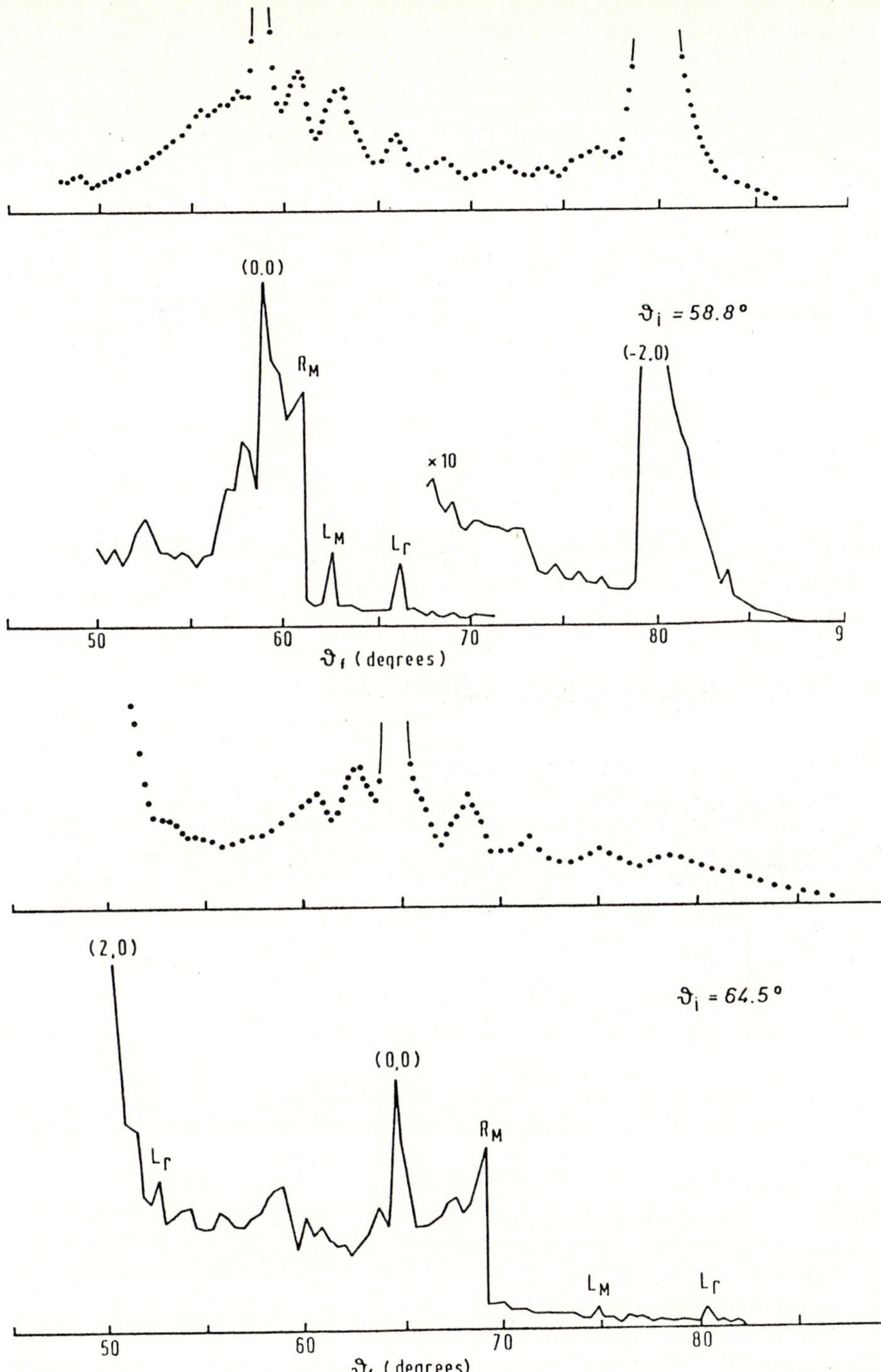

Fig.9.2. (cont.)

Table 9.1. Frequencies of the zone-boundary Rayleigh (R_M) and Lucas (L_M) modes and of the zone centre Lucas mode (L_Γ) derived from the alignments of the Ne-scattering kinematic focussing peaks near the van Hove critical points of the surface phonon dispersion curves. The actual phonon wave vector in units of $G/2$ (G = reciprocal lattice vector) is given in column 2. The frequencies are compared with the original slab shell model calculation of *Chen* et al. [9.28] and with more recent Green's function breathing shell model calculations [9.17]

set of aligned points	$K(G/2)$ fitted	fitted	Ω [10^{13} rad/s] calculated	
			slab	GF–BSM
R_M	0.96	3.91	3.81	3.85
L_Γ	0.04	5.42	5.31	5.50
L_M	0.81	5.95	5.90	5.75

conditions for a new high-resolution TOF helium spectrometer, and then to realize it.

9.3 The Experimental Breakthrough

The first measurement of the Rayleigh wave dispersion curve in LiF(001), by *Brusdeylins* et al. [9.46–48], revived great interest in the theory of surface phonons. Previous calculations for alkali halide surfaces, based on either the breathing shell model (BSM) and the GF method [9.14, 15] or the shell model and the slab method [x.28, 29], could now be tested and were found to be in overall good agreement with scattering data [9.49] (Fig.9.3). Some disagreement at the zone boundary suggested various theoretical improvements including elastic relaxation, surface changes in the electronic structure and many-body effects [9.49, 30]. Besides sharp peaks corresponding to Rayleigh waves, TOF spectra showed continuous regions attributed to bulk phonon densities (Fig.9.4).

Unlike neutron scattering — where the wave vector is conserved in three dimensions and only a finite number of bulk phonons are selected, yielding delta functions in the TOF spectra — in atom scattering only the parallel wave vector is conserved and surface-projected bulk-phonon bands occur in addition to surface phonon peaks. In order to extract the rich information contained in helium TOF spectra and to convert energy loss spectra into phonon densities one has to know the differential cross section. Then the comparison with the calculated phonon spectra permits one to draw some conclusions about dynamical models, surface structure, interatomic forces, etc. Green's function surface phonon densities combined with hard-wall inelastic scattering cross sections gave TOF spectra in quite good agreement with the experiment (Fig.9.4). The

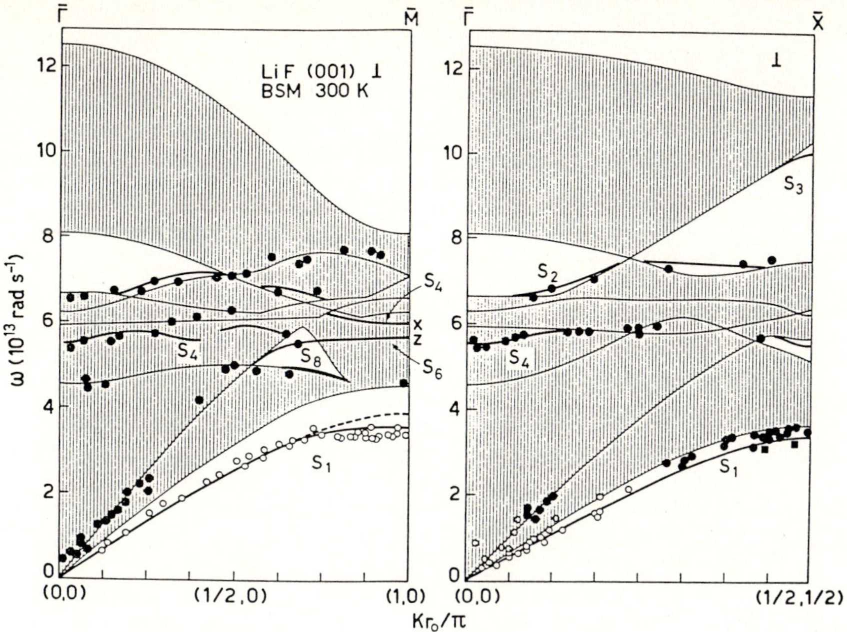

Fig.9.3. Surface phonon dispersion curves of LiF(001) calculated by the Green's function method and breathing shell model for sagittal polarization. Heavy lines are surface modes; shaded areas correspond to bulk bands projected onto the surface. He scattering data: open circles from [9.48]; full circles from [9.52]

experiment was in turn corroborated by such agreement, which demonstrated the minor role of multiphonon scattering processes (at least under the present physical conditions). Important deviations, consisting in a strong enhancement of some phonon peaks for certain kinematic conditions, could then be ascribed to bound-state inelastic resonances [9.43, 50].

The first He scattering TOF studies of surface phonons on LiF(001) gave no evidence of surface optical modes because neutral atoms have only a weak coupling to the *antiphase* (dipolar) motion of ions. Peaks ascribed to optical modes were found first for NaF(001) in time-of-flight spectra obtained with a more energetic He beam. The cost of raising the beam energy is, however, a multiphonon background and a severe loss in resolution. Since the multiphonon background should be almost structureless, subtracting from the actual spectrum the same spectrum averaged over a large number of channels [9.50] yields the structures of the one-phonon spectrum. As shown in Fig.9.5, the comparison with the calculated energy-loss spectrum confirms that the observed peaks correspond to one-phonon processes, involving Rayleigh as well as optical surface phonons. The weaker optical modes were often identified by tuning into a resonance in the region were they were expected to occur until a sharp resonance enhancement was observed. The dispersion curves of all sagittal surface modes in NaF(001) were measured in this way and found to be in very good agreement

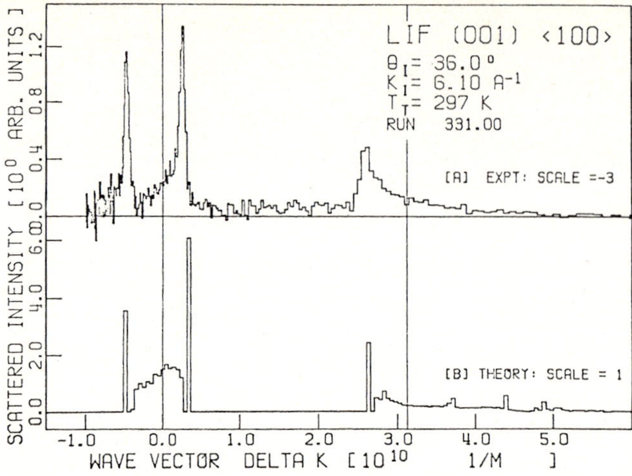

Fig.9.4. Inelastic scattering intensity of He from LiF(001) along the ΓM direction for the kinematics indicated in the inset (above) and theoretical scattering intensity for a hard-corrugated surface model

with previous calculations (Fig.9.6) [9.51]. The Genoa group, which first investigated inelastic resonances, inaugurated their brand new high-resolution TOF spectrometer with a more difficult experiment: The determination of the full set of surface optical branches in LiF(001) using the resonance enhancement method (Fig.9.3) [9.52].

The old question of kinematic focussing, started with the early Genoa experiments, was also solved at Göttingen: Angular distributions from NaF(001), with the scattering plane slightly tilted out of a symmetry direction in order to suppress the resonant features, gave clear evidence of kinematic focussing. The effect was then confirmed by TOF spectra and used for determining the Rayleigh wave dispersion curve from the kinematic focussing angles [9.53, 54] (Fig.9.7). But the question of whether resonance enhancement may assist kinematic focussing from optical modes (and therefore account for the alignments seen in Fig.9.2) remains open.

Recently, *Skofronick* and coworkers, at the Florida State University, have found and thoroughly analyzed further features of surface dynamics in alkali halides such as crossing resonances [9.49, 55–58], surface gap modes [9.57] and the relaxation-induced modes above the bulk maximum frequency [9.30, 57, 59, 60]. These observations have completed the assessment period for the powerful He scattering TOF spectroscopy, on one hand, and for the foundations of surface dynamics and atom–surface scattering theory, on the other.

216

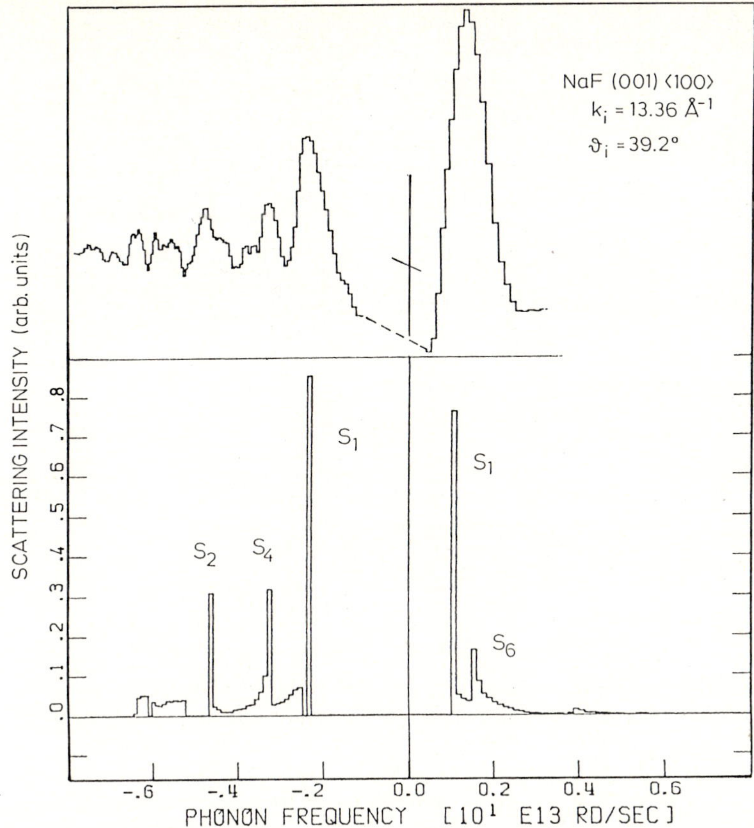

Fig.9.5. Experimental (above) and theoretical (below) energy loss spectrum of He from NaF(001) along the ΓM direction for a room-temperature nozzle He beam. The peaks labelled S correspond to Rayleigh waves, S_2 and S_4 to optical surface modes

9.4 New Perspectives in Surface Phonon Physics

The studies of *Toennies* and coworkers on different classes of materials have posed new problems and triggered a great many theoretical investigations on surface phonons. In the small world of ionic crystals and rare gas solids, where the electron system remains rather inert during surface vibrations, experiments faithfully obeyed theoretical predictions. In contrast, in other classes of crystals, such as metals, semiconductors, graphite and layered chalcogenides, where electrons participate actively in the nuclear motion, each new experiment showed something anomalous, at odds with the predictions of ideal surface models.

The philosophy can be expressed as follows: electron–phonon interaction mechanisms, which are hindered or forbidden in the bulk because of symmetry selection rules, may be activated or enhanced at the surface by symmetry reduction and may manifest themselves in visible anomalies. Many metals, al-

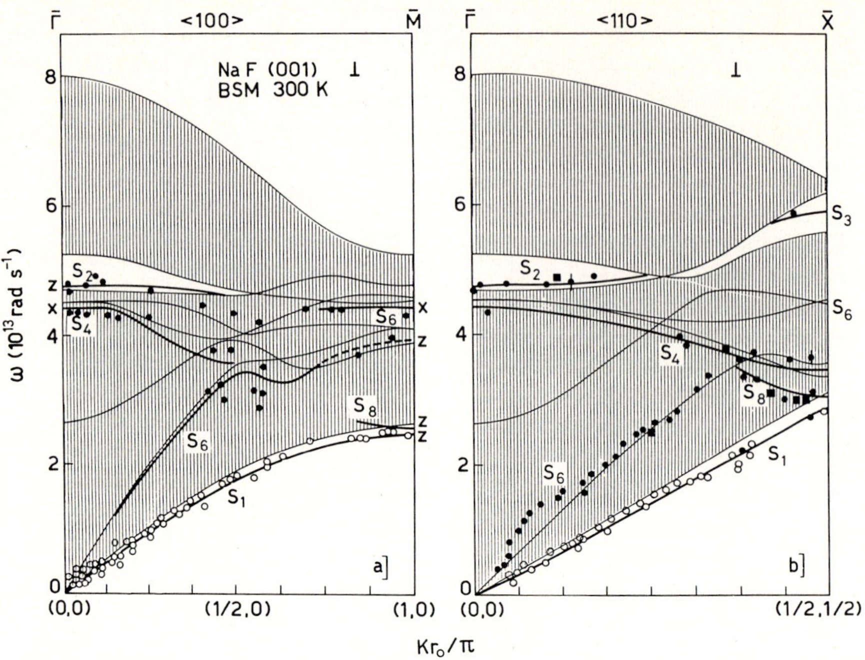

Fig.9.6. Calculated surface phonon dispersion curves of NaF(001) and inelastic He scattering data obtained with room-temperature (full circles) and low-temperature (open circles) nozzle beams

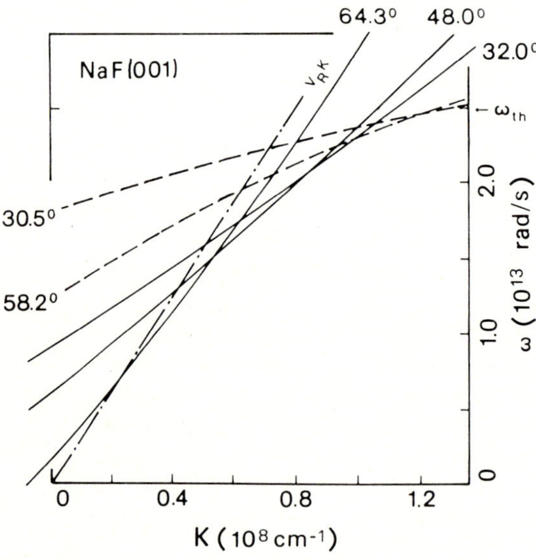

Fig.9.7. The dispersion curve of Rayleigh waves in NaF(001) along the ΓM direction derived as an envelope of scan curves at the kinematic focussing angles [9.53]

though they show no anomaly in bulk phonon dispersion curves, often display
anomalous surface phonon branches. This is the case of noble metals, where
experiments invariably give an unexpected acoustic resonance below the LA
edge and above the Rayleigh curve [9.61] (Fig.9.8).

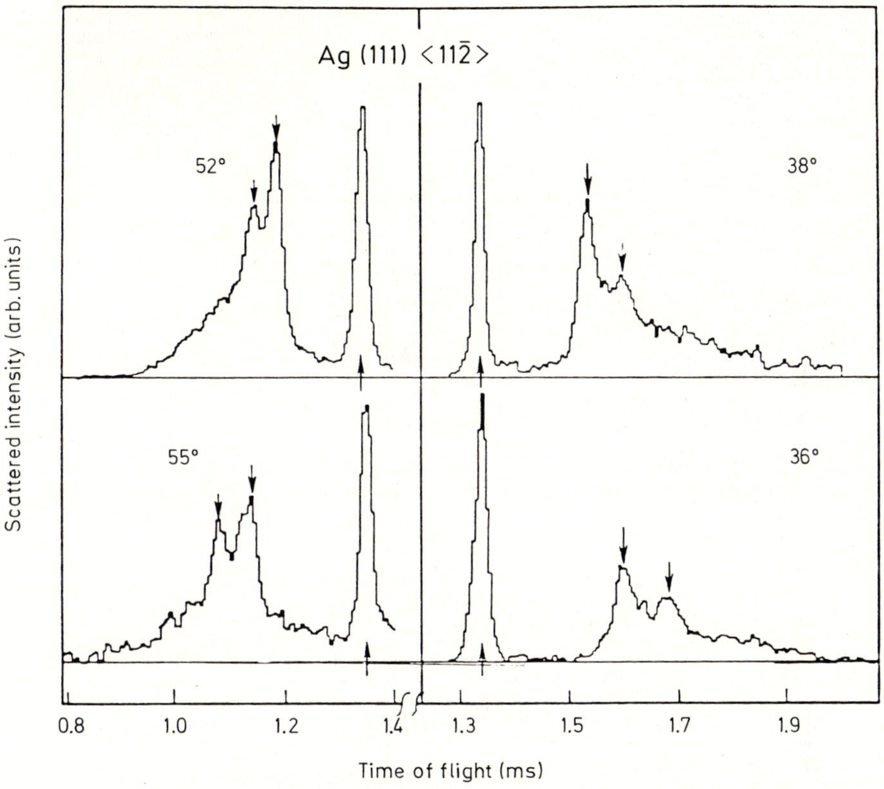

Fig.9.8. Time-of-flight spectra of He scattered from Ag(111) along the ΓM symmetry direction at four different incidence angles (from ref. [9.61] Downward arrows: surface modes;
upward arrows: elastic diffuse peak.

No ideal surface calculation accounts for such resonance [9.62]. Born–von
Kármán models allowing for force constant fitting predicted a substantial softening of in-plane surface force constants [9.63]. Actually, dynamical models
including electron many-body effects either in the form of multipoles [9.64, 65]
(Figs.9.9 and 9.10) or in the embedded-atom picture [9.66, 67], nicely reproduce
all surface phonon branches. In general terms, the LA surface resonance appears to be a direct consequence of surface charge redistribution, and the radial
softening can be viewed either as an increase of ion–ion coulomb attraction due
to the interposition of more electronic charge [9.64], or as an increased dipo-

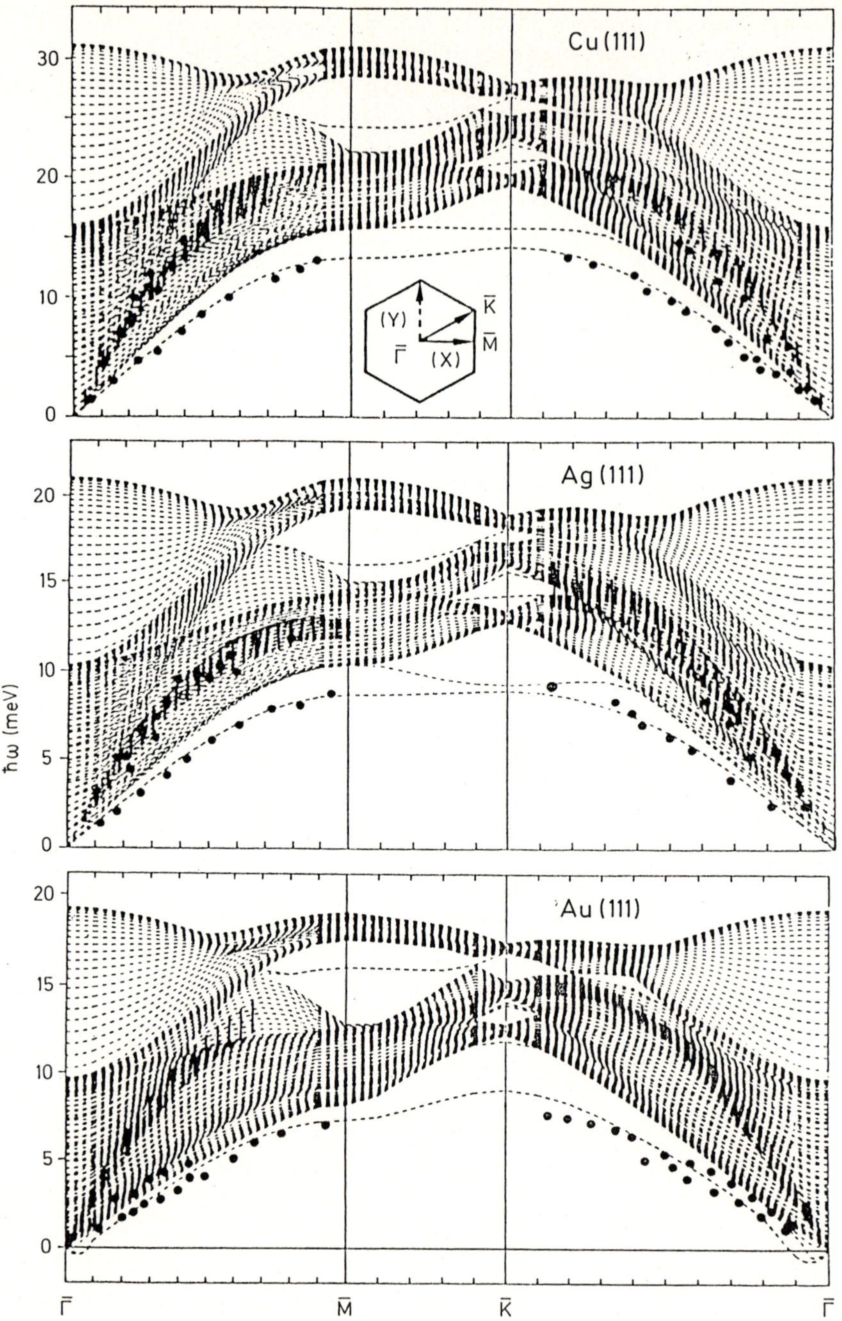

Fig.9.9. Dispersion curves of surface phonons in the (111) surface of noble metals. Experimental points are due to *Doak* et al. [9.61]. Slab calculation results are based on the multipolar expansion model [9.64]

lar polarizability of ions due to a larger sd–p hybridization [9.65]. This point raised an interesting discussion at a recent Solvay Symposium after the review presented by *Toennies* [9.68]: the calculations of *Euceda* et al. for Cu(111) surface electronic structure and *Kleinman*'s remarks pinpointed the larger sd–p hybridization as the leading mechanism.

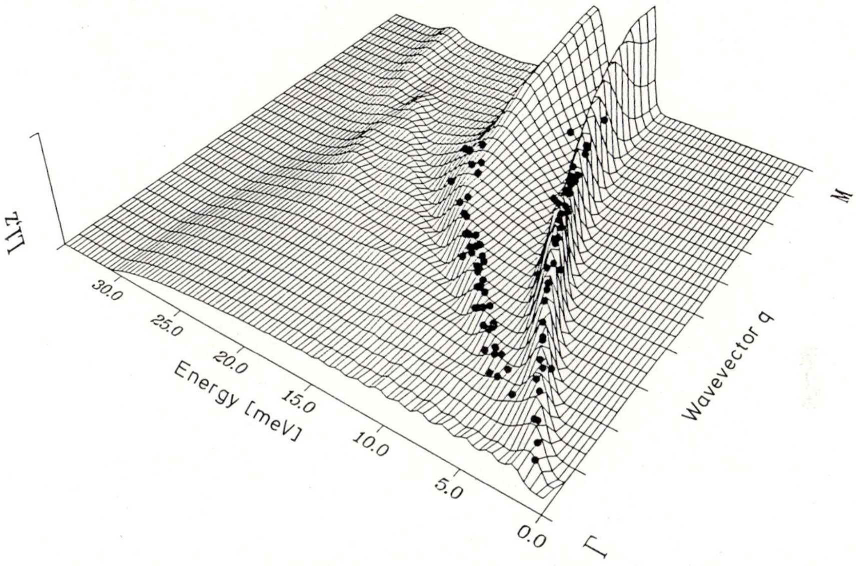

Fig.9.10. Multipolar model calculation of Cu(111) surface phonon projected densities for z-polarized component showing the resonant structure above the Rayleigh mode ridge, in excellent agreement with the experiment [9.65]

Investigations of semiconductor surfaces such as Si(111) 2×1 [9.69] and GaAs(110) [9.70, 71] led *Harten* et al. to the surprising discovery of a flat (optical) surface phonon branch at frequencies as low as about 10 meV. Although this branch roughly corresponds to a folded Rayleigh branch, due in both cases to the doubling of the surface unit cell with respect to the bulk one, its quantitative explanation has challenged theoreticians no less than the interpretation of *Ibach*'s mode in Si(111) at 55 meV. The two-fold problem of the silicon surface has been tackled with an arsenal of different techniques such as semi-empirical tight-binding [9.72, 73], bond-charge model [9.74, 75] and the powerful *Car–Parrinello* method [9.76].

Actually Si(111) 2×1 exhibits a surface phonon spectrum which deviates sigificantly from the predictions for the ideal surface [9.12, 35]. The sequence of five– and seven–atom rings, occurring in the π-bonded 2×1 reconstructed surface [9.77], determines a perturbation extending down to the fifth atomic layer, in such a way that one can speak of a buried interface between reconstructed and normal regions. With respect to the ideal surface spectrum one expects stiffer modes associated with the deformation of the five–atom rings, softer modes with the seven–fold rings. Moreover some modes may have more interface than surface character which leads to the conclusion that slab calculations should always be based on a sufficiently thick slab, regardless of which method is used.

A bond charge model (BCM) calculation of ours, in a 22 layer slab configuration (Fig.9.11), accounts quite well for both the 10 meV branch and the

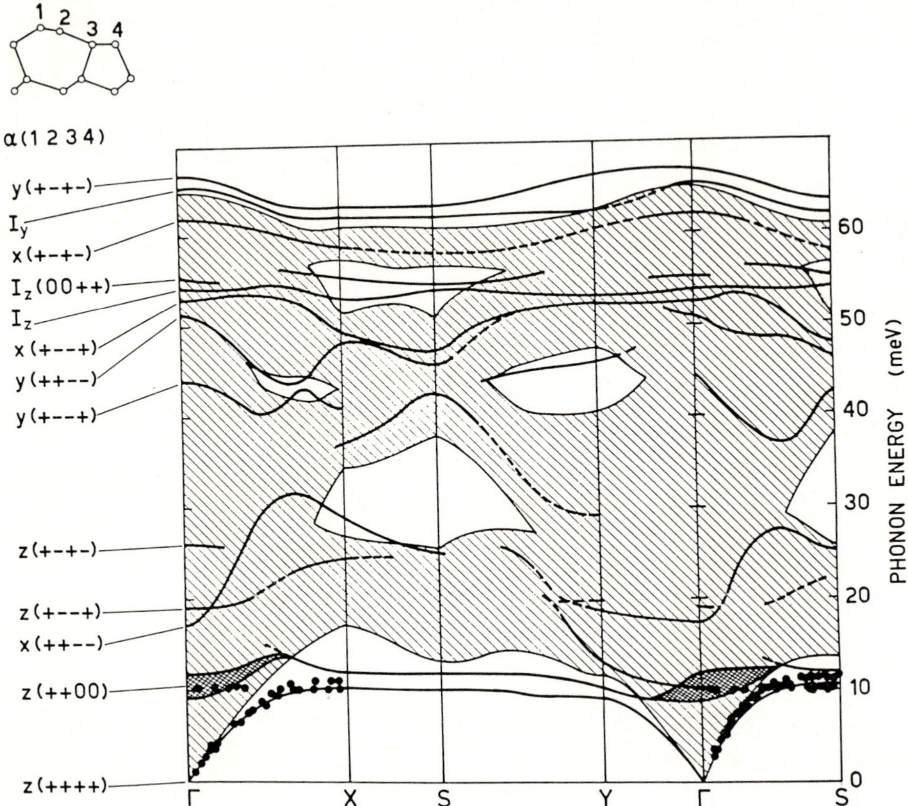

Fig.9.11. Bond-charge model calculation of surface phonons in the Si(111) 2×1 reconstructed surface (ref. [9.75]. The displacements of the surface chain atoms and polarizations for the zone center surface modes are indicated. The experimental points for the Rayleigh mode and the 10 meV resonance are due to *Harten* et al. [9.69]

optical mode at 55 meV [9.75]. The highly isotropic dipolar activity of the latter arises from the vertical motion of atoms in the interface region, which retain an appreciable intensity up to the surface. A remarkable feature of this approach is that no parameter fitting has been performed on the He-scattering data, so that the surface dynamics displays its intrinsic nature. The BCM analysis of experimental data also provides structural information: The small surface

222

phonon gap at the zone boundary folding points is a measure of the surface
π-bonded chain tilt. The measurements of *Harten* et al. suggest a dynamical
estimate of the tilt in agreement with density functional calculations [9.78], but
significantly smaller than the values deduced from fits to LEED data [9.79].

A more favorable situation occurs in GaAs(110), where the six–atom rings
are preserved at the surface, even if a tilted-chain configuration is present as
in the Si(111) 2×1 case. Here, semiempirical tight-binding [9.80] and BCM
calculations [9.71] agree in reproducing the flat low-lying branch again occurring at about 10 meV (Fig.9.12). The calculated surface phonon frequencies
are seen to depend strongly on the geometrical arrangement of surface atoms.
This situation allows a determination of the surface crystallography from an
accurate fit of the dispersion curves, in good agreement with previous LEED
data [9.81].

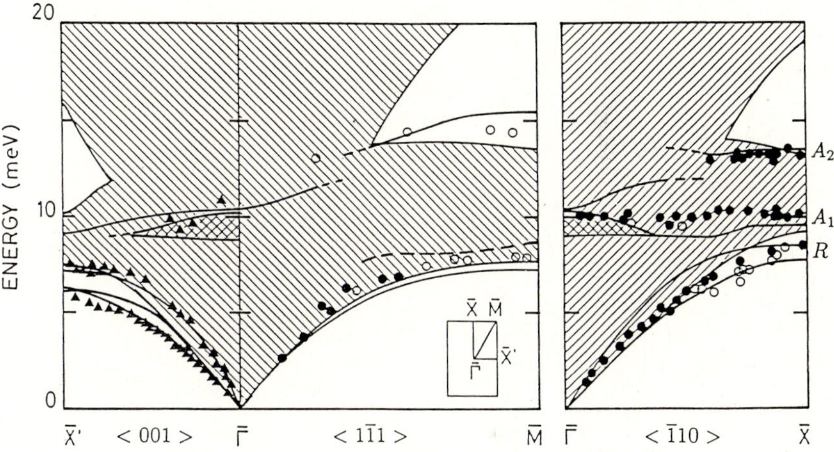

Fig.9.12. Bond-charge model calculation of surface phonons in the lower part of the spectrum
for the unreconstructed GaAs (110) surface (from [9.71]). Experimental points from [9.70,
71]

A careful He scattering study of the layered semiconductor GaSe gave much
less clear results, although the quality of the surface and the experimental
spectra were first class [9.82] (Figs.9.13, 14).

Layered crystals, displaying weak van der Waals interactions between layers and much stronger intra-layer bonds, are interesting objects with regard
to surface studies. Apart from the ease with which good surfaces can be prepared, one would expect the dynamical properties of the surface layer to be
very close to those of any bulk layer, owing to the weak inter-layer coupling.
Originally, He-scattering studies of layered crystals were considered a practical way for investigating bulk properties when thick monocrystals, suitable for
neutron studies, could not be easily grown. A good example is offered by the

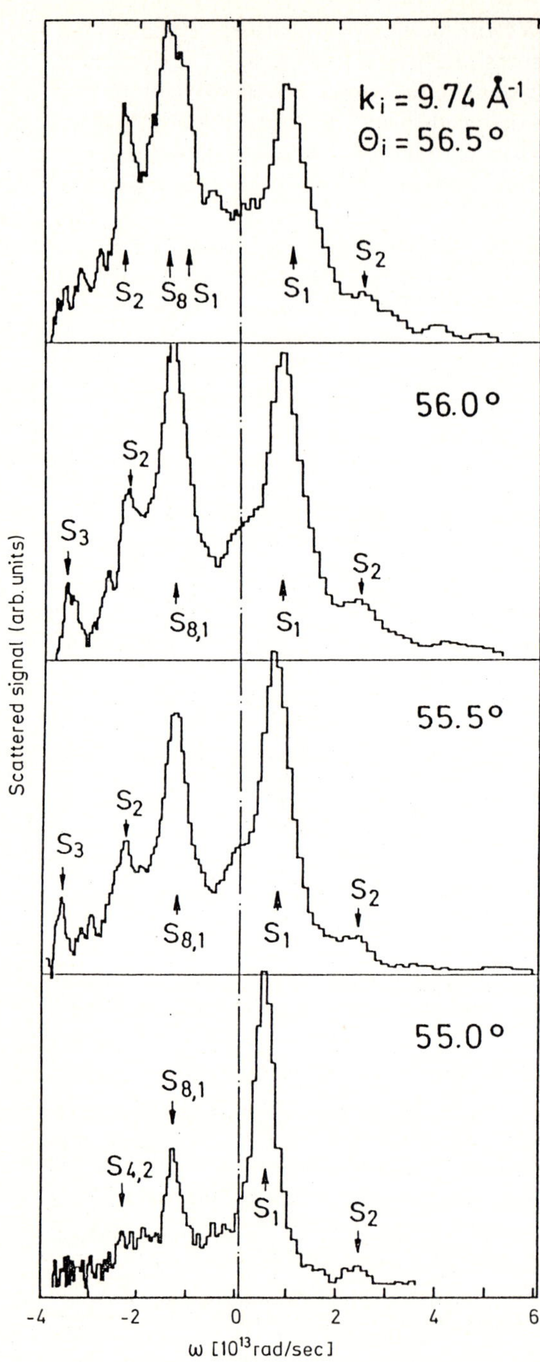

Fig.9.13. Inelastic He scattering spectra from GaSe(001) surface. Rayleigh (S_1) and optical surface modes (S_2 and S_3) are observed with excellent resolution [9.82]

224

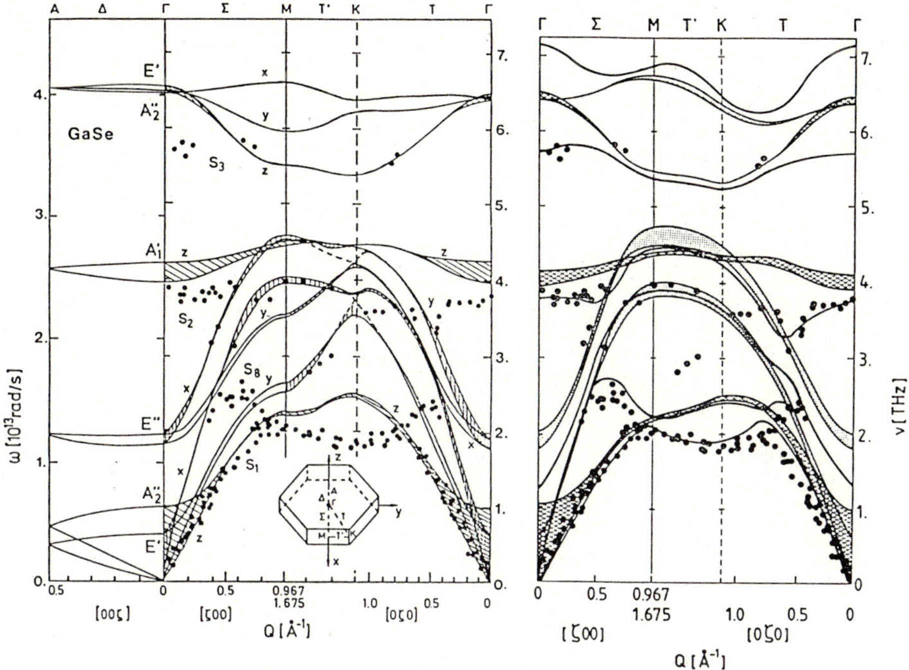

Fig.9.14. (a) Surface-projected bulk phonon bands with sagittal polarization (hatched regions) of GaSe(001) compared to He-scattering data. (b) A dispersive linear-chain fit of bulk and surface phonon dispersion curves of GaSe(001) [9.83]

first determination of the optical branches in $TaSe_2$ by means of He scattering (Fig.9.17, [9.83]).

This is clearly not the case in GaSe(001). The dispersion curves obtained from He scattering, presumably associated with surface modes, look rather different from the corresponding bulk phonon branches as obtained from neutron scattering. There is a general softening of the bulk branches in the surface region, which appears to be dramatic for certain Q-values. A dispersive linear chain (DLC) approach developed by one of us (LM), fitted to the He scattering data (Fig.9.14b); indicates an important softening of Se–Se interactions within the surface layer [9.84]. A local change of valence electron susceptibility, i.e., a larger Se polarizability at the surface, was surmised as a possible origin of such large anomalies.

The surface phonon density of GaSe(001) calculated from the fitted DLC model was found to be in surprisingly good agreement with the inelastic elec-

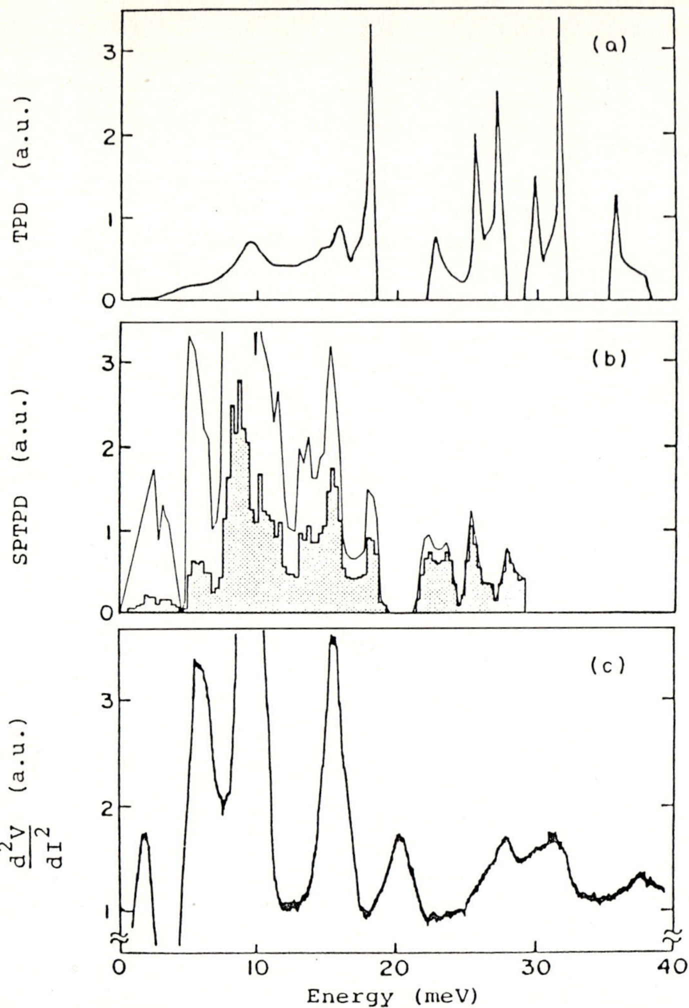

Fig.9.15. The surface phonon density of GaSe(001) calculated from the fitted dispersive linear-chain model (a) as compared to the bulk phonon density (b) and to the inelastic-electron-tunnelling spectrum of an ultrathin GaSe(001) slab between Pb superconducting contacts (c) (from [9.85])

tron tunneling spectra (IETS) of a GaSe ultrathin slab (10 – 20 layers) squeezed between two superconducting lead contacts [9.85] (Fig.9.15). Such a similarity suggests that the substantial softening of the intralayer force constants, possibly accompanied by a normal expansion of the layer itself, may be due to carrier accumulation (or injection in the IETS case) within the surface (film) layers. Such a working hypothesis has interesting consequences.

The first consequence concerns the tendency the of III–VI layered semiconductors, particularly InSe, to intercalate lithium and therefore to work as cath-

226

odes for solid state batteries and microionic devices [9.86]. Layer expansion and
the conductivity change associated with carrier injection are relevant properties
for layered cathodic materials, and their characterization by means of surface
phonon spectroscopic techniques is quite appealing. The second consequence
concerns another aspect of electron–phonon interaction at semiconductor surfaces. How would the surface phonon dispersion curves of heavily doped silicon
and GaAs be affected by the formation of *degenerate* accumulation layers?
Aside from the surface plasmon–phonon polariton interaction, widely investigated by means of optical techniques, may one expect appreciable changes of
surface force constants when the carrier confinement depth (at sufficiently low
temperature) becomes comparable to the depth of the extended reconstruction
region? This question might find an answer from new He scattering experiments
on heavily doped semiconductor surfaces.

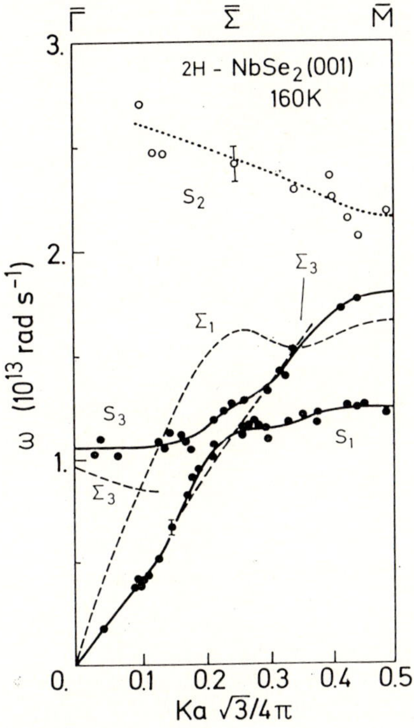

Fig.9.16. Surface phonon dispersion curves of 2H–NbSe$_2$(001) (black and open dots) [9.88].
Heavy and dotted lines are guidelines for surface phonons. Broken lines are bulk phonon
dispersion curves derived from neutron data [9.89]

9.5 Surface Phonon Anomalies

Conducting layered crystals such as $NbSe_2$ [9.87], manifest a contraction of
the van der Waals gap between the two topmost layers. An incomplete and
previously unpublished set of inelastic He scattering data on 2H–$NbSe_2$(001),
collected in 1984 and plotted in Fig.9.17 [9.88], is worth showing here. The

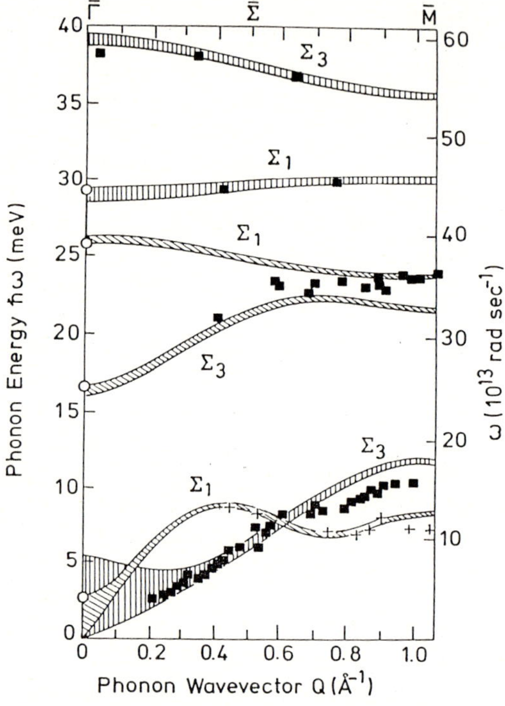

Fig.9.17. Experimental phonon dispersion curves in 2H-$TaSe_2$ (001) along ΓM at 140 K and
DLC calculated surface-projected bands [9.84]

experimental dispersion curves (black and open circles), measured at 160 K,
show interesting features. $NbSe_2$ has a charge density wave (CDW) transition
at 33 K and an important (presumably temperature dependent) anomaly at
about 2/3 of the reduced Brillouin zone in the LA branch of Σ_1 symmetry
[9.89]. In the acoustic region He data yield the Rayleigh wave branch (S_1) and
an almost dispersionless branch at about 1.2×10^{13} rad/s (S_3) with an evident
anti–crossing at one half of the zone. A few more, rather dispersed points give a
first determination of an optical branch in $NbSe_2$ (S_2). The two sections of the
Rayleigh branch are almost coincident with the lower edge of the bulk TA band
as obtained from room-temperature neutron scattering data [9.89] (Σ_3 broken
line). The lower section displays an upward curvature, which is a peculiarity of

228

weakly bound layered structures. The two sections of the flat S_3 branch reflect different origins:

— The left section, at small wave vectors, is lifted above the top edge (Σ_3) of the acoustic band, indicating a substantial stiffening of the radial force constant between the first two surface sandwiches. This is well consistent with the van der Waals gap contraction mentioned above.

— The right section of the flat branch closely reflects the shape of the LA branch in that region, with a shallow Kohn-type anomaly and a frequency of about 0.3×10^{13} rad/s below the LA branch. Such a difference between the branch at 160 K, measured by He scattering, and the branch at 300 K, measured by neutron scattering, may be interpreted as an anomalous softening occurring for T decreasing towards $T_c = 33$K, similar to that observed in bulk 2H–TaSe$_2$ with neutrons [9.89]. Assuming a linearity in the squared frequency, the frequency of the anomaly dip extrapolated at T_c would be 0.84×10^{13} rad/s. An interesting fact is that the S_1 and S_3 branches undergo a strong hybridization at 1/2 of the zone, and both modes show the weak anomaly at 3/5 of the zone, namely at a position shifted with respect to the anomaly of the Σ_1 mode.

The shift of the surface mode anomaly with respect to the corresponding bulk anomaly has been more clearly observed and carefully studied in 2H–TaSe$_2$ (001) [9.90] (Fig.9.18). Also here bulk neutron data show a deep Kohn anomaly in the LA branch at two-thirds of the irreducible zone, and the dip frequency decreases with temperature reaching a minimum at the CDW transition temperature $T_c = 123$K (Fig.9.18b). The Rayleigh branch from He-scattering data appears to be anomalous only in the narrow temperature range 80 K to 140 K and around one-half of the reduced Brillouin zone. At 70 K and 140 K a straight dispersion relation is observed, which we consider normal. At 120 K the anomaly appears as a sizeable softening. The lowest points, corresponding mostly to 110 K, permit us to locate the surface wave vector for the deepest surface anomaly at $Q_s = 0.53$Å^{-1}. A comparison with the bulk modes is shown in Fig.9.17b, where room-temperature bulk dispersion curves Σ_1 and Σ_3 measured by *Moncton* et al. [9.89] are shown along ΓM. The Rayleigh mode dispersion curve at 140 K (S_1) almost coincides with that of the transverse bulk mode Σ_3, except for some softening at the zone boundary. The bulk anomaly at 2/3 of the zone in the LA branch (Σ_1), despite its large temperature dependence, keeps its crossing with the LA branch away from the Rayleigh wave anomaly and cannot be responsible for the anomaly itself.

Furthermore, the Rayleigh wave frequency in the region of the anomaly is considerably lowered below the acoustic band edge, indicating a localization of the displacement pattern to within a few layers of the surface. Therefore the anomaly can be related to some change occurring in the first few layers. According to the Kohn-type anomaly mechanism, we may consider the positional shift of the anomaly as indicative of a surface electronic state cutting the Fermi surface at a wave vector different from that of the bulk Fermi wave vector. On the other hand the effect observed here has a striking similarity to that pre-

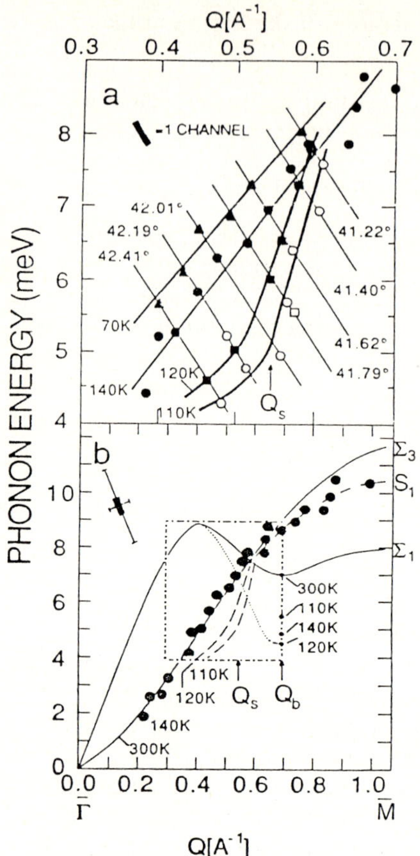

Fig.9.18. (a) The temperature dependent Rayleigh-mode dispersion of 2H–TaSe$_2$ over a limited wave vector range, revealing the anomalous softening at $Q_s = 0.53$Å^{-1}: filled circles=140 K; filled squares=120 K; open squares=115 K; open circles=110 K; lozenges=100 K; triangles=70 K. The interpolating thick solid lines are simply guides to the eye. The thin lines are the scan curves at different incident angles. (b) The experimental points (solid points) for the Rayleigh mode along ΓM at 140 K. The solid lines are TA (Σ_3)and LA (Σ_1)modes [9.88]. The dashed lines show the Rayleigh mode (S$_1$) dispersion curve and its T-dependent anomaly for 110 K and 120 K. For comparison the T-dependence of the bulk anomaly at $Q_b = 0.7$Å^{-1} is also shown [9.89]. The dotted line is our qualitative interpolation

dicted for the (001) surface of TiN [9.91] (Fig.9.19). In TiN the LA branch also exhibits a deep anomaly at the two-thirds of the reduced zone. This effect has been attributed to the change of the average range of the electron–phonon interaction due to the modified coordination at the surface. Hoverever, such a geometric mechanism does not seem to apply to a layered structure.

Another intriguing mechanism determining the position of certain anomalies has been proposed by *Bilz* and coworkers [9.92] for layered perovskites. According to this work the position of the anomaly is given by the wave vector

230

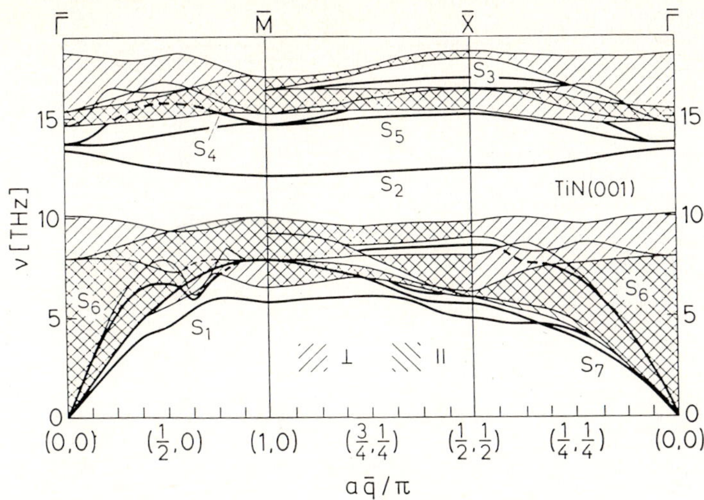

Fig.9.19. Phonon dispersion curves in TiN(001) calculated by the cluster deformability model and the Green's function method

of nonlinear periodic waves coupled to the phonons. Such waves originate from a nonlinearity in the electron–phonon interaction and their period is an integer fraction of a reciprocal lattice vector G determined by the exponent of the non-linear term. For example, a cubic term in the electron–phonon force yields an anomaly at $G/3$. In the absence of inversion symmetry, as occurs at the surface, even-power electron–phonon terms are induced and may cause anomalies at even fractions of G. However, only the ordinary 3×3 superstructure due to the bulk CDW is seen, and no superstructure with period four, corresponding to the half-zone anomaly, is detected by He-atom diffraction.

Generally speaking, shifted anomalies are a manifestation of competing symmetries, in this case not only between those of the lattice and the electron gas, but also between those of the surface and the bulk. Since the surface anomaly is peaked at the striped–incommensurate transition temperature, rather than at T_c, one could consider the intrinsic tendency of the surface to a different periodicity as responsible for the internal stress yielding discommensuration.

Below the CDW transition temperature phonons may be diffracted by the CDW superlattice and show replicas of the ordinary phonon branches at wave vectors displaced by any superlattice reciprocal vector g. This effect should be observable by He scattering in systems with large CDW amplitudes, and actually it has recently been observed in 1T–TaS$_2$(001) for its low temperature (120 K), commensurate $\sqrt{13} \times \sqrt{13}$ CDW phase [9.93]. The phonon dispersion curves of this system ($T_\mathrm{c} = 543$K) have been measured along both the crystallographic and CDW symmetry directions (Fig.9.20). The points are scattered over various regions of the frequency-wave-vector plane, but can be easily fitted by a single sinusoidal dispersion curve centered at the superstructure g-vectors contained

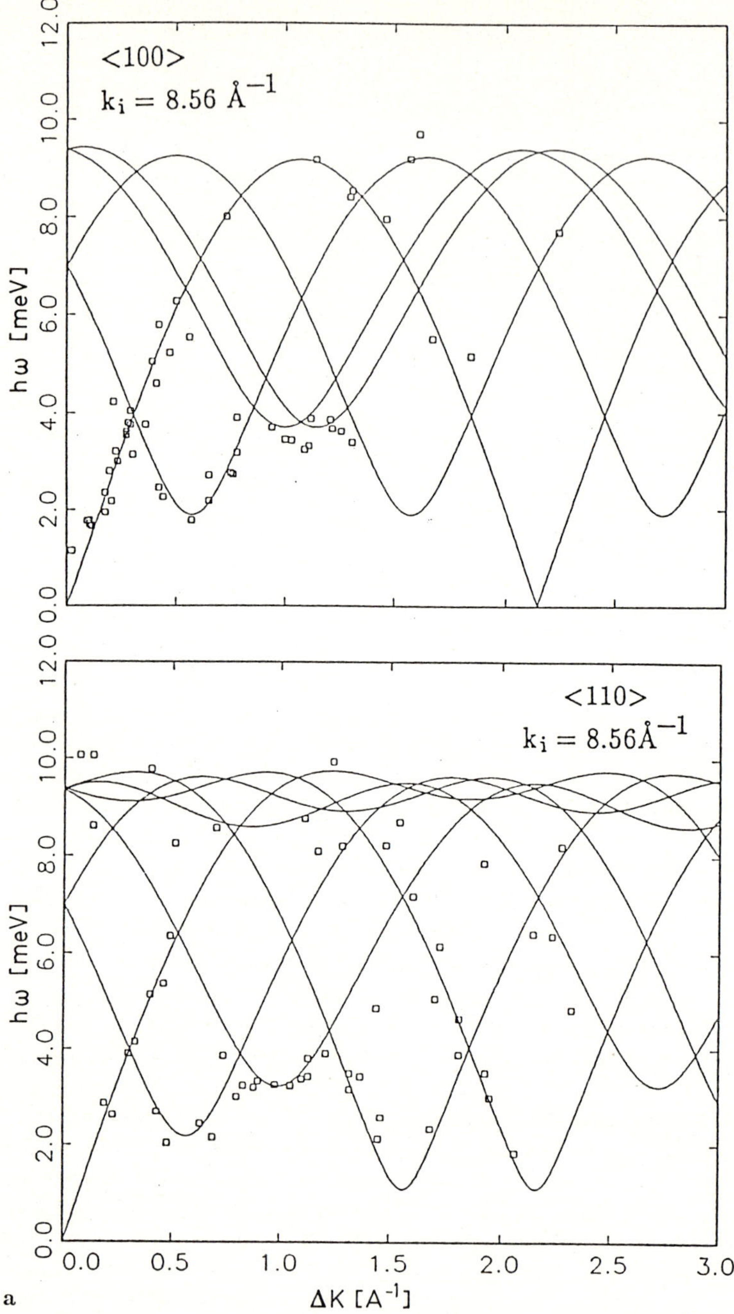

Fig.9.20. Dispersion curves of the Rayleigh mode of 1T–TaS$_2$ in the CDW phase at 120 K along the symmetry directions of the normal lattice (a) and of the CDW superlattice (b) as fitted by single sinusoidal components associated with a few superlattice reciprocal vectors

232

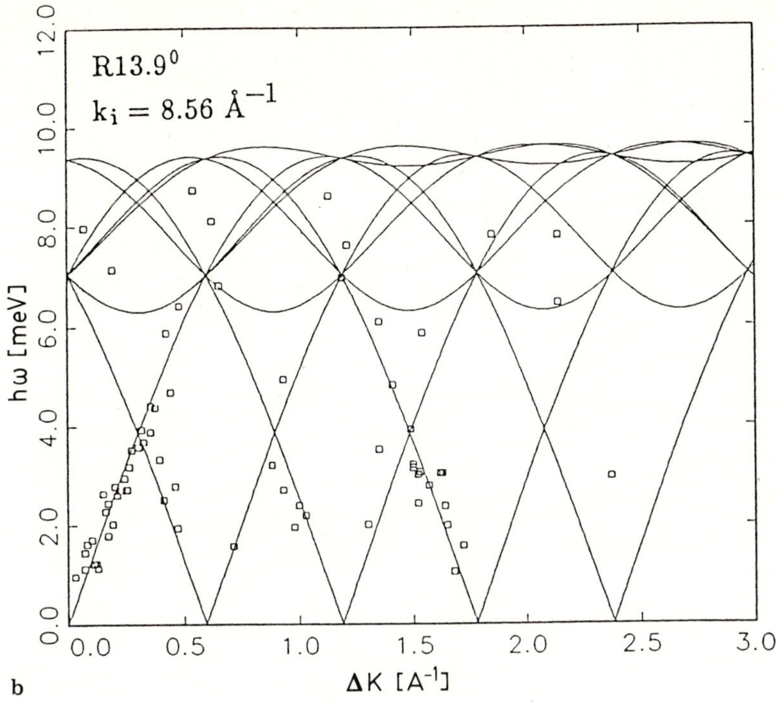

b

Fig.9.20. (cont.)

in the irreducible part of the lattice Brillouin zone. Out of a CDW symmetry direction the diffracted branches never vanish, but exhibit pronounced minima. CDW–phonon coupling yields anti–crossing between the ordinary ($\mathbf{g}=0$) and the diffracted branches. In this case, diffracted-branch minima may be interpreted as deep anomalies induced by the coupling itself. These data represent, to our knowledge, the first observation, made over the entire Brillouin zone, of phonons coupled to a charge density wave.

The dynamical coupling of phonons with hole-electron excitation yielding a CDW may be depicted as a harmonic interaction between the atomic lattice and an electron lattice of period corresponding to the CDW, the two lattices being in general incommensurate. It can be easily seen that such coupling yields anomalies in the dispersion relations of the atomic lattice of the type recently reported, e.g., in W(001) [9.94]. Consider a linear chain of equal masses M, interatomic spacing $a = 1$ and nearest neighbor force constant f. The massive chain is coupled to an incommensurate linear chain of electronic shells having a period $\alpha a = \pi/2k_F$ slightly larger than one (Fig.9.21a). Each shell is coupled to the closest atom by force constants g and to its neighboring shells by the force constant h.

Apart from perturbations due to the coarse lattice of discommensurations, we may write two coupled equations (in the adiabatic approximation)

233

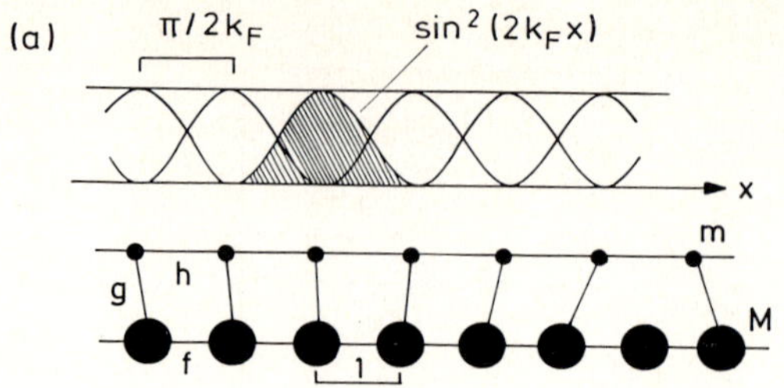

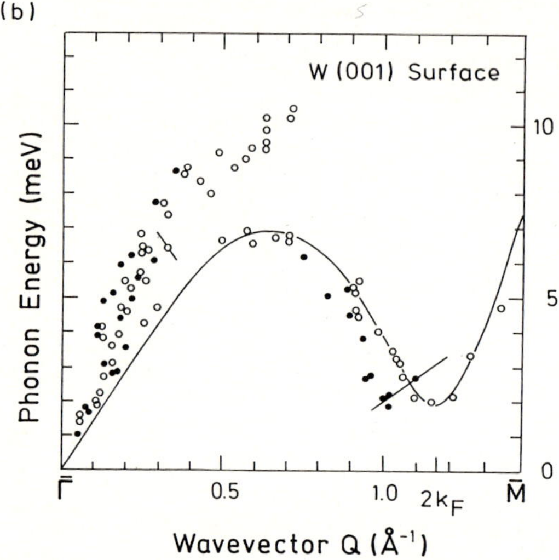

Fig.9.21. (a) Double linear chain model describing the coupling between atomic and electronic oscillations. The electron density is represented by a superposition of sine-squared shells. (b) Experimental dispersion curves of surface phonons in W(001) [9.94]. The points corresponding to the Rayleigh wave are above and show the cut-off at one-half of the zone. The lower branch is quasi-longitudinal and shows the deep, T-dependent anomaly induced by the surface charge density wave, precursor of the c(2 × 2) reconstruction. The full line displays the calculated dispersion curve of the anomalous acoustic surface mode

$$M\Omega^2 u_N \; = \; f(2u_N \, - \, u_{N+1} \, - \, u_{N-1}) \, + \, g(u_N \, - \, w_L) \tag{9.1a}$$

$$0 \; = \; g(w_L \, - \, u_N) \, + \, h(2w_L \, - \, w_{L-1} \, - \, w_{L+1}) \quad , \tag{9.1b}$$

and try the Bloch-wave solutions

$$u_N \; = \; u\exp(ikN) \; , \quad w_L \; = \; w\exp(i\alpha kL) \quad . \tag{9.2}$$

The resulting secular determinant is solved for

$$M\Omega^2 \; = \; 4f\sin^2(k/2) \, - \, g/[1 + (g/4h)\mathrm{cosec}^2(\pi k/4k_f)]. \tag{9.3}$$

Equation (9.3) gives an acoustic branch with a deep anomaly at $2k_{\mathrm{F}}$ (Fig.9.21b). The deep anomaly found by *Ernst* et al. [9.94] in the lowest surface

234

branch $A(SP_\parallel)$, longitudinal in character, of W(001) is nicely reproduced with $f/M = 38$ meV^2, $g/M = -40.6$ meV^2 and $h/M = 7.8$ meV^2 (Fig.9.21b).

By giving M the tungsten mass one finds $h = 5500$ dyne/cm for the intershell force constant. It is interesting to adjust the parabolic edge of the electron–hole excitation spectrum to the sinusoidal dispersion of a linear chain

$$\Omega = \frac{\hbar}{2m}k(2k_F - k) \approx \frac{\hbar k_F^2}{\pi m}\sin(\pi k/k_F). \tag{9.4}$$

In this way we view the electron–hole excitations as the oscillations of the electronic linear chain of force constant $h = \hbar^2 k_F^4/4\pi^2 m$. For the fitted value of h, and with m equal to the electronic mass, we have $2k_F \approx 2$Å^{-1}, which is folded back into the first zone at the anomaly position.

The incommensurate CDW also exerts a weak perturbation on the Rayleigh wave (RW) due to the fact that the RW atomic motion is not exactly transverse, normal to the surface, but has a small longitudinal component. The CDW induces a weak mixture of each RW of parallel wave vector K with modes of wave vector $K \pm 2k_F$ and equal frequency, when such a coupling is permitted by spectral conditions. This means that the RW is diffracted by the CDW potential, provided the final channels are permitted by the dynamics of the semi-infinite solids. In 1T–TaS$_2$ the CDW has a large amplitude and extends over the entire three-dimensional lattice; therefore the replica modes for $K \pm g$ are eigenmodes of the crystal (below T_c) and are open channels for diffraction.

In contrast to this, the CDW of W(001) is weak and practically restricted to the surface layer, so that the replicas of RWs are expected to be weak and those of bulk modes vanishing. On the other hand, the bulk bands of W, unlike those of layered compounds, have a large bandwidth (due to their large dispersion) and the diffracted RW easily falls inside the bulk continuum and ends up in a regular bulk mode channel. In this case the He scattering from the RW loses intensity in favor of the diffracted bulk channels, according to the kinematics described in Fig.9.22, yielding an apparent cut-off at sufficiently large momentum and energy transfer, as actually seen in experiment. A consequence of this mechanism would be the appearance of resonant enhancement of the bulk modes working as diffracted channels of the RW. This would be hardly visible in He scattering spectra. However, the recent molecular-dynamical study of the T-dependent surface phonon dispersion of W(001), performed by *Wang* et al. [9.95], also gives evidence of the RW cut-off and its apparent continuation in the bulk mode diffracted channel, see e.g. [Ref. 9.95, Fig.3].

The elastic scattering from the CDW may be viewed as an anharmonic process between a Rayleigh wave, a bulk mode, and a hole–electron excitation across the Fermi surface of zero energy and momentum $\pm 2k_F$. According to the diagrams depicted in Fig.9.22: the process can be driven by quadratic electron–phonon interaction to second order (a), or by linear electron–phonon interaction to fourth order (b and c). If we restrict ourselves to coupling within one single Brillouin zone (e.g., between K and $K - 2k_F$), we consider the matrix element $V(K, K - 2k_F)$ of the CDW potential (the one yielding, according to Fig.9.22,

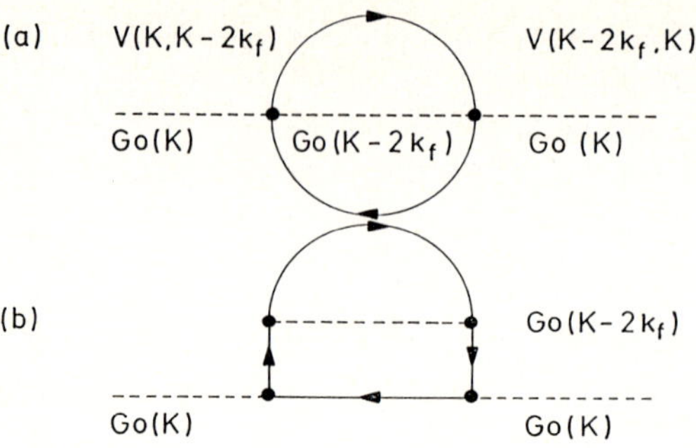

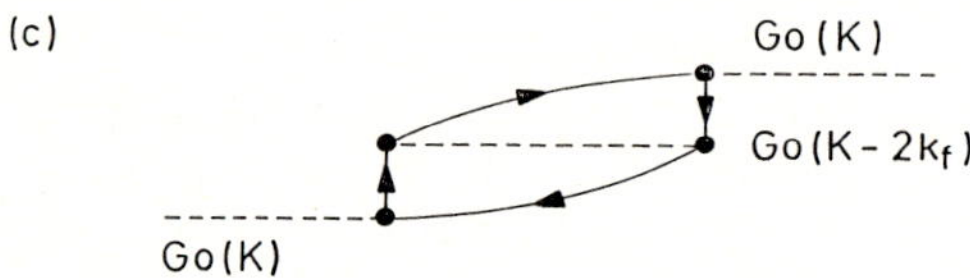

Fig.9.22. (a) Diagrams showing quadratic electron–phonon interaction to second order and (b, c) linear electron–phonon interaction to fourth order

the first cut-off of a forward propagating RW), and the processes arising from diagram (a). Then we write the perturbed Green's function $G(z)$ of the lattice in terms of the unperturbed one $G_0(z)$ as

$$G = (I - G_0 V)^{-1} G_0 \tag{9.5}$$

or, explicitly,

$$G(\boldsymbol{K}) = \begin{pmatrix} 1 & G_0(\boldsymbol{K})V(\boldsymbol{K}, \boldsymbol{K} - 2\boldsymbol{k}_F) \\ G_0(\boldsymbol{K} - 2\boldsymbol{k}_F)V(\boldsymbol{K} - 2\boldsymbol{k}_F, \boldsymbol{K}) & 1 \end{pmatrix}^{-1}$$

$$\times \begin{pmatrix} G_0 & 0 \\ 0 & G_0(\boldsymbol{K} - 2\boldsymbol{k}_F) \end{pmatrix} . \tag{9.6}$$

Thus the diagonal element $G(\boldsymbol{K})$ for the Rayleigh wave of wave vector $\boldsymbol{K}$ is

$$G(\boldsymbol{K}) = \frac{G_0(\boldsymbol{K})}{1 - G_0(\boldsymbol{K}) V(\boldsymbol{K}, \boldsymbol{K} - 2\boldsymbol{k}_F)^2 G_0(\boldsymbol{K} - 2\boldsymbol{k}_F)} ,$$

having assumed that $V(\boldsymbol{K} - 2\boldsymbol{k}_\mathrm{F}, \boldsymbol{K}) = V^*(\boldsymbol{K}, \boldsymbol{K} - 2\boldsymbol{k}_\mathrm{F})$. Since we are interested in the coupling between the RW and the bulk modes of wavevector $\boldsymbol{K} - 2\boldsymbol{k}_\mathrm{F}$, $G_0(\boldsymbol{K} - 2\boldsymbol{k}_\mathrm{F})$ is the complex-valued bulk-phonon Green's function

$$G_0(\boldsymbol{K} - 2\boldsymbol{k}_\mathrm{F}; \Omega^2) =$$

$$\int R(\boldsymbol{K} - 2\boldsymbol{k}_\mathrm{F}; z)(\Omega^2 - z)^{-1}\mathrm{d}z \; + \; \mathrm{i}\pi R(\boldsymbol{K} - 2\boldsymbol{k}_\mathrm{F}; \Omega^2), \tag{9.7}$$

where $R(\boldsymbol{K}; \Omega^2)$ is the surface projected bulk phonon density for the wave vector $\boldsymbol{K}$. The RW unperturbed Green's function projected onto the first layer can be approximated by

$$G_0(\boldsymbol{K}; \Omega^2) \; = \; 2\alpha c_0 \boldsymbol{K} \; / \; (\Omega_{K,R}^2 \; - \; \Omega^2 \; - \; 2\mathrm{i}0^+) \tag{9.8}$$

where the numerator is the amplitude at the topmost layer, c_0 the interplanar distance and α the constant ratio between RW wavelength and penetration depth. Thus the spectral response from the RW can be calculated as the imaginary part of the corresponding perturbed Green's function element

$$\mathrm{Im}\{G(\boldsymbol{K}; \Omega^2)\} \; = \; \frac{2\pi\alpha c_0 \boldsymbol{K} \Gamma_K}{(\Omega_{K,R}^2 - \Omega^2 + \Sigma_K)^2 \; + \; \Gamma_K^2} \tag{9.9}$$

where

$$\Sigma_K \; = \; -2\alpha c_0 \boldsymbol{K} \, V^2 \int \mathrm{d}z \, R(\boldsymbol{K} - 2\boldsymbol{k}_\mathrm{F}; z)/(\Omega^2 - z) \tag{9.10}$$

is the shift of the RW squared frequency induced by the CDW perturbation, $V^2 = V(\boldsymbol{K}, \boldsymbol{K} - 2\boldsymbol{k}_\mathrm{F})^2$, and

$$\Gamma_K \; = \; 2\pi c_0 \boldsymbol{K} V^2 R(\boldsymbol{K} - 2\boldsymbol{k}_\mathrm{F}; \Omega^2) \tag{9.11}$$

is the spectral width. This is obviously 0^+ when Ω is outside the phonon spectrum for $\boldsymbol{K} - 2\boldsymbol{k}_\mathrm{F}$, but it is finite when $\boldsymbol{K} - 2\boldsymbol{k}_\mathrm{F}$ brings Ω inside the spectrum. Such a broadening of the central RW peak due to the abrupt opening of bulk diffraction channels should account for the experimental cut-off.

9.6 Conclusion

The He scattering experiments performed in Göttingen, combined with an extensive modelling of surface dynamics along the lines illustrated here, have revived and solved old problems and opened new lines of research in surface phonon physics. In a way similar to neutron scattering spectroscopy, which in the fifties stimulated new thinking in lattice dynamics and the invention of microscopic models, He scattering data are challenging first-principles methods. We wish here to mention — but the list is largely incomplete — the early work by *Hanke* and *Muramatsu* on Si [9.96, 111], the pseudopotential perturbation theory for simple metal surfaces by *Calandra* and *Beatrice* [9.97],

the tight-binding method, e.g., of *Mazur* and *Pollmann* [9.73], the embedded
atom method of *Daw* and *Baskes* [9.66, 67] with encouraging applications to
noble metal surfaces, the ab initio interplanar force constant calculations of
Ho and *Bohnen* [9.98, 99] and *Wang* and *Weber* [9.100], the density-response
method successfully adopted by *Eguiluz* et al. for nearly-free electron metals
[9.101] (particularly aluminum [9.102]) and the above mentioned *Car–Parinello*
method [9.76].

Furthermore, the breakthrough made in Göttingen has triggered a great deal
of new He inelastic scattering studies in many laboratories all over the world,
which we cannot quote here due to space limitations. Together these have made
a further substantial contribution to the progress of surface phonon spectroscopy. A number of experimental studies have now been made of adsorbed
phases, epitaxially grown films, surface phase transitions, surface reconstruction and solitons, etc. The original competition of He inelastic scattering with
electron energy loss spectroscopy (EELS) has rapidly evolved into a fruitful co-
operation of two powerful complementary techniques: The former boasting an
excellent resolution at low phonon frequencies, the latter easily achieving the
high-frequency spectrum, where atoms still meet difficulties. The reader may
find an excellent overview of present problems and achievements, including a
direct comparison with EELS, in the article that J. Peter Toennies has written
for a recent Solvay symposium [9.68]. There, the reader will gain richer information and deeper insight than can be offered by the present chapter. Indeed,
the authors ask for indulgence for their omission of many excellent pieces of
work. This is just a little tribute to a great scientist and friend, whose youth-
ful enthusiasm and tough dedication have nourished our longstanding scientific
collaboration.

Acknowledgements

One of us (GB) gratefully acknowledges the support of the Alexander-von-
Humboldt-Stiftung, Germany, under the Research Award programme. We
thank Dr. E. Hulpke for the useful discussions on W(100) and for the accurate editing of the manuscript.

References

9.1   Lord Rayleigh: Proc. London Math. Soc. **17**, 4 (1887)
9.2   A.A. Maradudin: In *Nonequilibrium Phonon Dynamics*, ed. by W. E. Bron (New York
      1985) pp. 395–600
9.3   I.M. Lifshitz, L.M. Rozenzweig: Zh. Eksp. Teor. Fiz. **18**, 1012 (1948); for an English
      version see I.M. Lifshitz: Nuovo Cimento Suppl. **3**, 732 (1956)
9.4   A.A. Maradudin, E.W. Montroll, G.H. Weiss, I.P. Ipatova: *Theory of Lattice Dynamics
      in the Harmonic Approximation*, Solid State Physics, Suppl. 3 (Academic Press, New
      York 1971)

9.5   A.A. Maradudin, R.F. Wallis, L. Dobrzynski:*Handbook of Surfaces and Interfaces*, Vol
       3 (Garland, New York 1980)
9.6   F. Garcia-Moliner: Ann. Phys. (Paris) **2**, 179 (1977)
9.7   G. Armand: Phys. Rev. B **14**, 2218 (1976); see also G. Armand, P. Masri: Surface Sci.
       130, 89 (1983)
9.8   R.A. Allen: Surface Sci. **76**, 91 (1978)
9.9   A.A. Maradudin, J. Melngailis: Phys. Rev. **133**, A1188 (1964)
9.10   L. Dobrzynski, G. Leman: J. Phys. (Paris) **30**, 116 (1990)
9.11   S.W. Musser, K.H. Rieder: Phys. Rev. B **2**, 3034 (1970)
9.12   W. Goldhammer, W. Ludwig, W. Zierau, C. Falter: Surface Sci. **141**, 139 (1984)
9.13   J.E. Black, B. Lacks, D.L. Mills: Phys. Rev. B **22**, 1818 (1980)
9.14   G. Benedek: Phys. Status Sol. B **58**, 661 (1973)
9.15   G. Benedek: Surface Sci. **61**, 603 (1973)
9.16   G. Benedek, L. Miglio: In *Ab-Initio Calculation of Phonon Spectra*, ed. by J.T. De-
       vreese, V.E. van Doren, P.E. Van Camp (Plenum, New York 1983)
9.17   G. Benedek, G.P. Brivio, L. Miglio, V.R. Velasco: Phys. Rev. **B26**, 497 (1982)
9.18   L. Miglio, G. Benedek: In *Structure and Dynamics of Surfaces* II, ed. by W. Schom-
       mers, P. von Blanckenhagen (Springer, Heidelberg 1987) p. 35
9.19   G. Benedek, L. Miglio: In *Surface Phonons*, ed. by F.W. de Wette, W. Kress, Springer
       Ser. Surf. Sci., Vol. 27 (Springer, Berlin, Heidelberg 1991)
9.20   G. Benedek, M. Miura, W. Kress, H. Bilz: Phys. Rev. Lett. **52**, 1907 (1984)
9.21   T.E. Feuchtwang: Phys. Rev. **155**, 731 (1967)
9.22   V. Bortolani, F. Nizzoli, G. Santoro: In *Lattice Dynamics*, ed. by M. Balkanski (Flam-
       marion, Paris 1978)
9.23   The method is rewieved by R.F. Wallis in *Atomic Structure and Properties of Solids*,
       ed. by E. Burstein (Academic, New York 1972)
9.24   A.A. Lucas: J. Chem. Phys. **48**, 3156 (1968)
9.25   R. Fuchs, K.L. Kliever: Phys. Rev. **140**, A2076 (1965) and K.L. Kliever, R. Fuchs:
       Phys. Rev. **144**, 495 (1966); **150**, 573 (1966)
9.26   A. D. Boardman (ed.): *Electromagnetic Surface Modes* (Wiley, New York 1982)
9.27   R.E. Allen, G.P. Alldredge, F.W. de Wette: Phys. Rev. B **4**, 1648, 1661, 1682 (1971)
9.28   T.S. Chen, G.P. Alldredge, F.W. de Wette: Solid State Commun. **10**, 941 (1972)
9.29   T.S. Chen, F.W. de Wette, G.P. Alldredge: Phys. Rev. B **15**, 1167 (1977)
9.30   F.W de Wette, W. Kress (eds.): *Surface Phonons*, Springer Ser. Surf. Sci., Vol. 21
       (Springer, Berlin, Heidelberg 1991)
9.31   G.P. Alldredge: Phys. Lett. **41A**, 281 (1972)
9.32   H. Ibach: Phys. Rev. Lett. **24**, 1416 (1970)
9.33   A. Otto: Z. Phys. **216**, 398 (1968)
9.34   H. Ibach: Phys. Rev. Lett. **27**, 253 (1971)
9.35   W. Ludwig: Japan J. Appl. Phys., Suppl. **2**, Pt. 2, 879 (1974)
9.36   N. Cabrera, V. Celli, R. Manson: Phys. Rev. Lett. **22**, 346, (1969)
9.37   R. Manson, V. Celli: Surface Sci. **24**, 495 (1971)
9.38   G. Benedek, G. Seriani: Japan J. Appl. Phys., Suppl. **2**, Pt.2, 545 (1974)
9.39   B.R. Williams: J. Chem. Phys. **55**, 3220 (1971); **56**, 1895 (1972)
9.40   B.F. Mason, B.R.Williams: Japan J. Appl. Phys., Suppl. **2**, Pt. 2, 557 (1974)
9.41   G. Boato, P. Cantini: In *Dynamic Aspects of Surface Physics*, ed. by F.O. Goodman
       (Compositori, Bologna 1974) p. 707
9.42   G. Benedek: Phys. Rev. Lett. **35**, 234 (1975)
9.43   P. Cantini, G.P. Felcher, R. Tatarek: Phys. Rev. Lett. **37**, 606 (1976)
9.44   G. Benedek, G. Boato: Europhys. News **8**, 5 (1977)
9.45   J.P. Toennies, K. Winckelmann: J. Chem Phys. **66**, 3965 (1977)
9.46   G. Brusdeylins, R.B. Doak, J.P. Toennies: Phys. Rev. Lett. **44**, 1417 (1980)
9.47   G. Brusdeylins, R.B. Doak, J.P. Toennies: Phys. Rev. Lett. **46**, 437 (1981)
9.48   G. Brusdeylins, R.B. Doak, J.P. Toennies: Phys. Rev. B **27**, 3662 (1983)
9.49   G. Brusdeylins, J.P. Toennies, R.B. Doak: Phys. Rev. B **28**, 7277 (1983)

9.50 D. Evans, V. Celli, G. Benedek, R.B. Doak, J.P. Toennies: Phys. Rev. Lett., **50**, 1854 (1983)

9.51 G. Brusdeylins, R. Rechsteiner, J.G. Skofronick, J.P.Toennies, G. Benedek, L. Miglio: Phys. Rev. Lett., **54**, 466 (1985)

9.52 G. Bracco, M. D'Avanzo, C. Salvo, R. Tatarek, F. Tommasini: Surface Sci. **189/190**, 684 (1987)

9.53 G. Benedek, G. Brusdeylins, R.B. Doak, J.P. Toennies: Phys. Rev. B **27**, 2488 (1983)

9.54 G. Benedek, G. Brusdeylins, R.B. Doak, J..C. Skofronick, J.P. Toennies: Phys. Rev. B **28**, 2104 (1983)

9.55 G. Benedek, L. Miglio, G. Brusdeylins, J.G. Skofronick, J.P. Toennies: Phys. Rev. B **35**, 6593 (1987)

9.56 G. Chern, W.P. Brug, S.A. Safron, J.G. Skofronick: J. Vac. Sci. Technol. A **7**, 2094 (1989)

9.57 G. Chern, J.G, Skofronick, W.P. Brug, S.A. Safron: Phys. Rev. B **39**, 12828, 12838 (1989)

9.58 S.A. Safron, G. Chern, W.P. Brug, J.G. Skofronick, G. Benedek: Phys. Rev. B **41**, 10146, (1990)

9.59 W. Kress, F.W. de Wette, A.D. Kulkarni, U. Schröder: Phys. Rev. B **35**, 5783 (1987)

9.60 F.W. de Wette, A.D. Kulkarni, U. Schröder, W. Kress: Phys. Rev. B **35**, 2476 (1987)

9.61 R.B. Doak, U. Harten, J.P. Toennies: Phys. Rev. Lett. **51**, 578 (1983)

9.62 G. Armand: Solid State Commun. **48**, 261 (1983)

9.63 V. Bortolani, A. Franchini, F. Nizzoli, G. Santoro: Phys. Rev. Lett. **52**, 429 (1984)

9.64 C.S. Jayanthi, H. Bilz, W. Kress, G. Benedek: Phys. Rev. Lett. **59**, 795 (1987)

9.65 C. Kaden, P. Ruggerone, J.P. Toennies, G. Benedek: Vuoto **20**, 38 (1990) and to be published

9.66 M.S. Daw, M.I. Baskes: Phys. Rev. B **29**, 6443 (1984)

9.67 J.S. Nelson, M.S. Daw, E.C. Sowa: Phys. Rev. B **38**, (1988)

9.68 J.P. Toennies: In *Solvay Conference onf Surface Science*, ed. by F.W. de Wette, Springer Ser. Surf. Sci. Vol. 14 (Springer, Berlin, Heidelberg 1989)

9.69 U. Harten, J.P. Toennies, Ch. Wöll: Phys. Rev. Lett. **57**, 2947 (1986)

9.70 U. Harten, J.P. Toennies: Europhys. Lett. **4**, 833 (1987)

9.71 P. Santini, L. Miglio, G. Benedek, U. Harten, P. Ruggerone, J.P. Toennies: Phys. Rev. B **42**, 11942 (1990)

9.72 O.L. Alerhand, E.J. Mele: Phys. Rev. Lett. **59**, 657 (1987); O.L. Alerhand, E.J. Mele: Phys. Rev. B **37**, 2536 (1988)

9.73 A. Mazur, J. Pollmann: In *Phonons 89*, ed. by S. Hunklinger, W. Ludwig, G. Weiss (World Scientific, Singapore 1990) p. 943

9.74 U. Harten, J.P. Toennies, Ch. Wöll, L. Miglio, P. Ruggerone, L. Colombo, G. Benedek: Phys. Rev. B **38**, 3305 (1988)

9.75 L. Miglio, P. Santini, P. Ruggerone, G. Benedek: Phys.Rev. Lett. **62**, 3070 (1989)

9.76 F. Ancillotto, W. Andreoni, A. Selloni, R. Car, M. Parrinello: in *Phonons 89*, ed. by S. Hunklinger, W. Ludwig, G. Weiss (World Scientific, Singapore 1990) p. 931

9.77 K.C. Pandey: Phys. Rev. Lett. **47**, 1913 (1981); **49**, 223 (1982)

9.78 J.E. Northrup, M.L. Cohen: Phys. Rev. Lett. **49**, 1349 (1982); J. Vac. Sci. Technol. **21**, 333 (1982)

9.79 F.J. Himpsel, P. M. Marcus, R. Tromp, I.P. Batra, M.R. Cook, F. Jona, H. Liu: Phys. Rev. B **30**, 2257 (1984)

9.80 Y.R. Wang, C.B. Duke: Surf. Sci. **205**, L755 (1988)

9.81 C.B. Duke, S.L. Richardson, A. Paton, A. Kahn: Surf. Sci. **127**, L135 (1983)

9.82 G. Brusdeylins, R. Rechsteiner, J.G. Skofronick, J.P. Toennies, G. Benedek, L. Miglio: Phys. Rev. B **34** 902 (1986)

9.83 G. Benedek, L. Miglio, G. Brusdeylins, C. Heimlich, J.G. Skofronick, J.P. Toennies: Europhys. Lett. **5**, 253 (1988)

9.84 L. Miglio, L. Colombo: Phys. Rev. B **37**, 3025 (1988)

9.85 L. Miglio, G. Benedek: Europhys. Lett. **3**, 619 (1987)

9.86 C. Julien, I. Samaras, M. Tsakiri, P. Dzwonkowski, M. Balkanski: Mater. Sci. Eng. B **3**, 25 (1989)

9.87 A. Kahn: Surface Sci. Rep. **3**, 193 (1983)

9.88 G. Brusdeylins, J.P. Toennies: private communication

9.89 D.E. Moncton, J.D. Axe, F.J. DiSalvo: Phys. Rev. B **16**, 801 (1977)

9.90 G. Benedek, G. Brusdeylins, C. Heimlich, L. Miglio, J.G. Skofronick, J.P. Toennies, R. Vollmer: Phys. Rev. Lett. **60**, 1037 (1988)

9.91 G. Benedek, M. Miura, W. Kress, H. Bilz: Phys. Rev. Lett. **52**, 1907 (1984)

9.92 H. Bilz, H. Büttner, A. Bussmann-Holder, W. Kress, U. Schroeder: Phys. Rev. Lett. **48**, 264 (1982)

9.93 G. Brusdeylins, F. Hofmann, P. Ruggerone, J.P. Toennies, R. Vollmer, G. Benedek, J.G. Skofronick: In *Phonons 89*, ed. by S. Hunklinger, W. Ludwig, G. Weiss (World Scientific, Singapore 1990) p. 892

9.94 H.-J. Ernst, E. Hulpke, J.P. Toennies: Phys. Rev. Lett. **58**, 1941 (1987)

9.95 C.Z. Wang, A. Fasolino, E. Tosatti: Europhys. Lett. **7**, 263 (1988)

9.96 A. Muramatsu, W. Hanke: Phys. Rev. B **27**, 2609 (1983) and a review article in Physics Reports (North-Holland, Amsterdam 1984)

9.97 C. Beatrice, C. Calandra: Phys. Rev. B **28**, 6130 (1983)

9.98 K.M. Ho, K.P. Bohnen: Phys. Rev. Lett. **56**, 934 (1986)

9.99 K.M. Ho, K.P. Bohnen: Phys. Rev. B **38**, 12897 (1988); Europhys. Lett. **4**, 345 (1987)

9.100 X.W. Wang, W. Weber: Phys. Rev. Lett. **58**, 1452 (1987)

9.101 A.G. Eguiluz, A.A. Maradudin, R.F. Wallis: Phys. Rev. Lett. **60**, 309 (1988)

9.102 A.G. Eguiluz: In *Phonons 89*, ed. by S. Hunklinger, W. Ludwig, G. Weiss (World Scientific, Singapore 1990) p. 851

10. Physisorbed Rare Gas Adlayers Studied with Helium Scattering

G. Comsa, K. Kern, B. Poelsema

Rare gas films physisorbed on single crystal surfaces provide an experimental realization of quasi two-dimensional systems. Due to the competing adsorbate–adsorbate and adsorbate–substrate interactions these systems exhibit an amazingly large number of interesting and quite subtle phenomena, such as commensurate–incommensurate transitions, continuous melting or orientational epitaxy. Going from 2D to 3D, i.e. growing thicker films, epitaxial growth and wetting phenomena can be studied.

Thermal energy helium scattering appears to be particularly appropriate to investigate the properties of these layers. Being a rare gas at thermal energy it is nondestructive and is sensitive to the outermost surface layer; the He wavelength (1–0.5 Å) is well suited for diffraction and its energy (10–100 meV) allows high resolution measurements of energy losses and gains (phonons); the He diffuse scattering, due to the large scattering cross-section of isolated adatoms and defects, is very sensitive to disorder.

10.1 Background

The study of physisorbed rare gases has been shown to be a powerful tool in the understanding of elementary processes in surface physics, such as adsorption, desorption, melting or epitaxy [10.1]. In addition, physisorbed adlayers have acquired model character in the study of two-dimensional (2D) phases and their mutual transitions [10.2]. In analogy to bulk matter, these adsorbed layers can form quasi-2D gas, liquid or solid phases. Of particular interest are the solid phases; the substrate provides a periodic potential relief which interferes with the lattice structure of the monolayer, inducing modulations in the latter. In addition, since the adatoms belong to a different species than the substrate atoms, the strength of the lateral interactions within the adlayer differs in general from the strength of the adsorbate–substrate interaction. Depending on the delicate balance between these forces and on structural relationships, the adsorbed monolayer can form ordered solid structures which are commensurate (in registry) or incommensurate (out of registry) with the substrate (Fig.10.1).

Springer Series in Surface Sciences, Vol. 27   **Helium Atom Scattering from Surfaces**
Editor: E. Hulpke   © Springer-Verlag Berlin, Heidelberg 1992

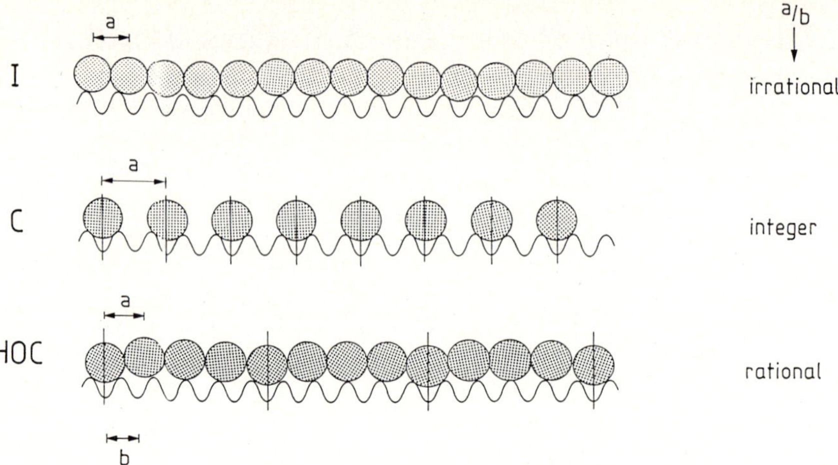

Fig.10.1. One-dimensional model of a physisorbed monolayer. The substrate is represented by a sinusoidal potential of period b and the adlayer by a chain of atoms with lattice constant a

When the periodic modulation of the adsorption potential in the directions parallel to the substrate surface (the so called corrugation of the substrate surface) is dominant, the adsorbed adatoms only occupy the energetically most favored sites and form a structure commensurate (C) with the underlying substrate. At the other extreme, when the adsorbate–adsorbate interaction dominates the periodic variation in the adsorbate–substrate interaction, the adlayer forms a structure which is determined by the dense packing of adatoms largely ignoring the substrate structure. This adlayer phase is termed incommensurate (I). Intermediate situations also occur when only a certain fraction of the adatoms occupies preferred adsorption sites; these are termed high order commensurate (HOC). The existence of these HOC phases has only recently been demonstrated experimentally. Only when the lateral adatom interaction and the substrate corrugation are comparable, can an ordered commensurate (C or HOC) adlayer undergo a transition into an incommensurate (I) phase as a function of coverage or temperature.

A variety of probe particles, such as electrons, neutrons, X-ray photons and thermal energy atoms are used as tools to investigate the structure of physisorbed films (cf. Ref. [10.2]). These methods of course all have their pros and cons. Low energy electron diffraction (LEED), for instance, is the most accessible technique, but suffers from the consequences of the strong interaction of the electrons with the adsorbed layer (multiple scattering, electron stimulated desorption etc.). Synchrotron X-ray scattering, on the other hand, has an excellent momentum resolution, but is hampered in application by its modest surface sensitivity.

The important characteristics of the He beam as a surface analytical tool are connected with the nature of the He–surface interaction potential. At large distances the He atom is subject to the weak attraction of the dispersion forces.

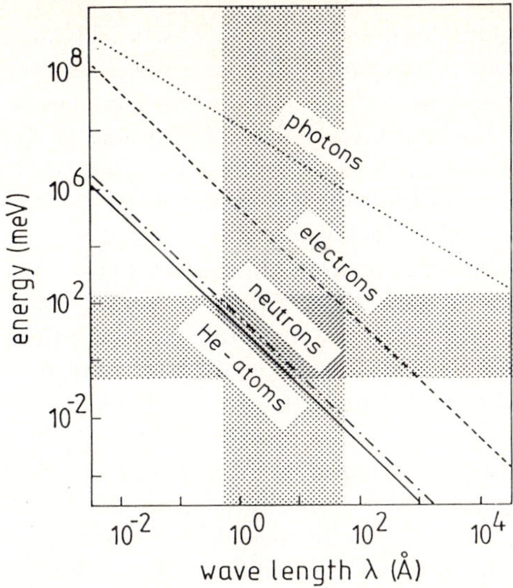

Fig.10.2. Energy-wavelength dispersion of common probe particles in surface scattering experiments. The shaded regions mark characteristic energy- and length-scales in surface processes

Upon closer approach, the electronic densities of the He atom and of the surface atoms overlap, giving rise to a strong repulsion. The classical turning point for thermal energy He is a few angstroms in front of the outermost surface layer. It is this interaction mechanism which makes the He atom sensitive exclusively to the outermost layer. The low energy of the He atoms and their inert nature ensures that He scattering is a completely nondestructive surface probe. This is particularly important when delicate phases, like physisorbed layers, are investigated. The de Broglie wavelength of thermal He atoms is comparable with the interatomic distances in adsorbed overlayers. Thus from measurements of the angular positions of the diffraction peaks the size and orientation of the two-dimensional (2D) unit cell, i.e., the structure of the outermost layer can be straightforwardly determined [10.3]. Analysis of the peak intensities yields the potential corrugation, which usually reflects the geometrical arrangement of the atoms within the 2D unit cell.

The energy of thermal He atoms is comparable with the energies of collective excitations (phonons) in overlayers. Thus, in a scattering experiment the He atoms may exchange an appreciable part of their energy with the surface [10.4]. This energy change can be measured in time-of-flight experiments with a resolution ≈ 0.3 meV; the resolution can even reach ≈ 0.1 meV (when using very low beam energies, $\simeq 8$ meV). Thus, surface phonon dispersion curves of rare-gas layers can be mapped out by measuring energy-loss spectra at various momentum transfers in different crystallographic directions. This is a definite advantage of inelastic He scattering over inelastic neutron scattering. (In view of the random orientation of powdered samples, which have to be used in surface investigations with neutron scattering, only average phonon density of states, but not dispersion curves can be obtained.) The range of energy transfer that can be covered by thermal He atoms is limited at the low end by the present maximum resolution of ≈ 0.1 meV and at the high end by the nature of the scattering mechanism. The thermal He-beam–surface interaction time being non-negligible, the upper limit for the observable phonon modes is ≈ 40 meV.

Besides the inelastic component, a certain contribution of elastically scattered He atoms is also found in between the coherent diffraction peaks. We will refer to this scattering as diffuse elastic scattering. This diffuse intensity is attributed to scattering from defects and impurities. Accordingly, this diffuse elastic scattering provides valuable information on the degree and nature of surface disorder. It can be used, for example, to study the growth of thin films [10.5] or to deduce information on the size, nature, and orientation of surface defects [10.6]. Very recently the TOF-peak width analysis of the diffuse elastic component has also been used to study the diffusive motion of adatoms [10.7].

Another remarkable way to use He scattering for the study of adsorbed layers is based on the large total cross section Σ for diffuse He scattering by isolated adsorbates (e.g., in the case of adsorbed Xe, $\Sigma_{\mathrm{Xe}}^{\mathrm{He}} \approx 110\mathrm{\AA}^2$ for $E_{\mathrm{He}} = 18$ meV) [10.8]. This large cross section is attributed to the long-range attractive interaction between adatom and the incident He atoms, which causes He atoms to be scattered out of the coherent beams. The large size of the cross section allows the extraction of important information concerning the lateral distribution of adsorbates, mutual interaction between adsorbates, dilute–condensed phase transitions in 2D, adatom mobilities, etc., simply by monitoring the attenuation of one of the coherently scattered beams, in particular of the specular beam [10.8]. This technique also allows the detection of impurities (including hydrogen!) in the ‰ range, a level hardly attainable with any other methods.

10.2 Experimental Aspects of Helium Scattering at Surfaces

Figure 10.3 shows schematically a He-atom–surface scattering experiment [10.9]: A highly monochromatic beam of thermal He atoms ($\Delta\lambda/\lambda \lesssim 0.01$) is generated in a high pressure supersonic expansion and collimated to a few tenths of a degree by a series of specially shaped collimators, the so called skimmers. Depending on source temperature (T_0), the wavelength of the He atoms ranges between $\lambda \approx 0.3$ and 2.0 Å; typical fluxes are of the order 10^{19} He atoms/s·sr:

$$E_{\mathrm{He}} = \frac{5}{2} k_{\mathrm{B}} T_0 \tag{101}$$

$$\lambda_{\mathrm{He}} = \frac{h}{\sqrt{2mE_{\mathrm{He}}}} \ . \tag{10.2}$$

The time-of-flight spectrometers are designed to measure structural and dynamical surface properties in one experiment, i.e. designed to measure the double differential scattering cross section:

$$\frac{d^2\sigma}{d\Omega dE_f} = \frac{1}{I_0} \cdot \frac{\delta Z}{\delta E_f} \quad , \tag{10.3}$$

where δZ is the number of He atoms in the energy interval δE_f scattered into the solid angle element $\delta\Omega$, and I_0 the intensity of the incoming He beam. Except for trivial factors, the cross section is determined by the energy exchange ($\hbar\omega$) and the momentum exchange ($\boldsymbol{Q}$):

$$\frac{d^2\sigma}{d\Omega dE_f} \simeq S(\boldsymbol{Q}, \ \hbar\omega)$$

$$\hbar\omega = E_f - E_i, \quad \boldsymbol{Q} = \boldsymbol{k}_f - \boldsymbol{k}_i \quad . \tag{10.4}$$

Since the initial parameters of the incoming He beam and the geometry of the experiment are known, the energy and momentum transfer during the scattering can be calculated straightforwardly.

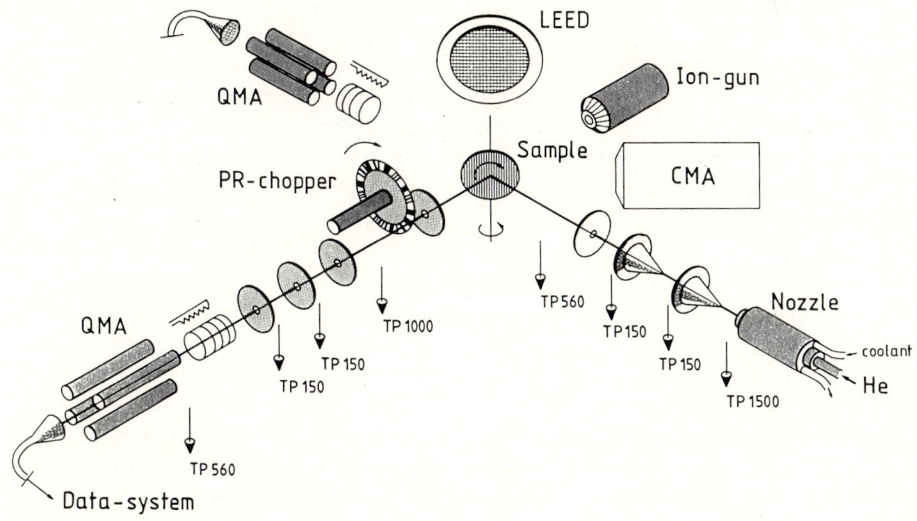

Fig.10.3. Schematic view of a high-resolution He–surface scattering spectrometer

The major experimental problems in high resolution He scattering are connected with the detection: large background and low detection efficiency. While with some technical effort the background can be efficiently reduced, no real breakthrough has been so far achieved in the attempt to increase the detection efficiency of He beams without impairing the time resolution. The method of choice is still electron bombardment ionization, followed by ion mass analysis.

The effective detection efficiency hardly surpasses 10^{-5}, with the time resolution still sufficient for time-of-flight He velocity analysis. Recent attempts to monitor the He metastables (also generated by electron bombardment) instead the He ions can be advantageous, when the He background pressure in the detector cannot be made low enough. In this type of detection, the relatively low efficiency is compensated by a favorable directionality effect: the He atoms in the beam are more efficiently detected than those in the isotropic background.

A relatively high He background is obviously inherent in all He beam experiments. Even a highly collimated beam puts a continuous, heavy He load on the pumps evacuating the sample chamber. In the case of the apparatus shown in Fig.10.3 the beam collimated to $0.2°$ (1.5×10^{-6}sr) supplies the sample chamber with about 1.5×10^{13} He/s (5.7×10^{-7} mbar· l/s). In view of some additional He load from the beam generator system and with a reasonably sized pumping system, the He background level in the sample chamber is rarely below 10^{-9} mbar. A detector located in this chamber would have a dynamical range of only $\sim 10^2$–10^3. This is acceptable when the information is inferred from the intense specular beam, but out of question when diffraction patterns of adlayers or inelastic scattering spectra are sought. The only practical way for a radical background reduction is the differential pumping. In the example in Fig.10.3, the scattering (sample) chamber is separated by three differential pumping chambers (one of them contains the chopper) which, together with the very efficient (high compression ratio) pumping of the detector chamber, leads to a He background pressure of $\sim 10^{-15}$ mbar in the latter. This results in an effective dynamical range of $\sim 10^6$. Very low intensity diffraction peaks (like second-order diffraction peaks from clean close packed metal surfaces) or weak inelastic resonances can be clearly resolved above the background.

The velocity distribution of atoms scattered from a surface is determined by measuring the time they need to cover a given distance (the so-called time-of-flight (TOF) method). The flight path is defined by the position of the chopper blade, which is preferably placed between sample and detector, and the center of the electron beam bombardment ionizer. For optimum resolution, the chopper opening function (gate function) should be in the ideal case a δ-function. The time interval between gate openings has to be large enough so that superpositions of slow and rapid atoms originating from successive openings can be neglected. The transmission is equal to the ratio between gate opening time and the interval between openings. Thus the width of the gate function, i.e. the time resolution, is a compromise between the ideal δ-function and an acceptable transmission which usually results in very low transmission values (of order 0.1%) for reasonable time resolutions.

An elegant and efficient alternative is the use of pseudorandom chopping. This method was first introduced in inelastic neutron scattering studies in the late 1960s. The gate function is a "pseudorandom" sequence of slots and bars. We use in our laboratory a binary shift register sequence of 512 slots and 511 bars, of 2.5 μs width each (at a rotor speed of 391 Hz). The distribution of arrival times is deconvoluted to the TOF distribution by cross-correlation. A detailed description of the pseudorandom TOF technique can be found in

[10.10–12]. The transmission is close to 50%, independent of time resolution. The stringent requirements of a highly constant rotation period of the chopper during the whole measuring time and a very low phase jitter (smooth rotation) of the chopper during each revolution are not easy to satisfy under UHV conditions. This has been realized, however, by using a chopper with magnetic bearings.

A matter of controversial discussion is the comparison of pseudorandom and conventional single slot chopping; for details see [10.12]. The comparison can be expressed quantitatively by introducing the gain factor $G(f_k)$, which represents the ratio of the variances resulting in channel k from the two kinds of chopping. For binary shift register sequences, the gain factor is given by

$$G(f_k) \;=\; \frac{\sigma^2_{\text{single-slot}}}{\sigma^2_{\text{pseudorandom}}} \;\simeq\; \frac{1}{2}\frac{f_k + \bar{u}}{\bar{f} + \bar{u}/n} \quad ,$$

where f_k is the number of counts of the time-dependent signal in channel k, $\bar{f}$ is its mean value, $\bar{u}$ is the mean number of counts per channel of the time-independent signal (the real background), and n the number of slots of the pseudorandom sequence. The pseudorandom chopping is favoured when $G(f_k) > 1$. This is always the case when $\bar{u} > 2\bar{f}$ because $n \gg 1$. The general interest, so far, is focussed on the position, height, and shape of the peaks in the TOF spectra and much less on the shape of a possibly present energy-dependent background. In this case a much less restrictive condition can be deduced: the pseudorandom chopping is favoured when in the "interesting" channels $f_k > 2\bar{f}$, irrespective of the value of $\bar{u}$. This condition is fulfilled for all significant peaks and thus the pseudorandom chopping is of advantage, even if technically more demanding.

The effective resolution of the He-scattering spectrometer as a whole is in fact the figure of primary interest. For the spectrometer used in the authors' laboratory the overall instrumental resolution (taking into account the scattering geometry, chopper gate function, energy spread of the incoming He-beam and the finite length of the ionization region) amounts to $\approx 7\mu s$ FWHM for a typical flight time of 800 μs corresponding to 0.32 meV FWHM at a 18 meV beam energy. The characteristic parameters of the high resolution He spectrometers HETOF–1 and –2 run in our laboratory in Jülich are summarized in Table 10.1. Today, there are about ten high resolution He spectrometers of similar characteristics in operation (Chicago–Sibener, Genua–Tatarek, Göttingen–Toennies, Murray Hill–Doak).

The capabilities of such a high resolution instrument are demonstrated in Figs.10.4 and 10.5. In Fig.10.4 we show as an example a He diffraction scan from a complete Xe monolayer adsorbed on Pt(111) ($\Theta_{Xe} \approx 0.42$ Xe atoms per Pt substrate atom); a well behaved diffraction pattern with sharp Bragg peaks is observed. This diffraction scan has been measured in a fixed scattering geometry $\theta_f + \theta_i = 90°$ by rotating the Pt crystal around an axis perpendicular to the scattering plane; the angle on the abscissa is transformed to the parallel momentum transfer scale by the relation $Q^{\parallel} = 5.723(\sin\theta_f - \cos\theta_f)$. The

Table 10.1. Characteristics of the high resolution He time-of-flight spectrometers operated in Jülich

Beam energy:	2–170 meV	
Beam energy spread:	$\Delta E/E = 2\Delta\lambda/\lambda =$	
	0.014	$(E_{\mathrm{He}} = 18$ meV$)$
	0.020	$(E_{\mathrm{He}} = 69$ meV$)$
Beam intensity:	$\sim 10^{19}$ He atoms/s·sr	
Detector sensitivity:	$\geq 3 \cdot 10^{-5}$	
Transfer width:	~ 400Å for $\theta_{\mathrm{i}} = \theta_{\mathrm{f}} = 45°$	
Spectrometer energy resolution:	0.32 meV	$(E_{\mathrm{He}} = 18$ meV$)$
	1.70 meV	$(E_{\mathrm{He}} = 69$ meV$)$

diffraction pattern characterizes a hexagonal close packed 2D Xe crystalline solid with a nearest neighbor distance of 4.33 Å. In the inset of Fig.10.4 the orientational structure (with respect to the Pt substrate) of the Xe monolayer is characterized in an azimuthal diffraction profile, which is obtain by rotating the Pt crystal around its surface normal at a fixed polar angle corresponding to a Bragg position. The symmetric peak-doublet centered along the $\bar{\Gamma}\bar{K}_{\mathrm{Pt}}$ direction of the substrate surface characterizes a Novaco–McTague rotated phase rotated $\pm 3.3°$ off the natural R30° orientation of submonolayer Xe films. The tiny satellite-peak at $Q \sim 0.3$ Å^{-1} with an intensity of about 10^{-5} of the incident beam has been identified as the mass-density-wave (MDW) satellite of the rotated Xe monolayer [10.13]. These diffraction satellites which arise due to a periodic deviation of the positions of monolayer atoms from their regular lattice sites in the rotated phase have been looked for since its original prediction in 1978 by *Novaco* and *McTague* [10.14] but were not observed with the classical diffraction techniques (electrons, neutrons, X-ray). It is indeed the high sensitivity of thermal He atoms to very small modulations of the adlayer topography which allows the detection of MDW satellites in a diffraction experiment.

The resolving power of inelastic He scattering is demonstrated in Fig.10.5. A series of TOF spectra taken from a full Ar monolayer on Pt(111) at different scattering angles are plotted over an energy-loss range from -2 to -6 meV. In these spectra the various features of the dynamical coupling between the adlayer and Pt substrate [10.15] can be distinguished. As a reference, let us first look at the spectrum (d) taken at $\theta_i = 35°$ which corresponds to the creation of an Ar monolayer phonon with wave vector $Q = 0.78$Å^{-1}, i.e., close to the edge of the 2D Ar Brillouin zone $[Q(\bar{\mathrm{M}}) = 0.98$Å$^{-1}]$. Since at these large Q values the Pt Rayleigh wave and the Pt bulk phonons have much higher energies than the Ar adlayer mode, no coupling of the Ar phonons to the Pt substrate is expected here. Indeed, the small linewidth $\Delta E = 0.32$ meV of the Ar loss peak at -4.8 meV is determined only by the instrumental resolution $\Delta E_{\mathrm{instr}}$ of the spectrometer without any measurable additional broadening. Spectrum (a)

250

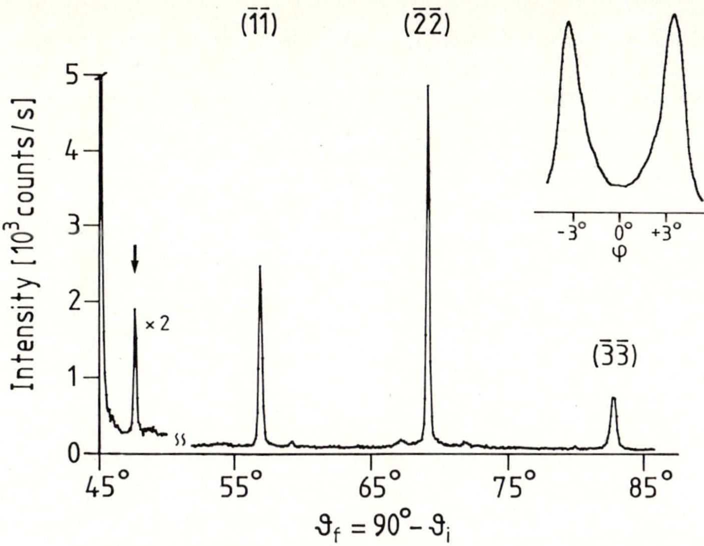

Fig.10.4. Polar and azimuthal (inset) He diffraction scan of a complete Xe monolayer on Pt(111); He wavelength $\lambda = 1.098$Å, surface temperature 25 K. The polar scan is taken along the $\bar{\Gamma}\bar{M}_{Xe}$-azimuth

exhibits two-phonon loss peaks. The feature at -2.9 meV is a signature of the Pt Rayleigh wave dynamically coupled to the Ar adlayer mode. Furthermore, the Ar loss at -5.0 meV is much broader ($\Delta E \simeq 0.41$ meV) than in spectrum (d). This difference is due to the fact that the Ar phonon in (a) is created at the center of the Brillouin zone $\bar{\Gamma}$ ($Q = 0.0/$Å^{-1}) where an effective damping of the Ar phonons due to the coupling to the projected bulk phonon bands of the Pt substrate occurs. This so called "radiative damping" of the adlayer phonon results in a linewidth broadening $\varepsilon \equiv \left[(\Delta E)^2 - (\delta E_{\mathrm{instr}})^2\right]^{1/2}$ of 0.2–0.3 meV. It is instructive to estimate the corresponding lifetime shortening due to this radiative damping. Using the Heisenberg uncertainty principle $\varepsilon \Delta t = h$ the lifetime becomes $\Delta t \sim 3 \times 10^{-12}$ s. Relating this value to the time scale of the Ar–Pt vibration $\hbar\omega = 5.0$ meV, i.e., $\nu = 1.2$ THz, a mean phonon lifetime of about three vibrational periods is obtained. Besides the radiative damping also the hybridization of the Ar adlayer mode and the Pt Rayleigh wave is observed. This is shown in spectrum (c) of Fig.10.5. At these scattering conditions ($\theta_i = 37.7°$) corresponding to a wave vector $Q = 0.45$Å^{-1} of the Ar phonon, the crossing of the adlayer mode and the Pt Rayleigh wave should occur. Instead, due to the hybridization of the two modes this crossing is avoided and a phonon doublet with an energy splitting of ~ 0.8 meV is observed (note also the comparable heights of the two peaks). Details of the dynamical coupling between an adsorbed layer and the substrate and of the phonon spectrum of thin physisorbed films (1–25 monolayers) can be found in [10.16–18]. The high energy resolution of the spectrometer at sufficiently high primary energies ($\Delta E = 0.32$ meV at $E_{\text{He}} = 18.3$ meV) recently enabled us to measure for the first time the intrinsic linewidth of a surface phonon and its temperature dependence [10.19].

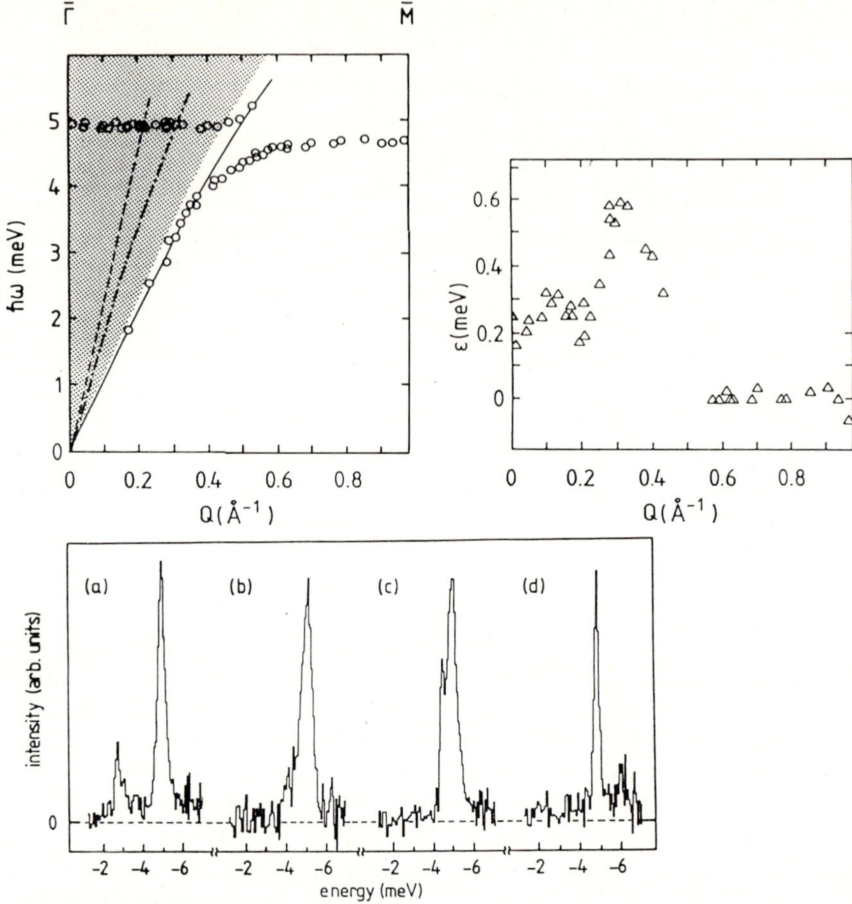

Fig.10.5. Experimental dispersion curves from a monolayer Ar on Pt(111) (perpendicular polarized mode) and He TOF spectra (inset) taken at different incident angles θ; along the $\bar{\Gamma}\bar{M}$ direction of the Ar unit cell: (a) 40.4°, (b) 38.5°, (c) 37.7°, (d) 35°. The energy of the primary He beam was 18.3 meV

10.3 Rare Gas Monolayer Phases on Pt(111)

10.3.1 Isolated Adatoms and the 2D Gas → 2D Gas + 2D Solid Phase Transition

For the perfect (111) surface of the fcc metal platinum, the corrugation sensed by a He atom of thermal energy is extremely small, giving rise to diffraction intensities of less than 10^{-4} of the specularly reflected intensity: a close-packed metal surface acts as an almost perfect mirror for thermal energy He-atoms. Adsorption on clean surfaces results in a dramatic decay of the specular intensity; the mirror becomes blind. From the initial slope of the adsorption curve (specular intensity versus coverage) it is possible to define a total scattering cross section for diffuse scattering Σ, associated with one adatom [10.20]:

$$\Sigma = -\frac{1}{n_s} \cdot \frac{d(I/I_0)}{d\Theta}\bigg|_{\Theta=0} \quad , \tag{10.5}$$

where I and I_0 are the intensities of the specular beam of the adsorbate covered and the clean surface, respectively, Θ is the adsorbate coverage, defined as the number of adatoms per Pt surface atom and n_s is the first layer Pt atom density. The experimental Σ values, determined through (10.5) are of the same order as gas phase total scattering cross sections ($\approx 100\text{Å}^2$). This large cross section is attributed to the long range attractive interaction between the adatom and the incident He atom, which causes the He atoms to be scattered out of the coherent beams. The remarkable size of the cross section, 4–6 times the geometrical size, A, of the adsorbate allows the extraction of important information concerning the lateral distribution of adsorbates, mutual interactions between adsorbates, dilute–condensed phase transitions in 2D, adatom mobilites, etc., simply by monitoring the attenuation of one of the coherently scattered beams.

As mentioned above the possibility to investigate the lateral distribution of adsorbates, in particular the dilute–condensed phase transition in 2D, is based upon the large difference between the cross section for diffuse scattering, Σ, and the geometrical size, A, of the adsorbates. The degree of overlap of the cross sections Σ at a certain adsorbate coverage Θ which determines the surface reflectivity, depends on the nature of the lateral distribution of the adsorbate. For instance, as long as the adsorbates form a lattice gas, the surface reflectivity for He atoms depends on Θ according to

$$\frac{I}{I_0} = (1-\Theta)^{\Sigma n_s} \tag{10.6}$$

$$\frac{I}{I_0} \simeq 1 - \Sigma n_s\, \Theta \quad \text{for } \theta << \Sigma n_s \quad . \tag{10.6.1}$$

On the other hand, when island formation starts, i.e. at the 2D gas → 2D solid transition, the reflectivity is determined by the much smaller geometrical size A of the adatoms in the 2D condensed phase:

$$\frac{I}{I_\mathrm{c}} \simeq 1 - An_\mathrm{s}\,(\Theta - \Theta_\mathrm{c}) \quad , \tag{10.7}$$

with I_c being the specular intensity at the critical coverage Θ_c, where condensation sets in. A comparison of (10.6) and (10.7) shows that, when condensation sets in, the slope of the specular He intensity I versus coverage Θ is expected to change dramatically: the ratio $\Sigma/A \simeq 4$–6. This is illustrated in Fig.10.6 for a Pt(111) surface at 54 K exposed to Kr at $p_\mathrm{Kr} = 2.1 \times 10^{-9}$ mbar.

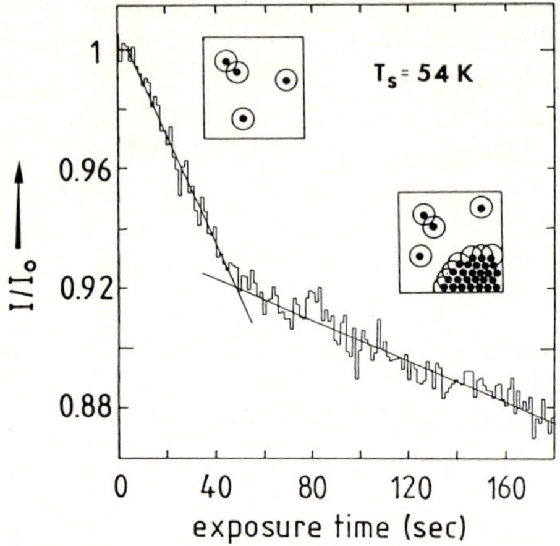

Fig.10.6. Specularly reflected He intensity from a Pt(111) surface at 54 K during exposure to $p_\mathrm{Kr} = 2.1 \times 10^{-9}$ mbar

The sudden slope change in Fig.10.6 obviously marks directly the onset of islanding. The critical coverage Θ_c corresponds to the 2D vapor pressure of Kr on Pt(111) at 54 K. From measurements of Θ_c at different temperatures, the 2D latent heat of vaporization is obtained under the assumption that the 2D gas is nearly perfect. In Table 10.2 the 2D latent heats of vaporization obtained in this way for Xe, Kr, and Ar on Pt(111) are listed. They actually represent the lateral interaction energy between the adlayer atoms.

Table 10.2. Lateral interaction energies for rare gases adsorbed on Pt(111)

adatom	Xe	Kr	Ar
ε_ℓ [meV]	43	26	17

254

10.3.2 Commensurate, High Order Commensurate and Incommensurate Adlayers

When the lateral adatom interaction and the substrate corrugation are comparable, the adsorbed monolayer may form various ordered structures, commensurate as well as incommensurate phases, allowing the investigation of the corresponding CI transition as a function of coverage or temperature. This CI transition is driven by the formation of line defects, so called misfit dislocations as was demonstrated first by *Frank* and *van der Merwe* [10.21]. These authors studied a linear chain of atoms with a lattice constant placed in a sinusoidal potential of amplitude V and periodicity b, representing the mutual interactions of the atoms in the chain by springs with spring constant K. The calculations reveal that for slightly different lattice parameters of chain and substrate, i.e. for a weakly incommensurate adlayer, the lowest energy state is obtained for a system which consists of large commensurate domains separated by regions of bad fit. The regions of poor lattice fit are dislocations with Burgers vectors parallel to the chain.

In two-dimensional systems these regions of bad fit (the domain walls) are lines. Because in a triangular lattice there are three equivalent directions, domain walls can cross. Using Landau theory, *Bak* et al.[10.22] have shown that it is the wall crossing energy Λ which determines the symmetry of the weakly incommensurate phase and the nature of the phase transition. For attractive walls, $\Lambda < 0$, a hexagonal network of domain walls (HI) will be formed at the CI transition because the number of wall crossings tends to be as large as possible. This C–HI transition is predicted to be first order. For repulsive walls, $\Lambda > 0$, the number of wall crossings tends to be as small as possible, i.e. a striped network of parallel walls (SI) will be formed in the incommensurate region. The C–SI transition is predicted to be continuous. The striped phase is expected to be stable only close to the CI phase boundary. At large incommensurabilities the hexagonal symmetry should be recovered in a first order SI–HI transition. In Fig.10.7 we summarize the possible domain wall structures. Superheavy and heavy walls are characteristic for those systems in which the incommensurate phase is packed more densely than the commensurate phase, while for light and superlight walls the opposite holds.

The best known examples of the CI transition in 2D adlayer systems are those which occur in the Kr monolayer on the basal (0001) plane of graphite [10.23] and in the Xe monolayer on the Pt(111) [10.2,24] surface. Here we will discuss briefly the physics of the Xe/Pt system (Fig.10.8). Below coverages of $\Theta_{Xe} \simeq 0.33$ and in the temperature range 60–99K, xenon condenses in a $(\sqrt{3} \times \sqrt{3})R30°$ commensurate solid phase. This phase has sharp diffraction peaks characteristic for coherent Xe domains which are about 800 Å in size. As the coverage is increased above 0.33 the relatively loosely packed Xe structure ($\approx 9\%$ larger lattice constant than in bulk Xe) undergoes a transition from the commensurate $\sqrt{3} \times \sqrt{3}$ structure to an incommensurate striped solid phase with superheavy walls. This weakly incommensurate solid is able to accommodate more Xe atoms than the commensurate phase by dividing into regions of

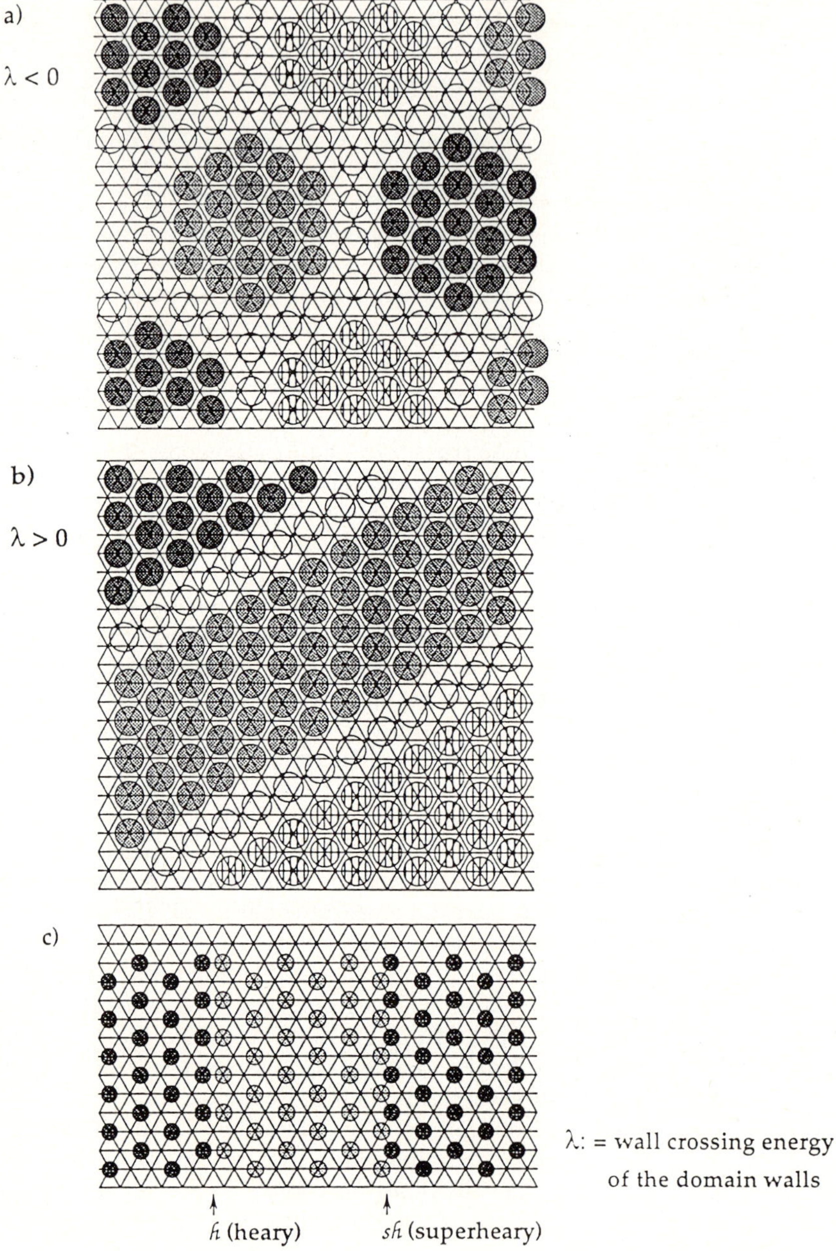

Fig.10.7. Domain wall systems for a $(\sqrt{3} \times \sqrt{3})R30°$ phase on a triangular substrate lattice; the walls are of the "light" type

256

commensurate domains separated by a regularly spaced array of striped denser domain walls. Increasing coverage causes the commensurate domains to shrink and brings the walls closer together. The domain walls are thus a direct consequence of the system's efforts to balance the competition between the lateral Xe–Xe and the Xe–Pt interactions. The C–SI transition can also be induced by decreasing the temperature below ~ 60K at constant coverage $\Theta_{Xe} < 0.33$; the driving force for this temperature-induced CI transition being anharmonic effects [10.25].

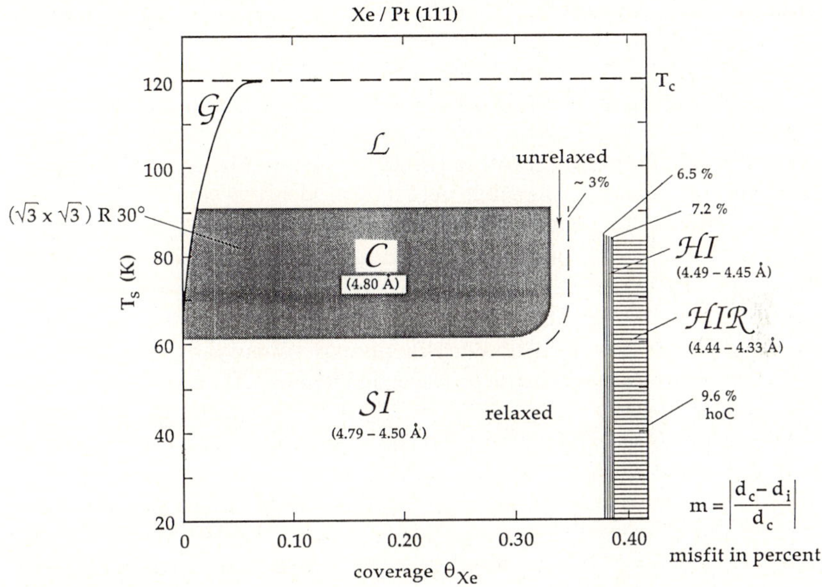

Fig.10.8. Schematic phase diagram of monolayer Xe on Pt(111). C, SI, HI, HIR denote the commensurate $(\sqrt{3} \times \sqrt{3})\text{R}30°$, the striped incommensurate, the hexagonal incommensurate and the hexagonal incommensurate rotated 2D solid phases. G and L denote the 2D gas and liquid, respectively

The usual measure for the incommensurability of an I phase is the misfit $m = (a_c - a_i)/a_c$, where a_c is the lattice parameter of the commensurate phase and a_i that of the incommensurate structure. For striped I phases, the misfit has of course uniaxial character, being defined only along the direction perpendicular to the domain walls. Quantitative measurements of the misfit during the C–SI transition of Xe on Pt(111) have revealed a power law of the form

$$m = \frac{1}{\ell} \propto (1 - T/T_c)^{0.51 \pm 0.04} \quad , \tag{10.8}$$

i.e. the distance ℓ between nearest neighbor walls scales with the inverse square root of the reduced temperature. This square-root dependence is the result of

an entropy-mediated repulsion between meandering nearest neighbor walls and
is in accord with theoretical predictions [10.26].

With increasing incommensurability the domain wall separation becomes
progressively smaller until at a critical misfit of $\sim 6.5\%$ the Xe domain wall
lattice rearranges from the striped to the hexagonal symmetry (Fig.10.7) in a
first order transition [10.27]. A further increase of the incommensurability by
adding more and more Xe eventually results in an adlayer rotation to misalign
itself with the substrate in order to minimize the increasing strain energy due to
the defect concentration [10.27]. This continuous transition to a rotated phase
(HIR) follows a power law $\varphi \propto (m - 0.072)^{1/2}$ starting at a critical separation
between nearest neighbor walls $\ell_c \simeq 10$ Xe-row distances (see also Fig.10.4).

It is generally accepted that on the same substrate the absolute magnitude
of the corrugation increases with the size of the rare gas adatom, while the
corrugation decreases relative to the binding energy of the adatom as well as
relative to the lateral adatom interaction [10.28]. Thus, within our simple model
of competing interactions (corrugation versus lateral attraction) we expect a
gradual transition from the floating Xe monolayer with its rich diversivity of
incommensurate domain wall phases to locked Kr or Ar layers which are dominated by the lock-in forces of the substrate, favoring HOC phases instead of
incommensurate domain wall phases.

Until recently, there has been no convincing experimental evidence for the
existence of high order commensurate physisorbed layers. This appeared to support the widespread belief that "experimentally it is impossible to distinguish
between a high-order C structure and an incommensurate structure" (Per Bak
in Ref.[10.29]).

This belief is certainly legitimate if the only accessible experimental information is the ratio of the adlayer and substrate lattice basis vectors. Indeed,
because one can always find a rational number within the confidence range
of any experimental irrational number, i.e. the basis vectors supplied by the
most refined experiment are always compatible with a high (enough) order
commensurate phase. There are, however, two other experimentally accessible parameters which allow a unequivocal distinction between a high order
commensurate "locked" and an incommensurate "floating" layer [10.30]. First:
the superstructure formed by the atoms located in equivalent, energetically favorable high symmetry sites. These more strongly bound atoms being located
"deeper" in the surface than the others, the adlayer is periodically buckled. Because of the extreme sensitivity of He scattering to the surface topography, this
superstructure, which characterizes high order commensurate layers, is directly
accessible to a high resolution He diffraction experiments. Second: the thermal expansion. Indeed, a "floating" layer is expected to thermally expand very
much like the corresponding rare gas bulk crystal, while a "locked" layer has by
definition to follow the substrate to which it is locked. The thermal expansion
of rare gas solids is at least ten times larger than that of substrates normally
employed, and so the distinction between high order commensurate "locked"
and incommensurate "floating" becomes straightforward. This very sharp cri-
terion requires that the "locking" is strong enough to withstand temperature

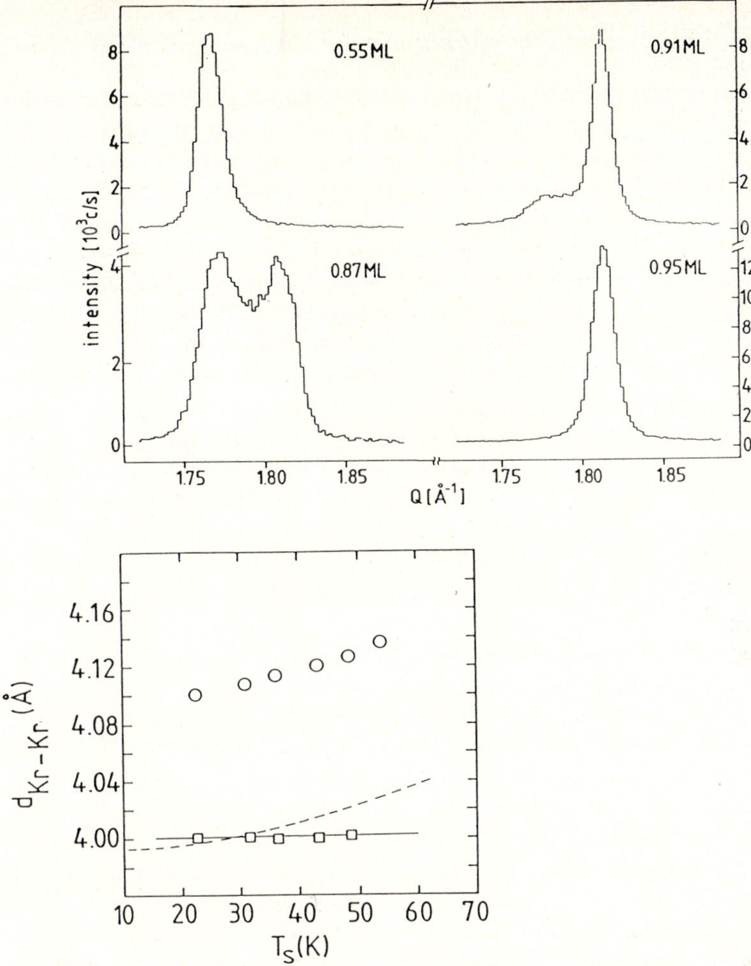

Fig.10.9. a) Polar He-diffraction scans of the $(1,1)_{Kr}$ -diffraction order from Kr adlayers on Pt(111) at various Kr submonolayer coverages at $T_s = 25$K. b) Kr-layer lattice spacing vs T_s for the (□) high (0.95 ML) and (○) low (0.5 ML) coverage phase; temperature dependence of the lattice spacing of (—) Pt-substrate, and of (- - -) bulk Kr

variations over a suffcently large range (≥ 10K) to allow for reliable thermal expansion measurements.

Figure 10.9a shows a series of polar He scans of the (1,1)Kr diffraction peak taken at 25 K along the $\bar{\Gamma}\bar{M}_{Kr}$ direction of the Kr monolayers adsorbed on a Pt(111) surface at coverages between 0.5 and 0.95 ML. The sequence is characteristic for a first order phase transition from a hexagonal solid phase with wavevector $Q = 1.769\text{Å}^{-1}$ ($d_{Kr} = 4.10$Å) to one with $Q = 1.814\text{Å}^{-1}$ ($d_{Kr} = 4.00$Å), below and above 0.8 ML, respectively. During the phase transition the intensity diffracted from one phase increases at the expense of the other.

The question concerning the incommensurate "floating" versus high order commensurate "locked" nature of the two Kr phases has been addressed by looking at their thermal expansion behavior and by searching for superstructure satellites. In Fig.10.9b the measured Kr–Kr interatomic spacing versus temperature is shown for submonolayer films of coverage 0.5 ML and 0.95 ML. The difference is striking. The low coverage phase shows a variation with temperature, very much like bulk Kr (dashed) and is thus an incommensurate "floating" phase. On the contrary, the lattice parameter of the high coverage phase is — like that of the Pt substrate (solid) — constant within experimental error in the same temperature interval; accordingly, this Kr-phase is high order commensurate "locked".

This assignment is supported by inspection of Fig.10.10, where polar scans (He energy 12 meV) in the $\bar{\Gamma}\bar{M}_{Kr}$-direction of the "floating" and of the "locked" Kr layer are shown. The scans differ substantially: the locked scan clearly evidences the presence of a superstructure, while the floating one does not. The superstructure peak at $Q = 0.532 \pm 0.022 \text{Å}^{-1}$ corresponds to 1/5 of the Pt substrate principal lattice vector. The origin of the superstructure peak is illustrated on the right hand of Fig.10.10 where the "locked" Kr phase on the Pt(111) surface is schematically shown. The Kr layer is rotated by 30° with respect to the substrate and its translational position is fixed by locating the central Kr atom in a preferred three fold hollow site (say fcc). Obviously, the Kr atoms in fcc sites (filled small circles) form a hexagonal $(5 \times 5)R0°$-superstructure which is responsible for the diffraction satellite at $Q_{Pt}^{(1,1)}$. Note that the superstructure is aligned with the substrate lattice while the Kr layer as a whole is rotated by 30°; this is the reason while the superstructure satellite is seen in the $\bar{\Gamma}\bar{M}_{Pt} = \bar{\Gamma}\bar{M}_{Kr} = \bar{\Gamma}\bar{M}_{superstructure}$-direction. The particular ratio between the lattice parameters of adlayer and substrate $\sqrt{3}d_{Pt}/d_{Kr} = 6/5$ produces an additional pecularity; the same number of Kr atoms are located in hcp and fcc hollow sites. Thus, the $(5 \times 5)R0°$ -superstructure has a two-atomic basis. A simple counting in Fig.10.10 shows that one sixth of the Kr atoms are locked in a hollow site (fcc or hcp). This fraction appears to be sufficient to prevent the Kr layer from expanding freely over more than 25 K.

10.3.3 Influence of Extrinsic Defects

Almost any theoretical study of adsorbed layers deals with perfect substrates, i.e. strictly periodic arrangements of surface atoms extending over infinite distances. In experiment, however, this situation is almost never met. Indeed, unavoidable slight misorientations of the substrate plane with respect to the low-index planes during crystal surface preparation result in a finite step density. Today's best prepared low-index metal surfaces have step densities of $\approx 0.1\%$, corresponding to average domain sizes of perfect coherence of ≈ 2000–3000Å. In addition, even under the most careful experimental ultrahigh vacuum conditions (base pressure in the low 10^{-11} mbar range), small amounts of adsorbing

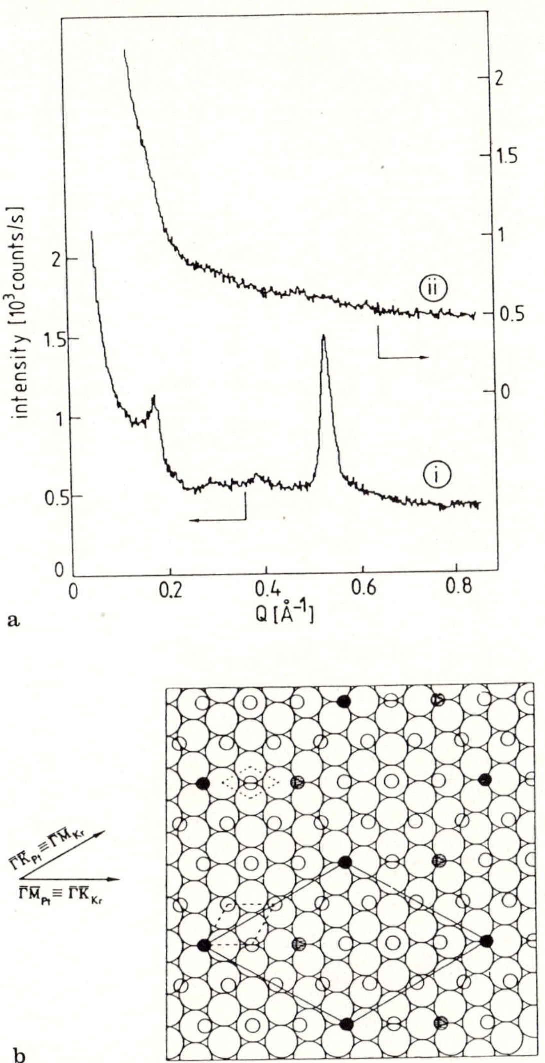

Fig.10.10. (a) Polar He diffraction scans of Kr monolayers in the vicinity of the specular peak ($Q = 0$Å^{-1}); (i) high (0.95 ML) and (ii) low (0.5 ML) coverage phase, taken along the $\bar{\Gamma}\bar{M}_{Kr}$ -azimuth. (b) Schematic representation of the high coverage phase of Kr on Pt(111); small circles represent Kr-atoms ($d_{Kr-Kr} = 4.00$Å) while large circles the Pt substrate atoms ($d_{Pt-Pt} = 2.77$Å).

impurities (H_2O, H_2, CO) cannot be completely avoided. These extrinsic defects can have a dramatic influence on phase transitions of adsorbed layers [10.31].

We have demonstrated recently that depending on the presence or absence
of minute amounts ($\approx 0.1\%$) of preadsorbed impurities, Kr monolayers may be
aligned (R0°) or rotated by 30°(R30°) with respect to the Pt(111) substrate,
respectively [10.32]. We have been able to demonstrate in a direct experiment
that when the Kr atoms are allowed to nucleate at the residual $\sim 0.1\%$ step
sites of the substrate (the binding energy at step sites is $\sim 25\%$ larger than
on terrace sites [10.33]) the R30° orientation is obtained; in contrast, when the
step sites are blocked by preadsorbed CO (or H, or even Xe) the growing Kr
layer cannot nucleate at step sites leading to an orientation aligned (R0°) with
the substrate.

To prepare Pt(111) surface with the CO-decorated steps we have taken
advantage of the cross-section overlap of admolecules (CO) and defects (steps).
If the CO molecules are located at a terrace they contribute with their full
diffuse cross section to the attenuation of the specular beam $\Sigma_{CO}^{eff,terrace} \approx \Sigma_{CO}$,
while CO molecules at step edges have a much lower cross section $\Sigma_{CO}^{eff,step} \ll \Sigma_{CO}$. This approach [10.34] is demonstrated in Fig.10.11

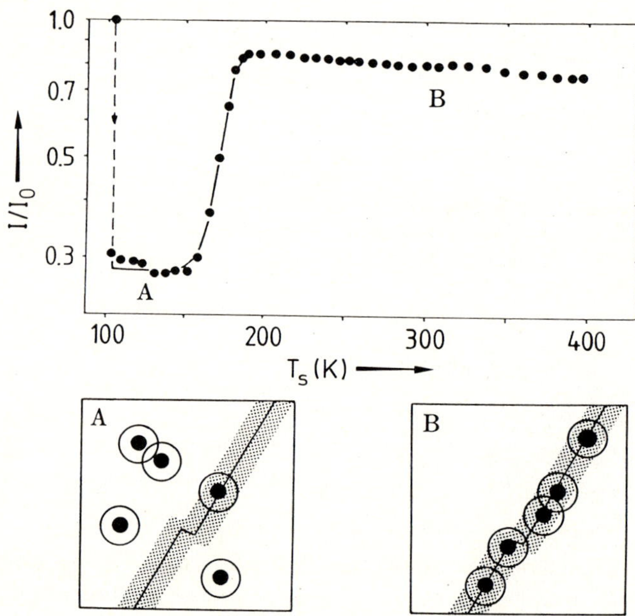

Fig.10.11. Relative He specular intensity I/I_0 versus surface temperature T_s. I is monitored
starting with the clean Pt(111) surface at $T_s = 107K$, then during CO adsorption and finally
during the linear heating (0.4 K/s) of the surface. The data are corrected for Debye-Waller
effects

First the surface is cooled to a low enough temperature ($T < 130K$) to ensure that the CO admolecules are immobile. A small quantity of CO molecules

262

is now adsorbed. Terrace and step sites are stochastically covered upon adsorption. Obviously, the CO molecules on terrace sites contribute mostly to the attenuation of the specular He beam; because of the strong overlap with the cross sections of the step atoms the effective contribution to attenuation of the CO molecules on step sites is minor. Finally, the surface is heated linearly and the intensity of the specular He beam monitored continuously. As soon as the CO molecules on the ideal terrace sites become mobile on the timescale of the heating rate, they migrate and are eventually trapped at step sites. The ideal terrace sites (the He mirror) become largely adsorbate free and thus the intensity of the specular He beam increases almost to the clean surface value. The overlap mechanism makes the migration from terrace to step sites directly visible. From the shape of the curve in Fig.10.11 we can also deduce an activation energy for CO diffusion on Pt(111); in the framework of a simple hopping model a value $E_{\mathrm{diff}} \approx 7$ kcal/mol is obtained [10.34].

References

10.1   S.K. Sinha (Ed.): *Ordering in Two Dimensions* (North Holland, Amsterdam 1980)

10.2   K. Kern, G. Comsa: In *Chemistry and Physics of Solid Surfaces* VII, ed. by R. Vanselow, R.F. Howe, (Springer, Berlin Heidelberg 1988) p.65

10.3   T. Engel, K.H. Rieder: *Structural Studies of Surfaces with Atomic and Molecular Beam Diffraction*, Springer Tracts Mod. Phys., Vol. 91 (Springer, Berlin, Heidelberg 1982)

10.4   J.P. Toennies: J. Vac. Sci. Technol. A **2**, 1055 (1984)

10.5   K. Kern, R. David, R.L. Palmer, G. Comsa: Phys. Rev. Lett. **56**, 2823 (1986)

10.6   A.M. Lahee, J.R. Manson, J.P. Toennies, Ch. Wöll: Phys. Rev. Lett. **57**, 471 (1986)

10.7   J.W. Frenken, J.P. Toennies, Ch. Wöll: Phys. Rev. Lett. **60**, 1727 (1988)

10.8   G. Comsa, B. Poelsema: Appl. Phys. A **38**, 153 (1985)

10.9   R. David, K. Kern, P. Zeppenfeld, G. Comsa: Rev. Sci. Instr. **57**, 2771 (1986)

10.10 G. Comsa, R. David, B.J. Schumacher: Rev. Sci. Instr. **52**, 789 (1981)

10.11 K. Sköld: Nucl. Instr. Meth. **63**, 114 (1968)

10.12 L.K. Verheij, P. Zeppenfeld: Rev. Sci. Instr. **58**, 2138 (1987)

10.13 K. Kern, P. Zeppenfeld, R. David, G. Comsa: J. Vac. Sci. Technol. A **6**, 639 (1988)

10.14 A.D. Novaco, J.P. McTague: Phys. Rev. B **19**, 5299 (1979)

10.15 B. Hall, D.L. Mills, J.E. Black: Phys. Rev. B **32**, 4932 (1985)

10.16 K. Kern, P. Zeppenfeld, R. David, G. Comsa: Phys. Rev. B **35**, 886 (1987)

10.17 B. Hall, D.L. Mills, P. Zeppenfeld, K. Kern, U. Becher, G. Comsa: Phys. Rev. B **40**, 6326 (1989)

10.18 P. Zeppenfeld, U. Becher, K. Kern, R. David, G. Comsa: Phys. Rev. B **41**, 8549 (1990)

10.19 K. Kern, U. Becher, P. Zeppenfeld, B. Hall, D.L. Mills: Chem. Phys. Lett. **167**, 362 (1990)

10.20 B. Poelsema, S.T. de Zwart, G. Comsa: Phys. Rev. Lett. **49**, 578 (1982); (E) **51**, 522 (1983).

10.21 F.C. Frank, J.H. van der Merwe: Proc. Roy. Soc. A **198**, 216 (1949)

10.22 P. Bak, D. Mukamel, J. Villain, K. Wentowska: Phys. Rev. B **19**, 1610 (1979)

10.23 E.D. Specht, A. Mak, C. Peters, M. Sutton, R.J. Birgeneau, K.L. D'Amico, D.E. Moncton, S.E. Nagler, P.M. Horn: Z. Phys. B **69**, 347 (1987); S.C. Fain, M.D. Chinn, R.D. Diehl: Phys. Rev. B **21**, 4170 (1980)

10.24 K. Kern, R. David, P. Zeppenfeld, R.L. Palmer, G. Comsa: Solid State Commun. **62**, 361 (1987)

10.25 M.B. Gordon, J. Villain: J. Phys. C **18**, 391 (1985)

10.26 V.L. Pokrovsky, A.L. Talapov: Sov. Phys. JETP **51**, 134 (1980)
10.27 K. Kern: Phys. Rev. B **35**, 8265 (1987)
10.28 G. Vidali, M.W. Cole: Phys. Rev. B **29**, 6736 (1984)
10.29 P. Bak: Rep. Prog. Phys. **45**, 587 (1982)
10.30 K. Kern, P. Zeppenfeld, R. David, G. Comsa: Phys. Rev. Lett. **59**, 79 (1987)
10.31 J. Villain: J. Phys. (Paris) Lett. **43**, 1551 (1982); M. Kardar, D.R. Nelson: Phys. Rev. Lett. **55**, 1157 (1985)
10.32 K. Kern, P. Zeppenfeld, R. David, G. Comsa: Phys. Rev. Lett. **57**, 3187 (1986)
10.33 R. Miranda, S. Daiser, K. Wandelt, G. Ertl: Surf. Sci. **131**, 61 (1983)
10.34 B. Poelsema, L.K. Verheij, G. Comsa: Phys. Rev. Lett. **49**, 1731 (1982)

11. Phonon Inelastic Scattering

D. Neuhaus

The quantum state of the motion of atoms within a solid with translational symmetry can be described by the phonon dispersion relation and the phonon density of states. At the surface the conditions of motion for the atoms are changed with respect to the bulk. The symmetry is reduced at the surface and the forces acting on the atoms are changed compared to the bulk. Due to these differences phonons localized at the surface (surface phonons) may exist which are not allowed in the bulk. Calculations by *de Wette* and *Benedek* revealed discrete phonon modes localized at the surface with dispersion relations in some cases separated from the broad bands of surface projected bulk dispersion curves. Some atomic motions which are inhibited in the bulk for symmetry reasons become possible at the surface by symmetry reduction [11.1,2]. Phonons behave like quasiparticles, characterized by energy and a vector very similar to the momentum vector called quasimomentum. The phonon energy shows a periodic dependence on quasimomemtum for crystals. This is a geometric property, independent of the interaction forces between the atoms and without the need to specify what kind of motion is described by the phonon.

Thermal neutron scattering has been successfully applied to measure phonon energies (typically in the range 0–40 meV) and to determine the bulk phonon dispersion relation but failed to measure the surface phonons, because of the low scattering cross sections. Scattering of photons, electrons and He atoms, which are sensitive to the surface, is used to probe surface phonons. To measure surface phonons with sufficient resolution, the dispersion relations of the phonons and the probing particles should be in the same range. This is not the case with light; here only surface phonons with very small momentum can be measured (Brillouin, Raman scattering).

Electrons of high energy (≈ 200 eV) have the required momentum to interact with surface phonons on the Brillouin zone boundary but in order to detect the very small energy changes resulting from the phonon interactions (≈ 2–40 meV) an extremely high energy resolution is required. The EELS technique developed by *Ibach* and coworkers has achieved 5 meV energy resolution [11.3].

Thermal energy He atoms appear to be the more suitable probe for surface studies of the atomic motions. Since their energy is low (< 70 meV) their interaction is completely non-destructive and because of the large cross sections they interact only with the outer layer (turning point for approaching He atoms

Springer Series in Surface Sciences, Vol. 27   **Helium Atom Scattering from Surfaces**
Editor: E. Hulpke   © Springer-Verlag Berlin, Heidelberg 1992

≈ 8 Å outside the surface) [11.4]. Helium used to study surface phonons is complementary to neutrons and bulk phonons.

Helium atoms became a useful probe to study surface phonons when it became possible to create high intensity monochromatic He beams. This was achieved by *Toennies* and coworkers in 1977, based on the early work of *Kantrowitz* et al. and *Becker* et al. [11.5,6]. They generate a He beam with a narrow velocity distribution by adiabatic expansion of high pressure helium gas into vacuum [11.7]. The quality of the helium beam for scattering experiments has been improved during recent years. Helium beams with 20 meV kinetic energy and a velocity resolution of $\Delta v/v \approx 1\%$, which corresponds to an energy resolution better than 0.1 meV, are now available.

The scattering process of helium atoms at a surface with translational symmetry is governed by the kinematic scattering conditions derived by *Benedek* [11.8]. Helium atoms are characterized in this process by their kinetic energy E and momentum $\hbar k$ (k is the wave vector). To facilitate the description k is split into components parallel and normal to the surface $k = (K, k_z)$. Conservation of energy is described by $E_f = E_i + \Delta E$ where E_i, E_f are the kinetic energy of the incident and the scattered (final) helium atoms; ΔE corresponds to the energy change of helium atoms caused by the interaction with the surface. Symmetry considerations allow only an interaction with surface atom vibrations polarized in the sagittal plane (defined by the flight direction of the incident helium atoms and the normal to the surface). Pure horizontally polarized vibrations are not detectable with helium scattering. In this case the one-phonon scattering cross section for incident helium atoms vanishes [11.9]. Due to the translational invariance parallel to the surface, momentum is conserved in this plane: $\hbar \Delta K = \hbar G + \hbar Q$, where $\hbar \Delta K$ describes the momentum change of the helium atoms by the scattering process parallel to the surface ($\hbar \Delta K = \hbar K_f - \hbar K_i$). The momentum change due to the interaction with the surface can be split into a wave vector Q within the first two-dimensional Brillouin zone and a surface reciprocal lattice vector G. ΔE and ΔK are determined in the scattering experiment. Time-of-flight (TOF) spectroscopy is used to measure ΔE. Single peaks in the TOF spectra, shifted on the energy scale with respect to the peak of the elastically scattered helium atoms by ΔE, are assigned to interaction of the helium atoms with the surface phonons. Both phonon creation ($\Delta E < 0$) and phonon annihilation ($\Delta E > 0$) are observed.

In Fig.11.1 a series of TOF spectra is shown. The momentum transfer $\hbar \Delta K$ can be calculated from ΔE in combination with the scattering angles. Prominent peaks in the TOF spectra ($\Delta E \neq 0$) are mostly related to single surface phonon interactions because this is the most probable interaction. Nevertheless, bulk phonons projected on the surface produce background signals in the TOF spectra which become a broad peak when a bulk phonon with a high density of states is involved. Frequently observed broad and shallow maxima underlying the spectra are related to multiphonon events. Finally one must check which reciprocal lattice vectors G contribute to ΔK. Extensive lattice dynamical calculations in combination with a full simulation of the expected TOF spectra make possible a detailed interpretation of the spectra. The intensity of a single

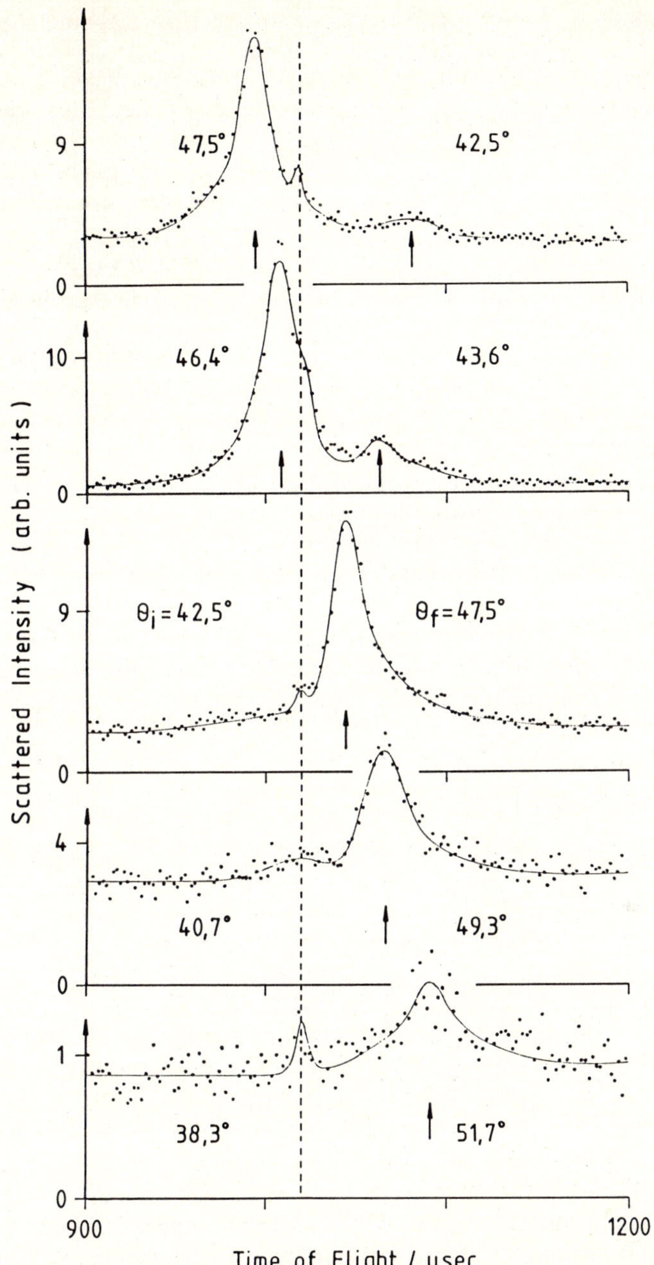

Fig.11.1. Time-of-flight spectra for helium atoms scattered from Pt(111) surface with ⟨110⟩ azimuth, crystal temperature 220K, beam energy 60 meV. The beam was scattered off the Pt surface under a fixed angle $\left(\Theta_i + \Theta_f = 90°\right)$. Angle of incidence Θ_i and scattering angle Θ_f are changed by sample rotation. Arrows indicate inelastic peaks, the dotted vertical line marks the position of elastically scattered He atoms. Thin lines: Maxwellian fit[11.29]

surface phonon interaction is proportional to the phonon density of states and the single phonon scattering cross section. The one-phonon distorted-wave Born approximation can be used to calculate the scattering cross sections [11.10]. Therefore, both dispersion relations and density of states of surface phonons can be determined with the helium scattering technique.

Helium diffraction ($\Delta E = 0$) can be used to measure the two-dimensional reciprocal lattice vectors of the surface. The intensities of the diffraction peaks are strongly related to the corrugation of the repulsive forces acting on arriving helium atoms in the surface plane. The repulsive forces on helium atoms are proportional to the electron density at the surface [11.11]. The electron density parallel surface at the helium atom turning point is nearly constant for densely packed metal surfaces due to the surface tension of the electron gas [11.12]. In this case helium atoms are mainly specularly scattered. The densely packed (111) surfaces of metals are nearly perfectly smooth mirrors for helium scattering experiments.

Adsorbates, especially chemisorbed atoms and molecules, on smooth substrate surfaces show a strong influence on the corrugation of the electron density at the surface. Helium scattering thus becomes very sensitive to adsorbates as shown by *Comsa* and co-workers. Adsorbate concentrations lower than 0.1% of a monolayer can be detected [11.13].

In the following, results of helium scattering experiments on noble metal surfaces are summarized. Measurements on the transition metal platinum will be discussed and the influence of chemisorbed and physisorbed adsorbates on metal surfaces, mainly platinum, will be studied.

11.1 The Noble Metals and Platinum

In 1983 *Doak* et al. reported surface phonon dispersion relations on Ag(111) measured with helium scattering. These showed, besides the expected Rayleigh mode, an additional clearly resolved higher frequency mode not predicted by theory (Fig.11.2) [11.14]. Other noble metals like copper and gold showed similar surface phonon branches at (111) surfaces along the $\bar{\Gamma}\bar{K}$ and $\bar{\Gamma}\bar{M}$ direction of the two dimensional Brillouin zone (Fig.11.2). In the case of gold, however, the two modes were not as clearly separated from each other as compared to silver and copper. To calculate phonon dispersion relations, lattice dynamics is predominatly described in force constant models. Two-body force constants, tangential and radial, and three-body angular force constants are considered [11.15,16]. As three-body forces are included, it is possible to account for the large violation of the Cauchy relation between the elastic constants in copper, silver, gold, platinum and other metals [11.17]. Interactions up to second neighbours were included to reproduce the measured bulk phonon dispersion relation by parametrization of force constants. Surfaces were introduced by slab models of a crystal with finite thickness and translational symmetry parallel to the surface [11.1]. Slab calculations, performed by utilizing force constants from bulk phonon calculations, show surface modes whose energy density is concentrated within a distance of the order of a wavelength below the surface. The measured

Rayleigh mode can be reproduced in these calculations. The Rayleigh wave is characterized by a propagating wave with an exponential decay of the displacement amplitude with depth beneath the surface and elliptical particle motion in the sagittal plane at each depth. To reproduce the measured additional surface mode above the Rayleigh mode in slab calculations it was necessary to change the force constants in the surface layer. Reduction of the nearest next neighbour radial force constant in the first layer of Ag(111) and Cu(111) yield a new surface vibration mode. [11.18]. In the case of silver the radial force constant had to be reduced by $\approx 50\%$ in order to fit the measured dispersion relation [11.9]. The eigenvectors associated with the new mode show a predominantly longitudinal displacement of the atoms in the first layer (longitidunal resonance).

Measured dispersion relations on the (111) surface of gold are much more difficult to interpret than those of the other noble metal surfaces of copper and silver, because the Au(111) surface is reconstructed. Helium diffraction experiments were used to study the structure of the first layer [11.19] The data are interpreted as showing regions of the surface with ABC stacking (expected for fcc lattice) and regions of ABA stacking, divided by transition regions. On average the surface is compressed. An attempt was made to explain the surface phonon dispersion relation on Au(111) in slab calculations ignoring the reconstruction of the surface. In these calculations, using the model of force constant parametrization, it was necessary to reduce the radial force constant by 70% and the angular force constant by 50%. in order to reproduce the measured dispersion curves [11.20]. Considerations of the eigenvectors in this study showed that the lower dispersion mode is related to a pseudo-Rayleigh wave and the upper mode to a longitidunal resonance. Pseudo-Rayleigh waves are characterized by a non-vanishing displacement of the atoms at infinite depth [11.21]. The apparent reduction in force constants at the surface has been qualitatively explained in the Thomas-Fermi total energy approximation of a metal. In this approximation the appearance of a surface results in a total energy decrease which explains the force constant reduction [11.20,22]. Besides the phenomenological force constant model to account for many-body interactions in a crystal, making use of extensive ad hoc softening of in-plane force constants to reproduce the observed surface phonon branches, an alternative approach to describe the lattice dynamics of noble metals was developed by *Jayanthi* et al. [11.23]. Many-body interactions in noble metals arise mainly from conduction s- and p-electron states located in the interstitial regions between the ion cores of predominantly d-character. In the new approach the s,p-charge density is described by a multipole expansion around the minimum of their density in the middle of the octahedral hole formed by the noble metal atoms in a fcc crystal. The charge density modulation around the minimum corresponds to a deformation of a coreless particle, known as a pseudoparticle. The total potential energy is a functional of the charge density which is described by ion cores and pseudoparticles. Expansion of the total energy along small displacements of the ion cores gives the equation of motion which in turn yields in the phonon dispersion relation. At the surface, the s,p-electron distribution and ion core po-

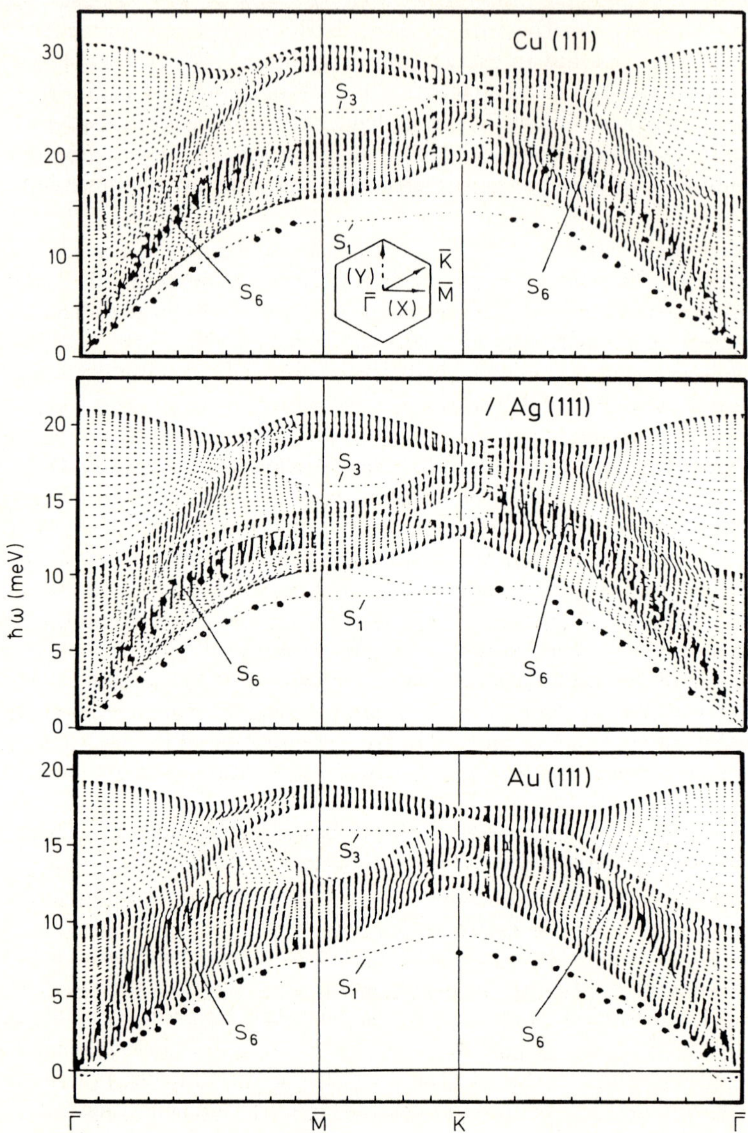

Fig.11.2. Calculated surface phonon dispersion curves of Cu(111), Ag(111), and Au(111). S_1, S_3, and S_6 represent the Rayleigh wave, the gap mode, and the surface longitudinal resonance, respectively. In a small region close to $\bar{\Gamma}$, we find for Au(111) a lattice instability, i.e., some of the phonon frquencies of the S_1 branch becomes imaginary (represented in the figure as negative ordinate values)[11.23]. Experimental points are depicted by filled circles[11.18]

270

sitions are different compared to the bulk. The work-function and s,p-electron charge distribution alone are sufficient to describe the surface parameters. In the cases of Cu(111) and Ag(111) the s,p-charge density is increased at the surface and the topmost layer of pseudoparticles is relaxed into the underlying ion core layer. The ion core–pseudoparticle attraction is increased by this relaxation and an effective reduction of the repulsive forces between the ion cores in the first layer is the result. The longitidunal resonance in Cu(111) and Ag(111) observed with helium scattering is explained by decreased repulsive forces between ion cores in the first layer due to redistributed s,p-electrons (Cu(111): $\approx 30\%$ reduction). In the case of Au(111), however, the reduction in core repulsive forces is not sufficient to explain the observed additional surface phonon mode close to the Rayleigh mode. *Jayanthy* et al. found a lattice instability at the surface, which gives rise to the observed reconstruction at the (111) surface of gold. The observed two surface phonon modes in Au(111) are attributed in this description to Rayleigh modes, which are split due to the anisotropy induced by the reconstruction at the surface [11.23].

Some metals show anomalies in the bulk phonon dispersion relation due to electron–phonon interactions. It is interesting to know whether surface vibrations of this metals show similar anomalies, in spite of electron redistribution, at the surface. One prominent representative of this kind is the transition metal platinum, which shows a phonon anomaly along the Γ, K direction of the three-dimensional Brillouin zone, measured by *Dutton* et al. with neutron scattering, in the T_1 phonon mode ($\langle 110 \rangle$ direction) appearing as a "dip" in the dispersion relation [11.24]. An essential factor for the observed anomalies is dielectric screening of lattice vibrations which can generate, along selected directions of the Brillouin zone, "dips" in the phonon dispersion relation. Large contributions to the dielectric screening function may arise from parallel sections of the Fermi surface or parallel electron–hole bands which cross the Fermi energy. On this understanding such phonon anomalies are images of the Fermi surface in the vibration spectra of a metal as predicted by *Kohn* in 1959 (Kohn anomaly) [11.25–27].

Surface phonon dispersion relation on Pt(111) were measured along $\bar{\Gamma}\bar{K}$ and $\bar{\Gamma}\bar{M}$ directions of the two-dimensional Brillouin zone by at least three groups; the results of two of them are presented in Figs.11.3,4 [11.29–31]. Similar to the dispersion relations of the noble metals copper and silver, two surface phonon branches were observed on platinum, visible in both of the selected directions of crystal symmetry ($\langle 110 \rangle$, $\langle 112 \rangle$). To fit the measured bulk phonon dispersion relation in the force constant model of a crystal (tangential, radial and angle bending force constants) fourth neighbour interactions have to be considered, however the Kohn anomaly in the T_1 mode is still not reproduced [11.27]. Interactions with 6th nearest neighbours are necessary in this model to reproduce the Kohn anomaly as measured with neutron scattering (14 force constants). Corresponding to Ag(111) the nearest neighbour radial force constant in the first layer has to be reduced in order to evolve the high energy surface phonon mode in the calculated dispersion relation and to fit it to the measured phonon energies on Pt(111). In this case the radial force constant is reduced by 40

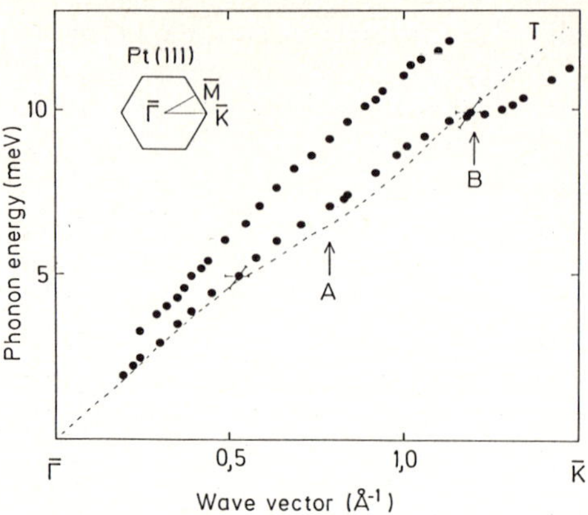

Fig.11.3. Reduced zone plot of surface phonon dispersion relations measured for Pt(111) with helium scattering at 220K sample temperature along the $\bar{\Gamma}\bar{K}$ azimuth. Dots: surface phonon dispersion relations. *A* and *B* mark the position of the anomalies in the lower (Rayleigh) branch. Dashed line: bulk phonon dispersion relation (T_1-mode as measured by *Dutton* et al. [11.24].), showing a bulk anomaly near *A*. Error bars are drawn in [11.29]

% relative to the bulk value. The measured low energy surface phonon mode is interpreted as a Rayleigh mode, while the high energy branch is attributed to a longitudinal resonance, analogous to the interpretation of the dispersion relations at the (111) surface of copper and silver. We might indeed expect platinum to behave in a way similar to the noble metals copper and silver, since platinum has only one less d-electron than a noble metal.

Measured surface phonon dispersion relations on Pt(111) show anomalies in the Rayleigh mode not reproduced in calculations with force constants models, even with the inclusion of up to 14 parameters. If particularities in the electron–phonon interaction are regarded as being responsible for the phonon anomalies, the dielectric function or the related generalized susceptibility, which describes the dielectric screening, must explain the observed anomalies. The generalized susceptibility was calculated by *Freeman* et al. for platinum [11.26]. The measured bulk phonon anomaly can be related to the calculated generalized susceptibility which shows a marked relative maximum in the same wave vector region where the phonon anomaly is observed. A distinct contribution to this maximum comes from electrons in the 5th band of platinum. The surface phonon anomaly, marked with (*A*) in Fig.11.3, which is also visible in Fig.11.4, coincides with the observed bulk anomaly, and is interpreted as a remainder of the latter observable through the finite penetration of the surface Rayleigh wave. The second anomaly, marked with (*B*) in Fig.11.2, found no confirmation in the surface phonon dispersion relation measured by the second group and depicted in Fig.11.3. In this case interpretation of anomaly (*B*) must be postponed. But it should be mentioned that anomaly (*B*) closely agrees with the less pronounced peak in the calculated generalized susceptibility at wave vector $Q = 1.2\,\text{Å}^{-1}$ and it is remarkable that anomaly (*B*) is measured in a wave vec-

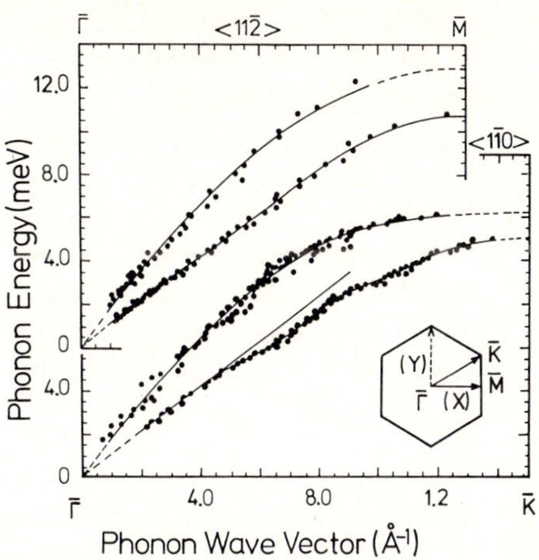

Fig.11.4. Surface-phonon dispersion curves for Pt(111) measured with helium scattering technique. The upper two curves are for the $\bar{\Gamma}\bar{M}$ direction and the bottom two curves for the $\bar{\Gamma}\bar{K}$ direction. The beam energies in both experiments were at fixed values between 10 and 32 meV. The target temperature in the $\langle 112 \rangle$ measurements was 400K, and in the $\langle 110 \rangle$ measurements 160K. The curves show the best-fit Fourier expansions. The solid line in the lowest set of data corresponds to a group velocity of 11.1 meV/$\mathring{A}$[11.30]

tor region where the surface Rayleigh mode crosses the bulk phonon T_1 mode. The temperature dependence of the Rayleigh mode was measured and shows a shift to higher surface phonon energies when the temperature is increased. This is in contrast to the decrease in phonon energy for the measured bulk phonon T_1 mode. On the other hand, the bulk phonon Kohn anomaly is more pronounced at low temperatures which corresponds to the observed temperature dependence of the surface phonon anomaly (A) [11.24,28], in accordance with the interpretation of anomaly (A) as a remnant of the bulk phonon anomaly.

11.2 Platinum with Chemisorbed Adsorbates

All the unusual features observed in the surface structure in the case of gold and in the surface dynamics for the noble metals copper, silver, gold and the transition metal platinum originate from a redistribution of electrons at the surface. It was shown by *Bortolani* et al. that a simple model, cutting the crystal at the surface but conserving the bulk binding conditions, could not reproduce the measured dynamical properties of the surface [11.20]. Consequently it is expected that directed changes in the electron distribution at the surface will modify the dynamical properties. By observing surface phonon dispersion relations on Pt(111) with chemisorbed (1×1) hydrogen or oxygen overlayers, one can study the influence of adsorbates in their capacity as electron acceptors or donors on the vibrational properties of the top layer platinum atoms. Hydrogen acts as an electron donor at the platinum surface, as apparent from the work function decrease with increasing hydrogen coverage [11.32]. The he-

lium surface potential corrugation on the (1×1)H/Pt(111) surface is increased
compared with the clean platinum surface, as evident from the more intense
diffracted helium beams when hydrogen is adsorbed on platinum [11.33].

Figure 11.5 shows the surface phonon dispersion relation at (1×1)H/Pt(111)
measured with helium scattering, compared to the result of lattice dynamics
calculations [11.28]. The measured phonon energies ($<$ 10 meV) are related
to platinum atom vibration, while vibrations of chemisorbed hydrogen atoms
are expected at much higher frequency. The effect of the additional hydrogen
mass on the platinum atom vibration can be neglected as the atomic mass ratio $m_{Pt}/m_H = 195$, so energy changes of less than 0.3 % at the Brillouin zone
boundary are expected. This was confirmed by deuterium substitutions, which
showed no change in phonon energy to within the experimental error [11.28]. In
contrast to the clean Pt(111) surface, only a single surface phonon branch was
detected on (1×1)H/Pt(111). The longitudinal resonance observed on the clean
platinum surface disappears with the (1×1) hydrogen overlayer. Compared to
the clean but dynamically modified platinum surface, the Rayleigh phonon energy is 1.5 meV lower at the hydrogenated surface. It was shown that when
bulk force constants are assumed at the surface, the calculated dispersion relation (force constant model) fits to the Rayleigh phonon mode [11.28]. It is
suggested from the calculated dispersion relation that the influence on binding

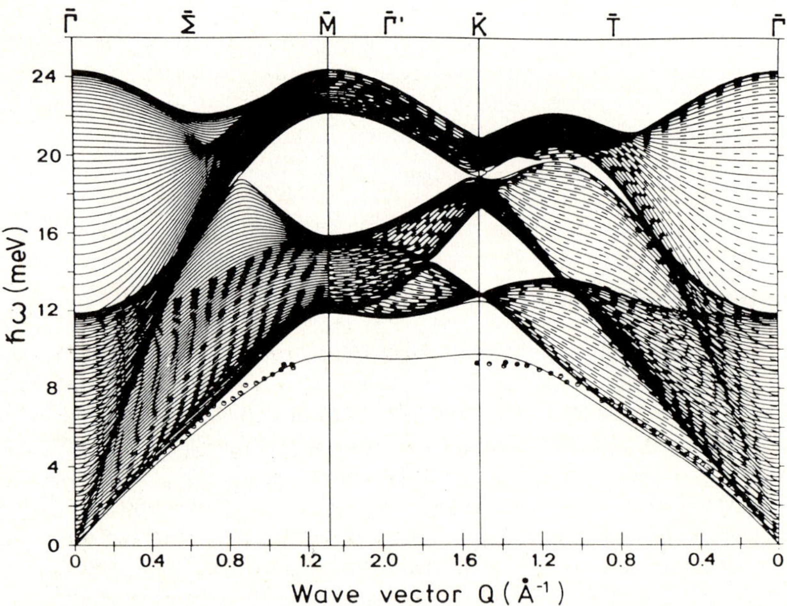

Fig.11.5. Comparison of the surface phonon dispersion curves from (1×1)H/Pt(111) measured with helium scattering technique with the results of a lattice-dynamical calculation
with unmodified bulk force constants for a slab of 60 atomic planes [11.28]. Measured surface
phonons are indicated by filled circles

274

forces between platinum atoms due to redistribution of electrons at the surface
is compensated by electrons from chemisorbed hydrogen atoms.

In contrast to hydrogen, chemisorbed oxygen affects the platinum surface as
an electron acceptor as indicated by the work function increase with growing
oxygen coverage at the surface [11.34]. The influence of chemisorbed oxygen
on the dynamical properties of Pt(111) is completely different from the hydrogenated surface. In Fig.11.6 measured surface phonon dispersion relations at

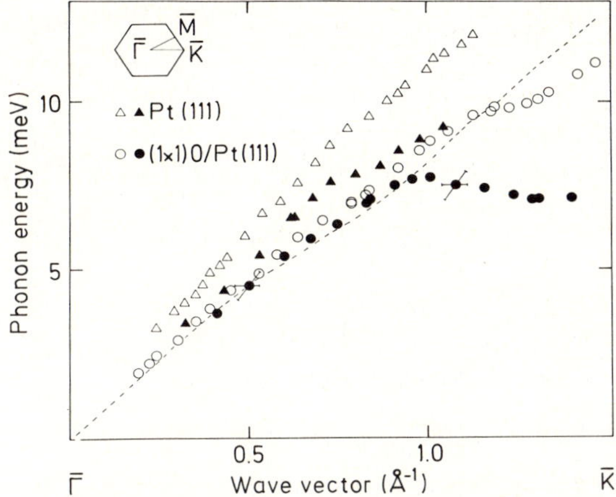

Fig.11.6. Reduced-zone plot of surface phonon dispersion relations measured with helium
scattering technique along the Γ̄K̄ azimuth. Open circles and triangles: Rayleigh mode and
longitudinal resonance on the clean Pt(111) surface at 220K sample temperature. Filled circles
and triangles: Rayleigh mode and longitudinal resonance on the (1×1)O/Pt(111) surface at
540K sample temperature. Dashed line: Bulk phonon dispersion relation (T_1 mode) which is
higher than or equal to the lower boundary of bulk phonons projected on the Pt(111) surface
[11.29,35]

the Pt(111) surface with (1×1) oxygen overlayer are compared to dispersion
relations for the clean surface [11.35]. The (1×1) oxygen overlayer on Pt(111)
with residual p(2×2) islands was formed at 540K sample temperature. The
structure of the adsorbed oxygen layer was checked by helium diffraction, as
shown in Fig.11.7 along different scan lines in the reciprocal lattice. At 300K
sample temperature a p(2×2) oxygen overlayer is observed, which changes to
a (1×1) overlayer as the sample temperature is increased. The residual p(2×2)
islands are indicated by broad and weak p(2×2) diffraction peaks along scan
lines [11.35]. Measurements on surfaces with a (1×1) adsorbate overlayer to
study the influence of adsorbates on the dynamical properties of surfaces have
the advantage that the interpretation of the dispersion relations obtained is not
influenced by additional Brillouin zone boundaries. On the clean platinum surface two surface phonon branches were observed, identified as Rayleigh mode
(lower branch) and longitudinal resonance (upper branch). The same interpretation is adopted for the (1×1) oxygen covered surface [11.35]. This means that

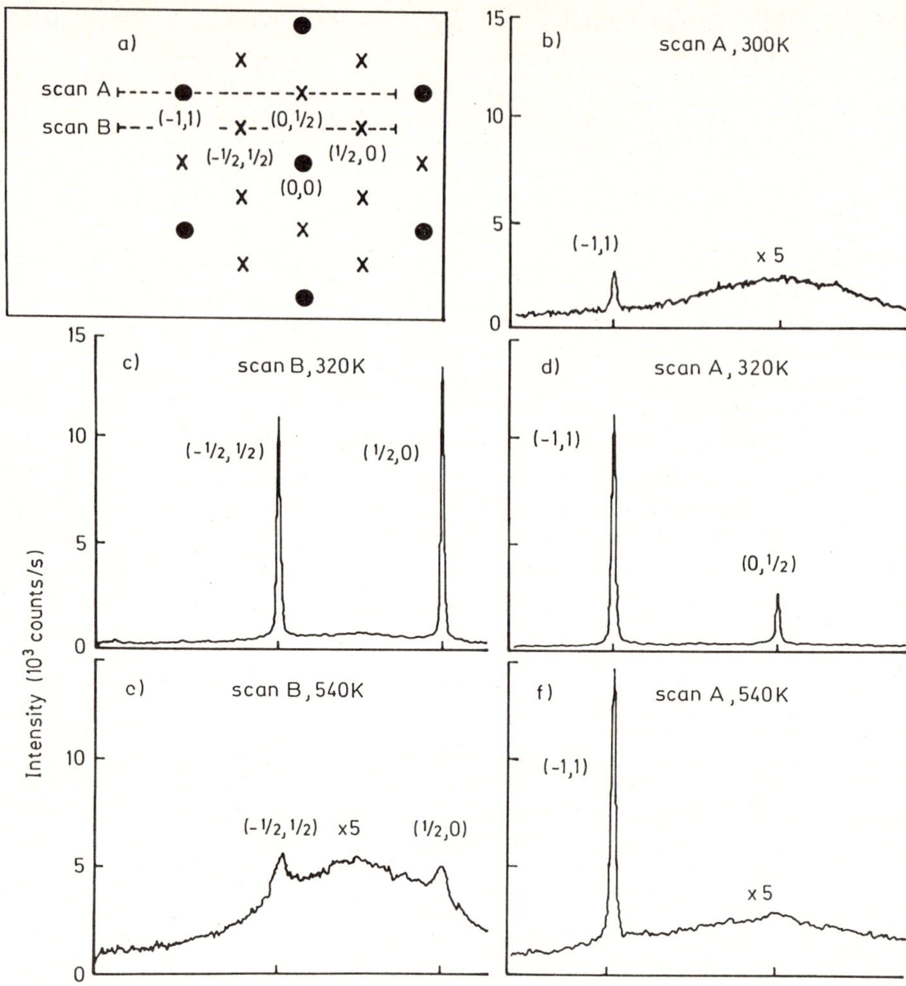

Fig.11.7. Helium diffraction pattern parallel to the $\bar{\Gamma}\bar{K}$ direction. (a) Scan lines in $\boldsymbol{k}$ space. (b) Scan A on the clean Pt(111) surface at 300K sample temperature; (c), (d), (e), (f) scan A or B at 320 or 540K sample temperature on oxygen-covered Pt(111) surface [11.35]. p(2×2)O/Pt(111) is observed at 320K sample temperature and (1×1)O/Pt(111) with residual p(2×2) islands at 540K sample temperature

the modes observed are modified platinum surface phonons rather than adsorbate vibrations, as the latter are expected to be widely split according to the Pt/O mass ratio of 12.2 and the measured Pt–O fundamental stretch mode of 59.5 meV [11.36,37]. The energies of the phonons at the oxygen covered platinum surface are reduced compared with the clean surface. The reduction in energy becomes more pronounced close to the Brillouin zone boundary. This is attributed to the decreasing penetration depth of surface modes with increas-

276

ing wave vector. Near the Brillouin zone boundary surface modes become more sensitive to surface properties, so changes in the binding forces at the surface have a stronger effect on the phonon energy. Measurements have confirmed that the observed phonon softening is not a temperature effect but solely due to the oxygen overlayer [11.35]. A qualitative model developed by *Heine* et al. for the binding forces in noble metals, but also applicable to platinum, will be used [11.38]. In this model the solid is described by d-electron shells, with strong repulsive pairwise interatomic forces, and an s,p-electron gas. The pairwise (d-shell) repulsion is balanced by multi-atom s,p-electron attraction. The latter acts as an ubiquitous "glue" between the d-shells. Surface phonon softening on (1×1)O/Pt(111) is explained by a reduction of s,p-electron density at the surface, as indicated by ultraviolet photoemission spectroscopy experiments, due to the oxygen adsorbate, which therefore should lead to reduced attractive interactions [11.39].

In conclusion it was shown that hydrogen and oxygen adsorbates on platinum, acting as electron donor and acceptor respectively, strongly affect the dynamical properties of the surface. The transition to ordered adsorbates less densely arranged than (1×1) overlayers introduces not only additional Brillouin zone boundaries but leads to specific dynamical properties, because binding forces between substrate surface atoms are changed in a more open grid. Concentrating on this aspect, surface vibrations at Pt(111) with p(2×2) chemisorbed oxygen overlayer were studied with the helium scattering technique. Measurements are described in the repeated Brillouin zone of the p(2×2)O/Pt(111) surface represented in Fig.11.8. The Brillouin zone of the clean Pt(111) surface is included. Surface phonon dispersion relations were studied in the ⟨112⟩ and ⟨110⟩ direction along $\bar{\Gamma}\bar{M}$ and $\bar{\Gamma}\bar{K}$ in the reciprocal space [11.40,41]. The points $\bar{\Gamma}$ and $\bar{M}$ in the two-dimensional Brillouin zone are equivalent, they differ only by a reciprocal lattice vector. Translational symmetry demands that the phonon dispersion relations along $\bar{\Gamma}\bar{M}$ are symmetric with respect to the point $\bar{M}'_2$ on the Brillouin zone boundary. The surface phonon energies measured with helium scattering on p(2×2)O/Pt(111) along $\bar{\Gamma}\bar{M}$ are shown in Fig.11.9 [11.42,40]. Three phonon branches are visible; they describe vibrations of platinum substrate atoms, as can be demonstrated with the same argument as in the case of the (1×1)O/Pt(111) surface. The low energy branch, symmetric about $\bar{M}'_2$, is interpreted as a Rayleigh mode. Both high energy branches are optical surface phonon modes, one of them being nearly dispersionless. The Rayleigh mode and the optical phonon branch with clear dispersion form a gap at $\bar{M}'_2$. The optical branch is explained as originating from the Rayleigh mode of the clean platinum surface. It was not possible to explain the measured band gap (≈ 0.85 meV) in a theoretical model by distinguishing between platinum atoms at the surface loaded with oxygen and not loaded, because in this case the calculated band gap was too small compared with the measurements [11.40]. The expected symmetry of the optical phonon branches with respect to $\bar{M}'_2$ at the Brillouin zone is not clearly resolved in the experiment. Calculations by *He* and *Rahman* showed three optical phonon branches, two of them being dispersionless with phonon energies at 9.5 meV and

10 meV [11.42]. With the assumption that the modes are visible alternatives in helium scattering experiments depending on the reciprocal lattice vectors involved, the observation can be explained. The theoretical study by *He* and *Rahman* on the dynamical properties of the p(2×2)O/Pt(111) surface is based on a model which retains the values of the force constants between the platinum atoms from the clean surface. Additional forces between platinum atoms at the surface are introduced by oxygen adsorbate atoms by means of binding forces between an oxygen atom and three platinum atoms when oxygen is assumed to sit above threefold hollow sites of the closed packed platinum surface. Platinum–oxygen force constants have been choosen to reproduce the measured 60 meV vertical vibration energy between the oxygen and the platinum atoms [11.37].

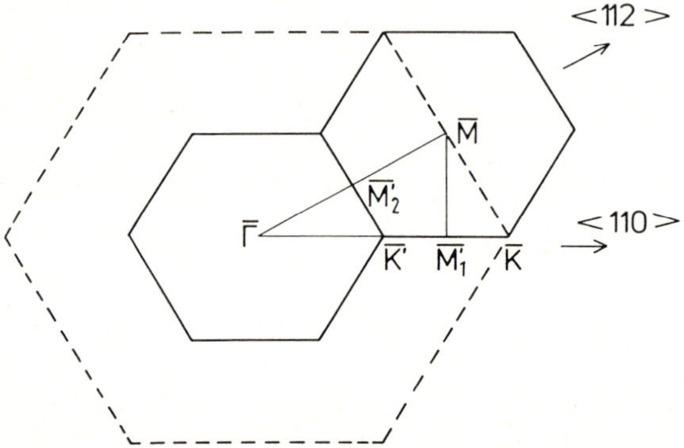

Fig.11.8. Brillouin zones of the p(2×2)O/Pt(111) system (solid lines). The two-dimensional Brillouin zone of the clean Pt(111) surface is represented with dashed lines [11.41]

Surface phonon dispersion relations for p(2×2)O/Pt(111) measured along the $\langle 110 \rangle$ direction is shown in Fig.11.10 [11.41]. Similar to the dispersion relations along the $\bar{\Gamma}\bar{M}$ direction in reciprocal space, three surface phonon modes are visible between $\bar{\Gamma}$ and $\bar{K}$, if the three phonons observed with energy higher than 10 meV are attributed to a phonon branch. Again surface vibrations of the platinum substrate are observed corresponding to the argument already mentioned. The measured lower branch is interpreted as a Rayleigh mode while the two upper branches are optical surface phonon modes. The energy of the Rayleigh phonons increases from $\bar{\Gamma}$ to $\bar{K}'$, and no difference from the measured Rayleigh phonon dispersion relation on the clean Pt(111) surface in the same azimuth is found [11.30]. As expected from the symmetry properties of the p(2×2)O/Pt(111) surface, the Rayleigh and the optical phonon branches from $\bar{K}'$ to $\bar{M}_1'$ are mirror symmetric to the branches from $\bar{M}_1'$ to $\bar{K}$ with the mirror plane through $\bar{M}$ and (Fig.11.8). The measured Rayleigh phonon energy at $\bar{M}_1'$ fits the energy of the Rayleigh phonon at the point $\bar{M}_2'$ as mentioned before [11.40]. The points $\bar{M}_1'$ and $\bar{M}_2'$ are equivalent in the two-dimensional repeated Brillouin zone. The energy gap of ≈ 0.85 meV between Rayleigh curve and opti-

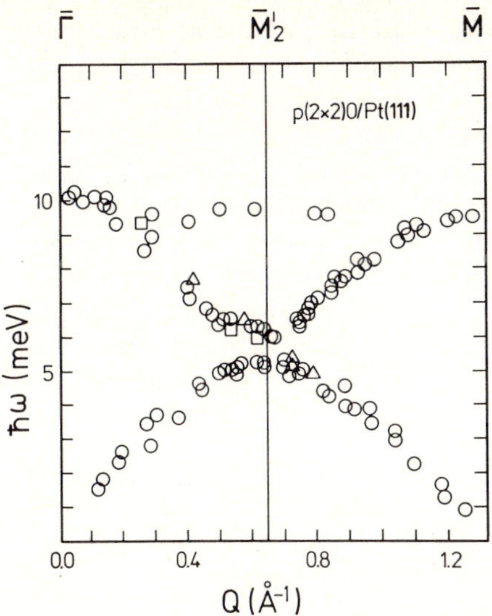

Fig.11.9. Experimental data points obtained in He scattering measurements along the $\bar{\Gamma}\bar{M}$ direction for the p(2×2) oxygen overlayer on Pt(111)[11.40,42]. Sample temperature 300K. The Brillouin zones are shown in Fig.11.8

cal phonon branch measured along $\bar{\Gamma}\bar{M}$ at $\bar{M}'_2$ was not resolved in the measurements along $\bar{\Gamma}\bar{K}$ at $\bar{M}'_1$, and the optical phonon energy at $\bar{M}'_1$, is approximately 1.7 meV lower than the energy of a phonon at $\bar{M}'_2$ achieved by an extrapolation of the measured optical phonons along $\bar{\Gamma}\bar{M}$ mentioned above. The large phonon energy difference can be explained as originating from different optical phonon modes. On the clean platinum surface, the Rayleigh phonon energy measured at $\bar{M}'_2$, which is 0.65 Å^{-1} along the $\bar{\Gamma}\bar{M}$ line, is only 0.5 meV lower than the energy of the Rayleigh phonon measured in the same distance from the $\bar{\Gamma}$ point at 0.65 Å^{-1} in the $\bar{\Gamma}\bar{K}$ direction [11.29,30]. This is completely different from the p(2×2)O/Pt(111) system, where the Rayleigh phonon energy at $\bar{M}'_1$ is 1.8 meV lower than the phonon energy at 0.65 Å^{-1} in the $\bar{\Gamma}\bar{K}$ direction ($\bar{M}'_1$ and $\bar{M}'_2$ are equivalent). This difference is more than three times the value that was observed on the clean Pt(111) surface. This points to a markedly increased anisotropy in the Rayleigh phonon dispersion relation of the p(2×2)O/Pt(111) system compared with the dispersion relation of the clean Pt(111) surface. The observed increase corresponds to larger oxygen-induced anisotropy in the elastic properties of the (111) surface of the platinum substrate.

Unfortunately, calculated surface phonon dispersion relations along the $\bar{\Gamma}\bar{K}$ direction of the p(2×2)O/Pt(111) system are not available at present, so more qualitative explanations, similar to the arguments applied to the (1×1)O/Pt(111) system, will be used. In this case, in contrast to the model developed by *He* et al. to describe the dynamical properties of the p(2×2)O/Pt (111) system, binding forces between the platinum atoms at the surface are diminished by an s,p-electron density reduction at the surface due to chemisorbed oxygen adsorbate. Reduction in binding forces at the surface can lead to a re-arrangement of the surface platinum atoms [11.43]. Analysis of EELS spectra from the p(2×2)O/Pt(111) surface shows oxygen adsorbed in threefold hollow sites on Pt(111) [11.37,44]. One probable rearrangement of the substrate atoms which would not affect the translational symmetry of the p(2×2) superstructure on Pt(111) is a small, symmetric, in-plane expansion of the platinum threefold hollow sites occupied by oxygen atoms, arising from the reduction of binding forces.

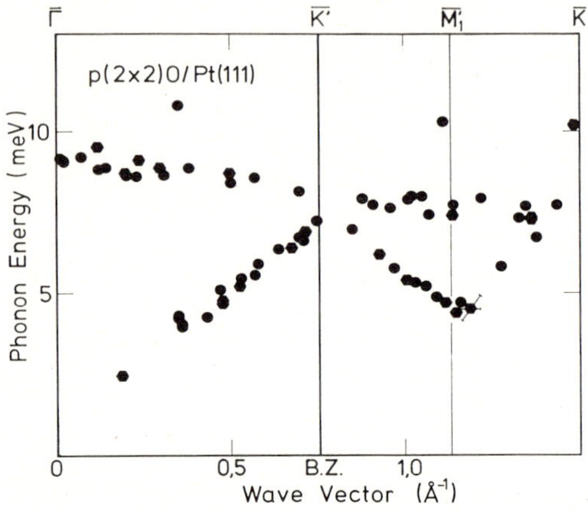

Fig.11.10. Extended-zone plot of the surface phonon dispersion relations measured with helium scattering on the p(2×2)O/Pt(111) system at 420K sample temperature. Solid circles and hexagons correspond to measurements on different crystals. The lower branches are Rayleigh phonon modes, the upper branches are explained as optical phonon modes [11.41]. The Brillouin zones are shown in Fig.11.8

In Fig.11.11 the nearest-neighbour distribution around the platinum surface atoms, not directly affected by the oxygen atoms of the p(2×2) superstructure, is shown. The platinum atoms not directly affected also form a p(2×2) superstructure. Open and solid arrows in Fig.11.10 indicate the direction of the displacement of platinum atoms in the first layer for oxygen adsorbates located at fcc sites or hcp sites, respectively. The strongest increase in anisotropy of the nearest-neighbour distribution is observed for oxygen adsorbed at hcp sites. In this case, the arrangement of the nearest-neighbour distribution confined to atoms of the second layer is adopted by the originally sixfold symmetry of the nearest-neighbour atoms in the first layer, which results in a marked threefold symmetry of the form of the nearest-neighbour atoms parallel to the surface. It was shown that nearest-neighbour interactions are decisive in the description of the dynamical properties of the Pt(111) surface by a semi-empirical force constant parametrization [11.28]. So it is to be expected that the strongest in-

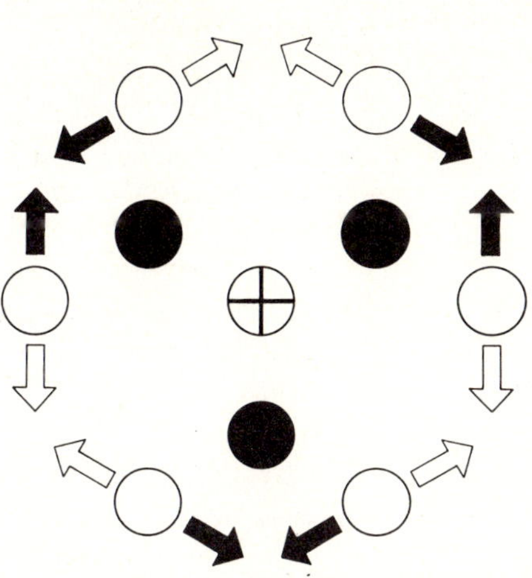

Fig.11.11. Nearest-neighbour distribution of a platinum atom in the first layer, described by an open circle with cross, which is not directly affected by oxygen atoms of the p(2×2) superstructure. Open circles describe platinum atoms in the first layer. Platinum atoms in the second layer are represented by solid circles. Displacement of atoms in the first layer with oxygen atoms adsorbed at "fcc sites" is described by open arrows. Solid arrows indicate the displacement for oxygen atoms adsorbed at "hcp sites" [11.41]

crease in the asymmetry of the elastic properties of the Pt(111) surface occurs for oxygen adsorbed at the hcp sites in the p(2×2)O/Pt(111) system [11.41,49].

11.3 Silver and Platinum with Physisorbed Adsorbates

Besides clean metal surfaces and metal surfaces with chemisorbed adsorbates, surfaces covered with physisorbed adsorbates represent a further system studied with the helium scattering technique. Normal to the surface, the physisorption potential is rather flat and broad and restoring forces are weak compared to forces acting in chemisorption systems. Adsorbate–substrate vibrations of physisorbed atoms are therefore low-frequency modes. Compared to chemisorbed adsorbates, scattering cross sections are increased for the incident helium atoms interacting with single phonon events [11.9]. (The vibrational frequencies of chemisorbed adsorbates are approximately one order of magnitude higher than the frequencies observed for physisorbed adsorbates.) The surface-phonon dispersion relation for krypton films on Ag(111) was measured with helium scattering by *Gibson* and *Sibener* [11.45]. Helium diffraction shows that krypton forms an azimuthally aligned but translationally incommensurate close packed lattice [Kr(111)] on the (111) surface of silver. The lattice constant for a krypton monolayer is slightly larger than that for bulk krypton ($\approx 1\%$) and approaches the bulk value for krypton in two- and three-layer films [11.45]. In Fig.11.12, measured dispersion relations of krypton overlayers on Ag(111) along $\bar{\Gamma}\bar{M}$ at 25K sample temperature are shown. The monolayer phonon branch is dispersionless to within the experimental error and is attributed to uncorrelated vibration of krypton atoms in the holding potential of the silver substrate (Einstein mode). Dynamical properties of the adsorbate are affected by substrate vibration modes, apparent from the increased phonon energy linewidth (decrease in phonon lifetime) as the Einstein mode overlaps substrate vibrational modes [11.45].

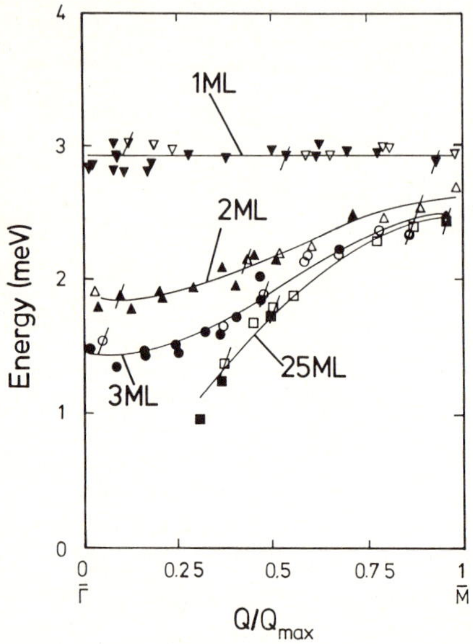

Fig.11.12. Surface phonon dispersion curves of the Kr overlayers on Ag(111) measured with helium scattering from $\bar{\Gamma}$ to $\bar{M}$. Open symbols are for energy loss of the beam, and filled symbols are for energy gain. Representative error bars are shown for energy and momentum loss features (with respect to the incident beam) which occurred in the first zone, i.e., for normal processes. Solid lines are polynomial fits to the data [11.45]. Dispersion curves from 1ML, 2ML, 3ML, and 25ML krypton adsorbate are shown (ML: monolayer). Sample temperature 25K

If the number of adsorbate monolayers on Ag(111) is increased, dispersion in the krypton adsorbate phonon energies appears. After ≈ 25 adsorbed monolayers the Rayleigh phonon dispersion relation of the (111) surface of a krypton crystal is obtained [11.45]. Vibrations from the substrate surface could not be observed. Obviously in this case vibrations from substrate atoms are effectively screened by krypton overlayers from interaction with incident helium atoms.

Dispersion relations similar to those for the system Kr/Ag(111) were obtained for xenon adsorbates on Pt(111) by *Kern* et al. [11.46]. The structure of a monolayer of xenon on Pt(111) at 25K sample temperature can be derived from the commensurate $\left(\sqrt{3} \times \sqrt{3}\right)$ R30° overlayer by ±3.3° rotation from R30°. Helium diffraction shows a nearly equal number of domains of R33.3° and R26.7°, with domain size of about 300 Å [11.46]. Figure 11.13 shows the surface phonon

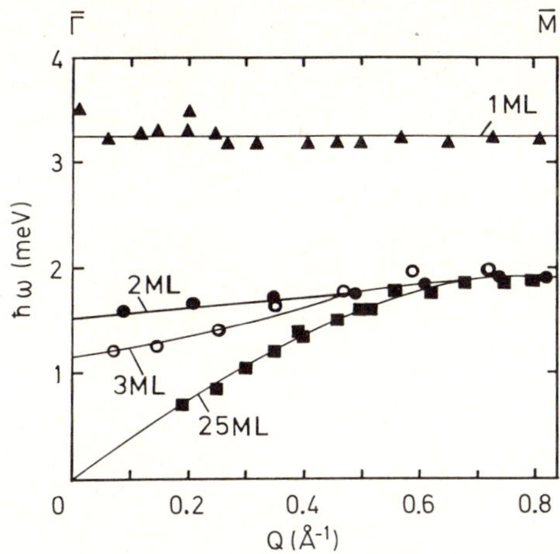

Fig.11.13. Surface phonon dispersion curves measured with helium scattering from xenon films on Pt(111) along the $\bar{\Gamma}\bar{K}$ azimuth [11.46]. Sample temperature 25K. Dispersion curves from 1ML, 2ML, 3ML, and 25ML xenon films on Pt(111) are shown (ML: monolayer)

dispersion relations for xenon films on Pt(111) at 25K sample temperature measured with helium scattering. Again, the phonon branch related to a single monolayer of physisorbed xenon on Pt(111) shows no dispersion and is interpreted as an Einstein mode. As additional xenon films are adsorbed at Pt(111) the dispersion of surface phonon modes increases, until after ≈ 25 adsorbed monolayers the surface phonon dispersion relation equivalent to the Rayleigh dispersion curve from a xenon crystal surface is measured [11.45,46]. In contrast to the systems Kr on Ag(111) and Xe on Pt(111), physisorbed krypton films on Pt(111) show strong adsorbate–substrate vibrational coupling [11.47]. On the clean Pt(111) substrate, helium diffraction indicates that krypton forms an incommensurate close-packed overlayer rotated by 30° with respect to the symmetry axes of the substrate [11.47]. In Fig.11.14, surface phonon dispersion relations of different numbers of krypton monolayers on Pt(111) are shown, measured with the helium scattering along the $\bar{\Gamma}\bar{K}$ direction. The dispersion relations are interpreted as describing the vibrational properties of physisorbed krypton atoms, which show a strong interaction with platinum substrate vibration modes in the region of the Brillouin zone where they overlap the platinum Rayleigh phonon dispersion curve. The interaction is visible even for three-layer krypton films on Pt(111). If ≈ 25 krypton monolayers are adsorbed, they form an adsorbate with the dynamical properties of a krypton crystal surface.

Adsorbate–substrate vibrational coupling leading to a strong hybridization between the monolayer mode and the Rayleigh wave, where the two dispersion curves cross, was predicted by *Hall* et al. in a theoretical study for physisorbed Ar, Kr and Xe overlayers on Ag(111) [11.48]. Phonon energy linewidth broadening, a consequence of the lifetime shortening induced by radiative damping

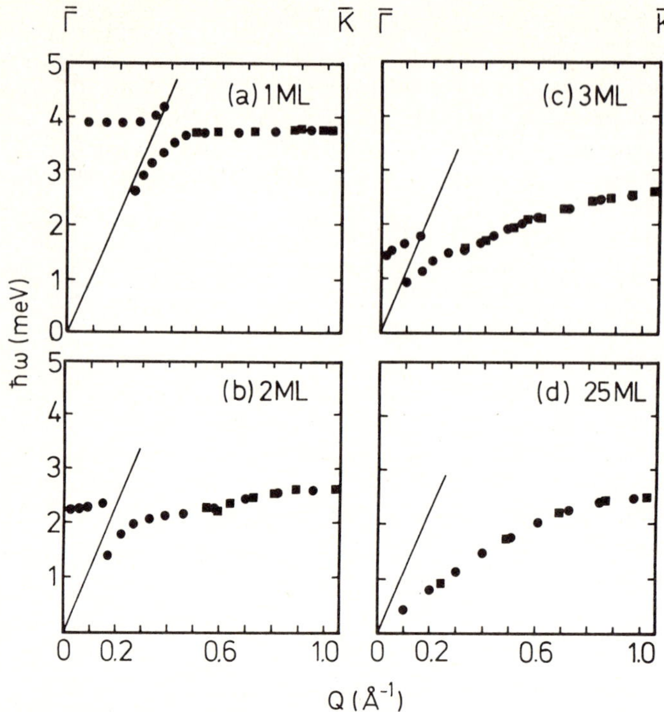

Fig.11.14. Experimental dispersion curves of single phonon processes of Kr films (a)1ML, (b) 2ML, (c) 3ML, and (d) 25ML on Pt(111) measured with helium scattering along the $(\bar{\Gamma}\bar{K})_{\mathrm{Kr}}$ azimuth. Sample temperature 25K. The circles and squares represent phonon creation and annihilation events, respectively. The solid line is the clean Pt(111) Rayleigh phonon dispersion curve [11.47]

due to the coupling to substrate bulk phonons in the region where the adsorbate dispersion relation overlaps the projected substrate bulk phonon dispersion relation was measured on krypton films adsorbed on Ag(111) and Pt(111). However, hybridization between adsorbate and substrate vibrational modes was observed only on krypton films on Pt(111). A real explanation of this interesting finding is, as far as the author knows, not yet available.

11.4 Summary and Conclusions

The (111) surfaces of the noble metals copper and silver and the $5d$ transition metal platinum show similar dynamical properties as described by their surface phonon dispersion relations. Besides the Rayleigh mode an additional high energy surface phonon mode was observed. The additional mode, attributed to a longitudinal resonance, was explained by a clear reduction of binding forces between atoms in the first layer. Compared to copper, silver, and platinum the (111) surface of gold is reconstructed which indicates dramatic changes in the binding forces at this surface.

Changes in binding forces between atoms in the Pt(111) surface could be introduced by chemisorbed adsorbates acting as electron donors or electron acceptors for the platinum surface. In the case of chemisorbed hydrogen (electron donor) the longitudinal resonance vanished on (1×1)H/Pt(111). Chemisorbed oxygen (electron acceptor) as a (1×1) overlayer on Pt(111) reduces the phonon energy of both the Rayleigh mode and the longitudinal resonance. On the platinum (111) surface with p(2×2) chemisorbed oxygen, an increased anisotropy in the Rayleigh phonon dispersion curves was observed.

Surface phonon dispersion relations from Ag(111) and Pt(111) with physisorbed krypton and xenon were measured as dispersionless Einstein modes, with the exception of physisorbed krypton on Pt(111) where a strong interaction (hybridization) between adsorbate and substrate vibration was observed. It is interesting to note that platinum, with an anomaly (Kohn anomaly) in the surface phonon dispersion relation of the (111) surface, showed this hybridization, not measured from physisorbed krypton films on Ag(111).

Several theoretical explanations suitable for parts of the experimental results were reported, but a comprehensive theoretical description covering the observed dynamical properties of "simple" surfaces with "simple" adsorbates has not yet been achieved.

Acknowledgement.

The author thanks B. Feuerbacher for many helpful discussions and a critical reading of the manuscript.

References

11.1 E. Allen, G.P. Alldrege, F. de Wette: Phys. Rev. B **4**, 1648 (1971)

11.2 G. Benedek: Surf. Sci. **61**, 603 (1976)

11.3 C. Stuhlmann, H. Ibach: Surf. Sci. **219**, 117 (1989)

11.4 V. Bortolani, A. Franchini, F. Garcia, F. Nizzoli, G. Santoro: Phys. Rev. B **28**, 7358 (1983)

11.5 A. Kantrowitz, J. Grey: Rev. Sci. Instrum. **22**, 328 (1951)

11.6 E.W. Becker, K. Bier: Z. Naturforschung **9a**, 975 (1954)

11.7 J.P. Toennies: J. Vac. Sci. Technol. A **2**, 1055 (1984)

11.8 G. Benedek: Phys. Rev. Lett. **35**, 234 (1975)

11.9 V. Bortolani, A. Franchini, F. Nizzoli, G. Santoro: Phys. Rev. Lett. **52**, 429 (1984)

11.10 N. Cabrera, V. Celli, F.O. Goodman, R. Manson: Surf. Sci. **19**, 67 (1970)

11.11 N. Esbjerg, J.K. Norskov: Phys. Rev. Lett. **45**, 807 (1980)

11.12 M.W. Finnis, V. Heine: J. Phys. F **4**, L37 (1974)

11.13 B. Poelsema, L.K. Verheij, G. Comsa: Phys. Rev. Lett. **51**, 2410 (1983)

11.14 R. Doak, U. Harten, J.P. Toennies: Phys. Rev. Lett. **51**, 578 (1983)

11.15 V. Bortolani, G. Santoro: Proceedings of the second international conference on phonon physics., August 26– 27 (1985), Budapest p.566

11.16 V. Bortolani, A. Franchini, F. Nizzoli, G. Santoro: *Dynamics of Gas-Surface Interaction*, ed. by G. Benedek, V. Valbusa, Springer Ser. Chem. Phys. Vol. 21 (Springer, Berlin, Heidelberg 1982) p.196

11.17 M. Born, K. Huang: *Dynamical theory of crystal lattices.* (Clarendon, Oxford 1954)

11.18 U. Harten, J.P. Toennies, C. Wöll: Faraday Discuss. Chem. Soc. **80**, 137 (1985)

11.19 U. Harten, A.M. Lahee, J.P. Toennies, Ch. Wöll: Phys. Rev. Lett. **54**, 2619 (1985)

11.20 V. Bortolani, G. Santoro, U. Harten, J.P. Toennies: Surf. Sci. **148**, 82 (1984)

11.21 G.W. Farnell: *Physical Acoustics* Vol.VI, ed. by W.P. Mason, R.N. Thurston (Academic, New York 1970)

11.22 D.E. Beck, V. Celli, G. Vecchio, A. Magnaterra: Il Nuovo Cimento **68B**, 230 (1970)

11.23 C.S. Jayanthi, H. Bilz, W. Kress, G. Benedek: Phys. Rev. Lett. **59**, 795 (1987)

11.24 D.H. Dutton, B.N. Brockhouse, A.P. Miller: Can. J. Phys. **50**, 2915 (1972)

11.25 L.M. Roth, H.J. Zeiger, T.A. Kaplan: Phys. Rev. **149**, 519 (1966)

11.26 A.J. Freeman, T.J. Watson-Yang, J. Rath: J. Magn. Mat. **12**, 140 (1979)

11.27 W. Kohn: Phys. Rev. Lett. **2**, 393 (1959)

11.28 V. Bortolani, A. Franchini, G. Santoro, J.P. Toennies, C. Wöll, G. Zhang: Phys. Rev. B **40**, 3524 (1989)

11.29 D. Neuhaus, F. Joo, B. Feuerbacher: Sci. **165**, L90 (1986)

11.30 U. Harten, J.P. Toennies, C. Wöll, G. Zhang: Phys. Rev. Lett. **55**, 2308 (1985)

11.31 K. Kern, R. David, L. Palmer, G. Comsa, T.S. Rahman: Phys. Rev. B **33**, 4334 (1986)

11.32 P.R. Norton, J.W. Goodale: Solid State Commun. **31**, 223 (1979)

11.33 J. Lee, J.P. Cowin, L. Wharton: Surf. Sci. **130**, 1 (1983)

11.34 G.N. Derry, P.N. Ross: J. Chem. Phys. **82**, 2772 (1985)

11.35 D. Neuhaus, F. Joo, B. Feuerbacher: Phys. Rev. Lett. **58**, 694 (1987)

11.36 G.P. Alldrege, R.E. Allen, F.W. de Wette: Phys. Rev. B **6**, 1682 (1971)

11.37 H. Steininger, S. Lehwald, H. Ibach: Surf. Sci. **123**, 1 (1982)

11.38 V. Heine, L.D. Marks: Surf. Sci. **165**, 65 (1986)

11.39 D.M. Collins, J.B. Lee, W.E. Spicer: Phys. Rev. Lett. **35**, 592 (1975)

11.40 K. Kern, R. David, R.L. Palmer, G. Comsa, J. He, T.S. Rahman: Phys. Rev. Lett. **56**, 2064 (1986)

11.41 D. Neuhaus: Phys. Rev. B **41**, 3397 (1990)

11.42 J. He, T.S. Rahman: Phys. Rev. B **34**, 5017 (1986)

11.43 K. Griffiths, T.E. Jackman, J.A. Davies, P.R. Norton: Surf. Sci. **138**, 113 (1984)

11.44 S. Lehwald, H. Ibach, H. Steininger: Surf. Sci. **117**, 342 (1982)

11.45 K.D. Gibson, S.J. Sibener: Phys. Rev. Lett. **55**, 1514 (1985)

11.46 K. Kern, R. David, R.L. Palmer, G. Comsa: Phys. Rev. Lett. **56**, 2823 (1986)

11.47 K. Kern, P. Zeppenfeld, R. David, G. Comsa: Phys. Rev. B **35**, 886 (1987)

11.48 B. Hall, D.L. Mills, J.E. Black: Phys. Rev. B. **32**, 4932 (1985)

11.49 F. Besenbacher and coworkers (Surf. Sci. **220** L701 (1989)) reported oxygen adsorbed at the fcc site in the p(2×2)O/Pt(111) system measured with transmission channeling technique, opposite to the interpretation of the measured strong anisotropy in the Rayleigh phonon dispersion relation

12. Quasielastic Helium Scattering Studies of Adatom Diffusion on Surfaces

J.W.M. Frenken and B.J. Hinch

Two-dimensional diffusion processes at surfaces can be studied on an atomic scale with quasielastic scattering of low-energy He atoms. In the last two decades He-atom scattering has matured into a powerful tool in the study of surfaces. Crucial contributions to both the development of the instrumentation for He-atom scattering and the exploration and interpretation of an extensive variety of surface phenomena which are accessible with this technique have come from Prof. J.P. Toennies and coworkers at the Max-Planck-Institut für Strömungsforschung in Göttingen (FRG).

In this chapter we describe a relatively new application of He-atom scattering. At present the energy resolution which can be attained by combining an optimally designed nozzle–skimmer source of He atoms with time-of-flight detection is just sufficient to measure the quasielastic scattering of He atoms from adatoms diffusing on a surface. The quasielastic scattering of neutrons is firmly established as an experimental probe of diffusive motion in bulk liquids and solids on an atomic scale [12.1, 2]. It is used to measure diffusion constants and to obtain detailed information the microscopic diffusion processes and mechanisms. Here we show that the same type of microscopic information on diffusion can be obtained for adatoms on surfaces with low-energy He-atom beams.

Other techniques to measure surface diffusion are either limited to extremely low diffusive mobilities (e.g. field-ion microscopy and field-emission fluctuation correlations) or provide only macroscopic information (e.g. mass-transfer and tracer-diffusion methods). An excellent overview of surface diffusion of adsorbates and the employed experimental techniques has been given by *Gomer* [12.3]. Recently, quasielastic neutron scattering has been used to investigate surface diffusion of adsorbed methane films on powders of MgO and graphite [12.4–6]. Like neutrons, He atoms are well suited for measurements of diffusive motion on a microscopic scale in systems with a large diffusion constant. Low-energy He beams have the additional advantage of extreme sensitivity to the surface and in particular to adatoms or other point defects.

287

The basic theory of quasielastic atom scattering is explained in the following section. After this we describe briefly the experimental procedure which we have adopted in the measurement and analysis of quasielastic atom-scattering data. So far we have applied the method to two systems with highly mobile adatoms, namely S adatoms on a Cu(111) substrate (heterodiffusion) and Pb adatoms on a melting Pb(110) surface (self-diffusion). The experimental results for these two systems are presented together with their interpretation in terms of the theory. We conclude this chapter with a short discussion of the strengths and limitations of this new technique.

12.1 Theory of Quasielastic Atom Scattering

12.1.1 Scattering Formalism

When a beam of He atoms is elastically scattered from a surface with diffusing adatoms, the energy distribution of diffusely scattered He atoms is, in fact, weakly inelastic. The broadening of the reflected energy distribution with respect to the incident energy distribution is brought about by small energy transfers, which are related to the diffusive motion of the adatoms. In this section we will derive the energy and intensity distribution of low-energy atoms quasielastically scattered from a flat surface, with area A, covered with a low concentration $\rho = N/A$ of laterally diffusing identical adatoms. We will closely follow a derivation of quasielastic atom scattering by *Levi* et al. [12.7] and apply it to several simple models of surface diffusion. The principle of quasielastic atom scattering can be formulated in analogy with the theory of quasielastic neutron scattering, which was first presented by *Van Hove* [12.8]. A few fundamental differences between atom and neutron scattering somewhat complicate the atom-scattering case, as we shall point out later in this section.

We model the surface by a hard wall. The time-dependent shape function $\zeta(\boldsymbol{R}, t)$ gives the local height of the surface at lateral position $\boldsymbol{R}$ and at time t. Within the eikonal approximation the differential scattering probability from this surface is [12.7]

$$
\begin{aligned}
\frac{\partial^2 P}{\partial \Omega \partial \omega} &= \frac{q(\boldsymbol{q}_i \cdot \boldsymbol{k})}{8\pi^3 A \mid q_{iz} \mid k_z^2} \int_{-\infty}^{+\infty} \exp(-\mathrm{i}\omega t) \int_A \int_A \exp[\mathrm{i}\boldsymbol{K} \cdot (\boldsymbol{R}^2 - \boldsymbol{R}^1)] \\
&\quad \times \langle \exp[-\mathrm{i}k_z \zeta(\boldsymbol{R}^1, 0)] \ \exp[\mathrm{i}k_z \zeta(\boldsymbol{R}^2, t)] \rangle \mathrm{d}t \mathrm{d}\boldsymbol{R}^1 \mathrm{d}\boldsymbol{R}^2 \\
&= \frac{q(\boldsymbol{q}_i \cdot \boldsymbol{k})}{8\pi^3 A \mid q_{iz} \mid k_z^2} \int_{-\infty}^{+\infty} \exp(-\mathrm{i}\omega t) J(\boldsymbol{k}, t) \mathrm{d}t \quad .
\end{aligned}
$$

$$(12.1)$$

Here $\hbar \boldsymbol{q}_i$ is the momentum of the incident He atom and $\hbar\boldsymbol{k} = \hbar(\boldsymbol{K}, k_z)$ and $\hbar\omega$ are the momentum and energy transfer upon scattering. The integration, $J(\boldsymbol{k}, t)$, is over the lateral positions $\boldsymbol{R}^1$ and $\boldsymbol{R}^2$ which run over the (large)

288

surface area A. The notation $\langle ... \rangle$ stands for the thermal average of the operator enclosed in the brackets. We model the shape function as follows.

$$\zeta(\boldsymbol{R}, t) = \sum_{n=1}^{N} Z[\boldsymbol{R} - \boldsymbol{R}_n(t)] \quad , \tag{12.2}$$

with $Z(\boldsymbol{R})$ describing the shape of an adatom and $\boldsymbol{R}_n(t)$ being the position of adatom n at time t. In the following we will assume that the profiles of the adatoms do not overlap, i.e. ζ is either zero or equal to the Z of the nearest adatom. This assumption is of course valid only for low adatom concentrations. The approximation of a *flat* substrate seems appropriate for weakly corrugated (e.g. metal) surfaces. Let us first perform the integrations over $\boldsymbol{R}^1$ and $\boldsymbol{R}^2$ in (12.1).

$$
\begin{aligned}
J(\boldsymbol{k}, t) = {} & \int_A\!\!\int_A \exp[\mathrm{i}\boldsymbol{K} \cdot (\boldsymbol{R}^2 - \boldsymbol{R}^1)] \\
& \times \langle \exp[-\mathrm{i}k_z\zeta(\boldsymbol{R}^1, 0)] \exp[\mathrm{i}k_z\zeta(\boldsymbol{R}^2, t)]\rangle \mathrm{d}\boldsymbol{R}^1 \mathrm{d}\boldsymbol{R}^2 \\
= {} & \int_A\!\!\int_A \exp[\mathrm{i}\boldsymbol{K} \cdot (\boldsymbol{R}^2 - \boldsymbol{R}^1)]\mathrm{d}\boldsymbol{R}^1 \mathrm{d}\boldsymbol{R}^2 \\
& + \langle \int_a\!\!\int_A \exp[\mathrm{i}\boldsymbol{K} \cdot (\boldsymbol{R}^2 - \boldsymbol{R}^1)]\{\exp[-\mathrm{i}k_z\zeta(\boldsymbol{R}^1, 0)] - 1\}\mathrm{d}\boldsymbol{R}^1 \mathrm{d}\boldsymbol{R}^2\rangle \\
& + \langle \int_A\!\!\int_a \exp[\mathrm{i}\boldsymbol{K} \cdot (\boldsymbol{R}^2 - \boldsymbol{R}^1)]\{\exp[\mathrm{i}k_z\zeta(\boldsymbol{R}^2, t)] - 1\}\mathrm{d}\boldsymbol{R}^1 \mathrm{d}\boldsymbol{R}^2\rangle \\
& + \langle \int_a\!\!\int_a \exp[\mathrm{i}\boldsymbol{K} \cdot (\boldsymbol{R}^2 - \boldsymbol{R}^1)]\{\exp[-\mathrm{i}k_z\zeta(\boldsymbol{R}^1, 0)] - 1\} \\
& \times \{\exp[\mathrm{i}k_z\zeta(\boldsymbol{R}^2, t)] - 1\}\mathrm{d}\boldsymbol{R}^1 \mathrm{d}\boldsymbol{R}^2\rangle \quad ,
\end{aligned}
\tag{12.3}
$$

where a stands for the surface region covered by adatoms, i.e. with nonzero ζ. Using the identities $\int_a \mathrm{d}\boldsymbol{R} \equiv Na_{\mathrm{at}}$ (where the area with non-zero height around each adatom is denoted by a_{at}), $\int_A \exp(\mathrm{i}\boldsymbol{K} \cdot \boldsymbol{R})\mathrm{d}\boldsymbol{R} \equiv 4\pi^2\delta(\boldsymbol{K})$ (for sufficiently large A) and defining the adatom form factor $F(\boldsymbol{k})$ by

$$F(\boldsymbol{k}) = F(\boldsymbol{K}, k_z) \equiv \int_{a_{\mathrm{at}}} \exp[\,\mathrm{i}\boldsymbol{K} \cdot \boldsymbol{R} + \mathrm{i}k_z Z(\boldsymbol{R})]\mathrm{d}\boldsymbol{R} \quad , \tag{12.4}$$

the integration in (12.3) can be carried out.

$$J(\boldsymbol{k},t) = 4\pi^2 A\delta(\boldsymbol{K})\{1 + \rho[F(0,k_z) + F^*(0,k_z) - 2a_{\mathrm{at}}]$$
$$+ A\rho\,|\,F(\boldsymbol{K},k_z) - F(\boldsymbol{K},0)\,|^2$$
$$\times \frac{1}{N}\langle \sum_{i,j=1}^{N} \exp[-\mathrm{i}\boldsymbol{K}\cdot\boldsymbol{R}_i(0)]\exp[\mathrm{i}\boldsymbol{K}\cdot\boldsymbol{R}_j(t)]\rangle \quad .$$

$$(12.5)$$

The differential scattering probability becomes [12.7]

$$\frac{\partial^2 P}{\partial\Omega\partial\omega} = \frac{q(\boldsymbol{q}_i\cdot\boldsymbol{k})}{|\,q_{iz}\,|\,k_z^2}\delta(\boldsymbol{K})\delta(\omega)\{1 + \rho[F(0,k_z)+F^*(0,k_z)-2a_{\mathrm{at}}]\}$$
$$+ \frac{q(\boldsymbol{q}_i\cdot\boldsymbol{k})}{8\pi^3\,|\,q_{iz}\,|\,k_z^2}\,\rho\,|\,F(\boldsymbol{K},k_z) - F(\boldsymbol{K},0)\,|^2 \int_{-\infty}^{\infty}\exp(-\mathrm{i}\omega t)$$
$$\times \frac{1}{N}\langle \sum_{i,j=1}^{N} \exp[-\mathrm{i}\boldsymbol{K}\cdot\boldsymbol{R}_i(0)]\exp[\mathrm{i}\boldsymbol{K}\cdot\boldsymbol{R}_j(t)]\rangle\mathrm{d}t \quad .$$

$$(12.6)$$

We see that apart from a purely elastic specular term [12.9], the scattering contains a contribution which depends on the form factor of the adatom and the scattering function [12.10]

$$S(\boldsymbol{K},\omega) = \frac{1}{2\pi}\int_{-\infty}^{\infty}\exp(-\mathrm{i}\omega t)I(\boldsymbol{K},t)\mathrm{d}t \quad , \qquad (12.7)$$

with the intermediate function $I(\boldsymbol{K},t)$ given by

$$I(\boldsymbol{K},t) = \frac{1}{N}\langle \sum_{i,j=1}^{N} \exp[-\mathrm{i}\boldsymbol{K}\cdot\boldsymbol{R}_i(0)]\exp[\mathrm{i}\boldsymbol{K}\cdot\boldsymbol{R}_j(t)]\rangle \quad . \qquad (12.8)$$

Note that in neutron scattering the form factor is unity as a result of the extremely short range of the interaction potential for neutrons. The scattering function $S(\boldsymbol{K},\omega)$ and the intermediate function $I(\boldsymbol{K},t)$ contain all statistical information on the dynamics (vibrations and diffusion) of the adatoms parallel to the surface. They are completely determined by the correlations in the relative positions of the adatom centers. This part of the scattering problem can be treated analogously to neutron-scattering theory. $I(\boldsymbol{K},t)$ is separated into a *self* and a *distinct* part.

$$I_{\mathrm{s}}(\boldsymbol{K},t) = \frac{1}{N}\langle \sum_{i=1}^{N}\exp[-\mathrm{i}\boldsymbol{K}\cdot\boldsymbol{R}_i(0)]\exp[\mathrm{i}\boldsymbol{K}\cdot\boldsymbol{R}_i(t)]\rangle \quad , \qquad (12.9a)$$

290

and

$$I_d(\boldsymbol{K},t) \;=\; \frac{1}{N}\langle\sum_{i=1}^{N}\sum_{j\neq i}^{N}\exp[-\mathrm{i}\boldsymbol{K}\cdot\boldsymbol{R}_i(0)]\exp[\mathrm{i}\boldsymbol{K}\cdot\boldsymbol{R}_j(t)]\rangle \quad . \tag{12.9b}$$

Following *Van Hove*'s treatment [12.1, 2, 8] we express these quantities in terms of the correlation functions G_s and G_d between pairs of atoms.

$$I_s(\boldsymbol{K},t) \;=\; \int G_s(\boldsymbol{R},t)\exp(\mathrm{i}\boldsymbol{K}\cdot\boldsymbol{R})\mathrm{d}\boldsymbol{R} \quad , \tag{12.10a}$$

$$I_d(\boldsymbol{K},t) \;=\; \int G_d(\boldsymbol{R},t)\exp(\mathrm{i}\boldsymbol{K}\cdot\boldsymbol{R})\mathrm{d}\boldsymbol{R} \quad . \tag{12.10b}$$

Classically, these correlation functions can be interpreted as the ensemble–averaged probability density of finding an atom at location $\boldsymbol{R}$ at time t, given that the same (G_s) or a different (G_d) atom was/will be at the origin at time $t=0$.

$$G_s(\boldsymbol{R},t) \;=\; \frac{1}{N}\langle\sum_{i=1}^{N}\delta[\boldsymbol{R}+\boldsymbol{R}_i(0)-\boldsymbol{R}_i(t)]\rangle \quad , \tag{12.11a}$$

$$G_d(\boldsymbol{R},t) \;=\; \frac{1}{N}\langle\sum_{i=1}^{N}\sum_{j\neq i}^{N}\delta[\boldsymbol{R}+\boldsymbol{R}_i(0)-\boldsymbol{R}_j(t)]\rangle \quad . \tag{12.11b}$$

We will first concentrate on the *self* part, since it describes the motion of individual adatoms and is therefore of primary interest in the characterization of diffusion mechanisms. Because $G_s(\boldsymbol{R},t)$ in fact describes both the (rapid) vibrational and the (slow) diffusive motions, it can be expressed as the convolution in space of the vibrational correlation function, G_s^V, with the diffusional correlation function, G_s^D [12.1].

$$G_s(\boldsymbol{R},t) \;=\; \int G_s^V(\boldsymbol{R}',t)G_s^D(\boldsymbol{R}-\boldsymbol{R}',t)\mathrm{d}\boldsymbol{R}' \quad . \tag{12.12}$$

Equation (12.12) assumes the vibrational motion of the scatterers to be uncorrelated with the diffusive motion, although it is known that vibrations may actually assist the diffusion process [12.11, 12]. From (12.12) it follows that

$$I_s(\boldsymbol{K},t) \;=\; I_s^V(\boldsymbol{K},t)I_s^D(\boldsymbol{K},t) \quad , \tag{12.13}$$

where I_s^V and I_s^D are the Fourier transforms in space of G_s^V and G_s^D, following (12.10a). This, in turn, leads to

$$S_{\mathrm{s}}(\boldsymbol{K},\omega) \;=\; \int_{-\infty}^{\infty} S_{\mathrm{s}}^{\mathrm{V}}(\boldsymbol{K},\omega')S_{\mathrm{s}}^{\mathrm{D}}(\boldsymbol{K},\omega-\omega')\mathrm{d}\omega' \quad , \tag{12.14}$$

with S_{s}, $S_{\mathrm{s}}^{\mathrm{V}}$ and $S_{\mathrm{s}}^{\mathrm{D}}$ the Fourier transforms in time of I_{s}, $I_{\mathrm{s}}^{\mathrm{V}}$ and $I_{\mathrm{s}}^{\mathrm{D}}$ respectively, according to (12.7). Equation (12.14) shows that the diffusive motion of the adatoms leads to a broadening of the energy distribution of both elastically scattered He atoms (elastic peak) and inelastically scattered He atoms (phonon creation and annihilation peaks). Here, we are interested in the elastic, or rather quasielastic peak. The zero-phonon (elastic) term in $S_{\mathrm{s}}^{\mathrm{V}}$ amounts to $\exp(-K^2\langle u_{\boldsymbol{K}}^2\rangle)\delta(\omega)$, with $\langle u_{\boldsymbol{K}}^2\rangle$ being the mean square vibrational displacement of the adatoms along the direction of $\boldsymbol{K}$. The quasielastic peak in S_{s} is obtained from (12.14) as

$$S_{\mathrm{s}}^{\mathrm{qe}}(\boldsymbol{K},\omega) \;=\; \exp(-K^2\langle u_{\boldsymbol{K}}^2\rangle)S_{\mathrm{s}}^{\mathrm{D}}(\boldsymbol{K},\omega) \quad . \tag{12.15}$$

It may seem surprising that (12.15) contains the Debye-Waller factor associated with only the lateral vibrational motion of the adatoms. However, this is a natural consequence of our simple choice of surface shape function in (12.2), which does not include the motion of the adatoms perpendicular to the surface. Uncorrelated perpendicular vibrational motion of the adatoms can be included in the model, which simply leads to the multiplication of the adatom form factor $F(0,k_z)$, $F^*(0,k_z)$ and $F(\boldsymbol{K},k_z)$, appearing in (12.5), (12.6) and (12.20) with an additional 'perpendicular' Debye-Waller factor of $\exp(-k_z^2\langle u_z^2\rangle/2)$, where $\langle u_z^2\rangle$ is the mean square vibrational displacement of the adatoms perpendicular to the surface [12.13].

In neutron scattering a natural distinction is made between what is called *coherent* scattering, $S(\boldsymbol{k},\omega)$, and *incoherent* scattering, $S_{\mathrm{s}}(\boldsymbol{k},\omega)$. Incoherence in the scattering in case of isotope mixtures or due to spin incoherence can result in a large incoherent cross section and a small coherent one. This is exploited in measurements of diffusion with quasielastic neutron scattering, where one usually studies the $G_{\mathrm{s}}(\boldsymbol{r},t)$ for strongly incoherent scatterers. With He atoms such an experimental insensitivity to the *distinct* correlation function $G_{\mathrm{d}}(\boldsymbol{R},t)$ cannot be obtained, since atom scattering is, in neutron-scattering language, always purely coherent.

In order to describe the *distinct* part of the scattering we employ *Vineyard*'s convolution approximation [12.1, 2, 14] to express the *distinct* correlation function G_{d} in terms of the *self* correlation function G_{s}.

$$G_{\mathrm{d}}(\boldsymbol{R},t) \;=\; \int g(\boldsymbol{R}')G_{\mathrm{s}}(\boldsymbol{R}-\boldsymbol{R}',t)\mathrm{d}\boldsymbol{R}' \quad . \tag{12.16}$$

Here, $g(\boldsymbol{R}) = G_{\mathrm{d}}(\boldsymbol{R},0)$ is the static pair correlation function. From this we obtain

$$
\begin{aligned}
G(\boldsymbol{R},t) &= G_{\mathrm{s}}(\boldsymbol{R},t) + G_{\mathrm{d}}(\boldsymbol{R},t) \\
&= \int [g(\boldsymbol{R}') + \delta(\boldsymbol{R}')] G_{\mathrm{s}}(\boldsymbol{R} - \boldsymbol{R}',t)\mathrm{d}\boldsymbol{R}' \quad,
\end{aligned}
\tag{12.17}
$$

with the intermediate function

$$
\begin{aligned}
I(\boldsymbol{K},t) &= I_{\mathrm{s}}(\boldsymbol{K},t) + I_{\mathrm{d}}(\boldsymbol{K},t) = [1 + I_{\mathrm{d}}(\boldsymbol{K},0)]I_{\mathrm{s}}(\boldsymbol{K},t) \\
&= I(\boldsymbol{K},0)I_{\mathrm{s}}(\boldsymbol{K},t) \quad,
\end{aligned}
\tag{12.18}
$$

and, finally, the scattering function

$$
S(\boldsymbol{K},\omega) = S_{\mathrm{s}}(\boldsymbol{K},\omega) + S_{\mathrm{d}}(\boldsymbol{K},\omega) = S(\boldsymbol{K})S_{\mathrm{s}}(\boldsymbol{K},\omega) \quad.
\tag{12.19}
$$

In (12.19) we have used the static structure factor $S(\boldsymbol{K}) \equiv I(\boldsymbol{K},0)$ of the adatom overlayer. Combining the second term of (12.6) and equations (12.7), (12.15) and (12.19) we obtain the following expression for the differential quasielastic scattering probability

$$
\begin{aligned}
\frac{\partial^2 P^{\mathrm{qe}}}{\partial\Omega\partial\omega} &= \frac{q(\boldsymbol{q}_i \cdot \boldsymbol{k})}{4\pi^2 \mid q_{iz} \mid k_z^2}\, \rho \mid \exp(-k_z^2 \langle u_z^2\rangle/2)F(\boldsymbol{K},k_z) - F(\boldsymbol{K},0) \mid^2 \\
&\quad \times \exp(-K^2\langle u_{\boldsymbol{K}}^2\rangle)S(\boldsymbol{K})S_{\mathrm{s}}^{\mathrm{D}}(\boldsymbol{K},\omega) \quad.
\end{aligned}
\tag{12.20}
$$

As discussed before, the reduction in the quasielastic intensity due to the vibrational motion of the adatoms perpendicular to the surface has been accounted for by multiplication of the form factor $F(\boldsymbol{K},k_z)$ in (12.20) with the 'perpendicular' Debye-Waller factor $\exp(-k_z^2\langle u_z^2\rangle/2)$ [12.12, 15].

The convolution approximation in (12.16) is known to be inaccurate [12.1, 2, 8], since it underestimates the collective aspects of the diffusive motion of nearby particles, at small times t. On the other hand, the *distinct* contribution to (12.20) contains an additional factor ρ with respect to the *self* contribution, and is therefore correspondingly small for low adatom concentrations. In addition, at a sufficiently low adatom concentration the correlations in the diffusion of different adatoms necessarily become unimportant.

Before discussing $S_{\mathrm{s}}^{\mathrm{D}}(\boldsymbol{K},\omega)$ for a selection of diffusion models we derive a simple expression for the static structure factor $S(\boldsymbol{K})$ of a low-density adatom overlayer with a lattice-gas structure. In that case

$$
S(\boldsymbol{K}) = \sum_{\boldsymbol{L}} P_{\boldsymbol{L}}\, \exp(\mathrm{i}\boldsymbol{K}\cdot\boldsymbol{L}) \quad;
\tag{12.21}
$$

$\boldsymbol{L}$ runs over all lattice vectors of the two-dimensional adatom lattice and $P_{\boldsymbol{L}}$ is the conditional probability of finding an occupied site at location $\boldsymbol{L}$ given that the origin is occupied. So $P_{\boldsymbol{L}=0} = 1$. At low adatom coverages, if the adatom-adatom interactions are weak and the adatom positions are not correlated, $P_{\boldsymbol{L}\neq 0} \approx \Theta$, where Θ is the fraction of occupied adatom sites. This results in

$$S(\boldsymbol{K}) = (1 - \Theta) + \Theta \sum_{\boldsymbol{L}} \exp(\mathrm{i}\boldsymbol{K} \cdot \boldsymbol{L})$$

$$= (1 - \Theta) + 4\pi^2 \rho \sum_{\boldsymbol{G}} \delta(\boldsymbol{K} - \boldsymbol{G}) \quad . \tag{12.22}$$

Apart from a set of discrete diffraction peaks at the reciprocal-lattice points $\boldsymbol{G}$, the structure factor is seen to be constant, namely $(1 - \Theta)$. Static-correlation effects would make $S(\boldsymbol{K})$ a slowly varying function of $\boldsymbol{K}$ which is discussed further in Sect.12.1.3. The dependence of $S_{\mathrm{s}}^{\mathrm{D}}(\boldsymbol{K},\omega)$ on ω remains to be interpreted in terms of model diffusion mechanisms.

12.1.2 Diffusion Models

The scattering function $S_{\mathrm{s}}^{\mathrm{D}}(\boldsymbol{K},\omega)$ in (12.20) depends on the particular diffusion model. The simplest model one can adopt is that of two-dimensional isotropic random continuous diffusion (given by *Fick*'s law). In that case the *self* correlation function is given by

$$G_{\mathrm{s}}^{\mathrm{D}}(\boldsymbol{R},t) = \frac{1}{4\pi D \mid t \mid} \exp\left(-\frac{R^2}{4D \mid t \mid}\right) \quad , \tag{12.23}$$

where D is the diffusion coefficient for individual adatoms parallel to the surface. Fourier transforming in space and in time we obtain respectively

$$I_{\mathrm{s}}^{\mathrm{D}}(\boldsymbol{K},t) = \exp(-K^2 D \mid t \mid) \quad , \tag{12.24}$$

and

$$S_{\mathrm{s}}^{\mathrm{D}}(\boldsymbol{K},\omega) = \frac{1}{\pi} \frac{DK^2}{\omega^2 + D^2 K^4} \quad . \tag{12.25}$$

For each $\boldsymbol{K}, S_{\mathrm{s}}^{\mathrm{D}}$ is a Lorentzian energy profile with an area of unity and a full width at half maximum (FWHM) of

$$\Delta E(\boldsymbol{K}) = \hbar \Delta \omega(\boldsymbol{K}) = 2\hbar D K^2 \quad . \tag{12.26}$$

If the substrate is not *flat* but provides the adatoms with a two-dimensional periodic array of binding sites, we may imagine the adatom diffusion to take place in the form of discrete jumps rather than continuously. On short length and time scales this leads to strong deviations of $G_{\mathrm{s}}^{\mathrm{D}}$ from the simple form of (12.23). Following the work of *Chudley* and *Elliott* [12.16] we describe the jump diffusion on a two-dimensional Bravais lattice with a rate equation.

$$\frac{\partial G_{\mathrm{s}}^{\mathrm{D}}(\boldsymbol{R},t)}{\partial t} = \sum_{\boldsymbol{j}} \frac{1}{\tau(\boldsymbol{j})} \left[G_{\mathrm{s}}^{\mathrm{D}}(\boldsymbol{R} + \boldsymbol{j},t) - G_{\mathrm{s}}^{\mathrm{D}}(\boldsymbol{R},t) \right] \quad . \tag{12.27}$$

Here $\tau(\boldsymbol{j})$ is the average time between successive jumps over jump vector $\boldsymbol{j}$ and the summation runs over all lattice vectors. In (12.27) we have tacitly assumed that the actual time involved in a single jump is very short compared with the time τ between successive jumps. The total jump rate is

$$\frac{1}{\tau} = \sum_j \frac{1}{\tau(j)} \quad , \tag{12.28}$$

and $\tau(-j) = \tau(j)$ as a result of the inversion symmetry of the Bravais lattice. The Fourier transform in space of equation (12.27) yields the rate equation for the intermediate function.

$$\begin{aligned}
\frac{\partial I_s^{D}(\boldsymbol{K},t)}{\partial t} &= I_s^{D}(\boldsymbol{K},t) \sum_j \frac{1}{\tau(j)} \left[\exp(-\mathrm{i}\boldsymbol{K} \cdot \boldsymbol{j}) - 1\right] \\
&= -I_s^{D}(\boldsymbol{K},t) \cdot 2 \sum_j \frac{1}{\tau(j)} \sin^2\left(\frac{\boldsymbol{K} \cdot \boldsymbol{j}}{2}\right) \quad ,
\end{aligned} \tag{12.29}$$

with the solution

$$I_s^{D}(\boldsymbol{K},t) = I_s^{D}(\boldsymbol{K},0) \exp[-\frac{1}{2}\Delta\omega(\boldsymbol{K})\,|\,t\,|] \quad , \tag{12.30}$$

where $\Delta\omega(\boldsymbol{K})$ is given by

$$\Delta\omega(\boldsymbol{K}) = 4 \sum_j \frac{1}{\tau(j)} \sin^2\left(\frac{\boldsymbol{K} \cdot \boldsymbol{j}}{2}\right) \quad . \tag{12.31}$$

From the boundary condition $G_s^{D}(\boldsymbol{R},0) = \delta(\boldsymbol{R})$ we know that $I_s^{D}(\boldsymbol{K},0) = 1$. From (12.30) we see that for each $\boldsymbol{K}$ the scattering function is still a Lorentzian energy profile,

$$S_s^{D}(\boldsymbol{K},\omega) = \frac{1}{\pi} \frac{\frac{1}{2}\Delta\omega(\boldsymbol{K})}{\omega^2 + [\frac{1}{2}\Delta\omega(\boldsymbol{K})]^2} \quad , \tag{12.32}$$

but now the FWHM amounts to

$$\Delta E(\boldsymbol{K}) = \hbar\Delta\omega(\boldsymbol{K}) = 4\hbar \sum_j \frac{1}{\tau(j)} \sin^2\left(\frac{\boldsymbol{K} \cdot \boldsymbol{j}}{2}\right) \quad . \tag{12.33}$$

We see that the periodicity of the available binding sites in real space makes the quasielastic energy width ΔE periodic in reciprocal space. Note that, at sufficiently small values of K, equation (12.33) always approaches the simple parabolic form of (12.26) with D given by *Einstein*'s relation

$$D_{K} = \frac{\langle j_{K}^2 \rangle}{2\tau} \quad , \tag{12.34}$$

where j_{K} is the component of $\boldsymbol{j}$ along the direction of $\boldsymbol{K}$, and D_{K} is the diffusion coefficient in that direction. In the thermal average $\langle j_{K}^2 \rangle$ the $\boldsymbol{j}$'s are weighted with their jump frequencies. This asymptotic behavior is to be expected, as probing the diffusion on a large length scale, or equivalently at small K-values, cannot reveal details of the microscopic diffusion mechanism. Equations (12.26) and (12.33) also imply that the quasielastic contribution to the specular scattering (at $\boldsymbol{K} = 0$) is always *purely* elastic ($\Delta E = 0$).

Table 12.1. Relation between quasielastic energy width ΔE and parallel momentum transfer $\hbar K$ for three one-dimensional jump-diffusion models (12.33). $\boldsymbol{K}$ is aligned with the direction of the diffusion. The average time between successive jumps is τ. For the third model the summation in (12.33) has been replaced by the corresponding integral. Also the diffusion constant is shown, expressed in the jump lengths and τ (equation (12.34)). For comparison also the parabolic $\Delta E(K)$ relation for random continuous diffusion is listed.

Diffusion model	$\Delta E(K)$	D
Jumps over $\pm L$	$\frac{4\hbar}{\tau}\sin^2\left(\frac{KL}{2}\right)$	$\frac{L^2}{2\tau}$
Jumps over $\pm L$ and $\pm 2L$	$\frac{2\hbar}{\tau}\left[\sin^2\left(\frac{KL}{2}\right) + \sin^2(KL)\right]$	$\frac{5L^2}{4\tau}$
Continuous distribution of jump lengths with a maximum length $j_{\max}$	$\frac{2\hbar}{\tau}\left[1 - \frac{\sin(Kj_{\max})}{Kj_{\max}}\right]$	$\frac{j_{\max}^2}{6\tau}$
Random continuous diffusion	$2\hbar DK^2$	—

Table 12.1 lists the expressions for $\Delta E(\boldsymbol{K})$ and D in terms of jump rates and jump distances for three simple one-dimensional jump diffusion models, with $\boldsymbol{K}$ aligned along the diffusion direction. The first two models in Table 12.1 show that when the jumps are restricted to lattice vectors the quasielastic energy width ΔE becomes a periodic function of K with nodes, $\Delta E = 0$, at the reciprocal-lattice points, $K = G = n2\pi/L$, with integer n.

In the case of jump diffusion on a two-dimensional, rectangular lattice, with lattice parameters L_x and L_y in the x- and y-directions, the diffusion coefficient and quasielastic energy width for a direction which makes an angle ϕ with the x-axis, are obtained as simple linear combinations of the corresponding quantities for one-dimensional diffusion along the principal directions.

$$D(\phi) = D_x \cos^2 \phi + D_y \sin^2 \phi \quad , \tag{12.35}$$

and

$$\Delta E(\boldsymbol{K}) = \Delta E_x(K \cos\phi) + \Delta E_y(K \sin\phi) \quad . \tag{12.36}$$

12.1.3 High Adatom Concentrations

The results in the previous two sections have been obtained for an adatom overlayer with a low concentration ρ. At higher coverages four of the approximations we have made may break down to some extent.

The eikonal approximation (12.1) does nothing to simulate multiple scattering effects. To some extent inter-adatom and adatom–substrate multiple scattering may still be accounted for within an effective form factor $F(\boldsymbol{k})$ for the isolated adatom. As the density of adatoms increases, however, intra-adatom multiple scattering cannot be accounted for in the purely kinematic approach of (12.1). At this limit the scattered intensity is no longer strictly expressible in terms of the Fourier transform of a pairwise correlation function.

In addition the additive form of the shape function ζ of the surface (12.2) becomes invalid as soon as the shape functions Z of the individual adatoms start to overlap. As a consequence, the factorization of the non-specular scattering probability into a part depending on the form factor of the adatom and a part depending on the correlations in space and in time of the adatom centers, (12.6) and (12.20), is no longer straightforward. For large $\boldsymbol{K}$, adatoms surrounded by other adatoms will generally scatter with a strongly reduced value of $\mid F(\boldsymbol{K}, k_z) - F(\boldsymbol{K}, 0) \mid^2$ compared to isolated adatoms. If the diffusion coefficients of surrounded and isolated adatoms do not differ one can define an effective adatom shape $Z'(\boldsymbol{R})$, which is less pronounced than $Z(\boldsymbol{R})$ and leads to a correctly reduced effective $\mid F'(\boldsymbol{K}, k_z) - F'(\boldsymbol{K}, 0) \mid^2$. However, when the diffusion coefficients do become different for high- and low-coverage regions on the surface, such as occurs with adsorbate islanding or terrace formation, one has to realize that the isolated adatoms scatter with a higher cross section than the surrounded adatoms.

The third approximation which may fail at higher adatom concentrations is $P_{L \neq 0} \approx \Theta$, which was used to arrive at the simple expression for the structure factor $S(\boldsymbol{K})$. Thus, at higher coverages (12.22) has to be replaced by

$$S(\boldsymbol{K}) = (1 - \Theta) + 4\pi^2 \rho \sum_{G} \delta(\boldsymbol{K} - \boldsymbol{G}) + \sum_{L \neq 0} (P_L - \Theta) \exp(\mathrm{i}\boldsymbol{K} \cdot \boldsymbol{L}).$$

$$(12.37)$$

The few terms at small $\mid \boldsymbol{L} \mid$ with $P_L \neq \Theta$ in the last sum of (12.37) introduce some angular variation in the magnitude of the structure factor. This alone produces no change in the energy widths of the quasielastic intensity distributions.

The most important approximation which probably breaks down at higher adatom coverages is the convolution approximation, (12.16). He-atom scattering measures the sum of $S_s(\boldsymbol{K}, \omega)$ and $S_d(\boldsymbol{K}, \omega)$ and is therefore really sensitive only to the sum of the correlation functions $G_s(\boldsymbol{R}, t)$ and $G_d(\boldsymbol{R}, t)$. Consequently the diffusion coefficient determined with atom-scattering measurements at small K is always the *transport diffusion coefficient*, i.e. the diffusion coefficient for lateral fluctuations in the total adatom density. The convolution approximation equates this diffusion coefficient with the so-called *tracer diffusion coefficient*, which is the diffusion coefficient for individual adatoms. Using this approximation we have therefore seen that it makes $S(\boldsymbol{K},\omega)$ equal to $S(\boldsymbol{K})S_{\mathrm{s}}(\boldsymbol{K},\omega)$, where $S(\boldsymbol{K})$ is a function of $\boldsymbol{K}$ only. The quasielastic atom-scattering results can then indeed be interpreted in terms of the *tracer diffusion coefficient*. One expects that increasing the adatom coverage will gradually make the convolution approximation (12.16) fail, at first at small times t. Eventually, the two diffusion coefficients become different, even if the microscopic diffusion mechanism remains unchanged. In fact, this takes place when the velocities of different atoms at different times start to become correlated [12.3]. It is easy to demonstrate that in the absence of adatom–adatom interactions the *transport diffusion coefficient* does not depend on the adatom coverage [12.17–19]. A change in quasielastic energy widths with increasing coverage would thus reflect directly the influence of such interactions on the diffusion process [12.20, 21]. In the two examples of surface diffusion, discussed in the following sections, such a coverage effect has not been observed, probably because the adatom coverages were still too low in those experiments.

12.2 Experimental Procedure

Both of the quasielastic He-atom scattering experiments described in the next section were performed with the apparatus shown schematically in Fig.12.1. This setup has been developed by the group of Prof. J.P. Toennies, at the Max-Planck-Institut für Strömungsforschung in Göttingen (FRG), for high-resolution studies of elastic atom scattering (e.g. surface diffraction) and inelastic atom scattering (e.g. surface phonons) [12.22–24].

Supersonic He beams were used with very low energies, between 7.4 and 8.0 meV for the Cu(111)\S measurements (Sect.12.3.1) and down to 2.2 meV for the Pb(110) measurements (Sect.12.3.2). To this end the He source, including the nozzle (10 μm diameter), was cooled to temperatures between 10 and 40 K. Typical He source pressures ranged from around 3 bar at the lower temperatures to 5 bar at higher temperatures. Scattered He atoms were detected at a fixed scattering angle of 90° with respect to the incident beam, by electron-impact ionization and He-ion counting. Energy spectra of scattered He atoms were obtained from the distributions of the flight time between chopper 1 and the detector. The energy resolution of the complete system, comprising the energy distribution of the incident He beam and the velocity resolution of the time-of-flight (TOF) measurement, amounted to $\sim$80 to $\sim$200 μeV for beam energies between 2.2 and 8.0 meV. This was determined both from scattering measurements with the surface at room temperature and from measurements of the purely elastic specular beam.

As the ratio of quasielastic signal to inelastic background was extremely low at high parallel momentum transfers and at high crystal temperatures, measurement times of up to 30 h were needed for a sufficiently precise determination of the quasielastic peak width at each angular setting. In order to

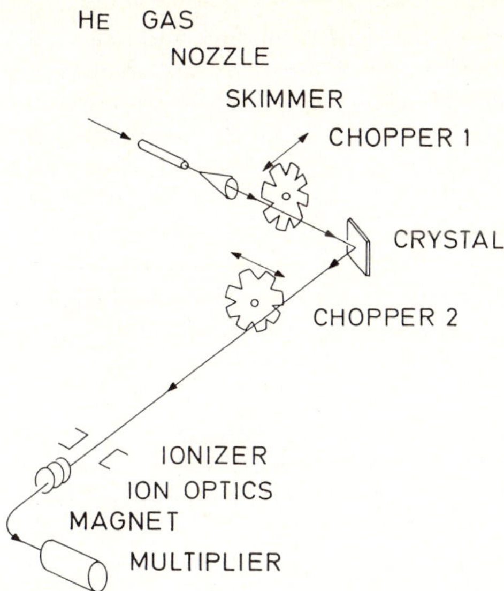

Fig.12.1. Schematic representation of the He-atom scattering apparatus

raise the signal-to-background ratio, a second chopper was installed between the crystal and the detector, which ran in phase, to within ± 2.5 μs, with chopper 1 [12.25]. Chopper 2 serves to select a specific time window Δt in the TOF spectrum. This time window is much narrower than the width ΔT of the complete TOF distribution (both Δt and ΔT are evaluated at the detector). The remaining fraction of the TOF distribution, $(\Delta T - \Delta t)/\Delta T$, is not transmitted by chopper 2. Consequently, more He pulses may be started at chopper 1 without signal overlap. The rate at which the selected part of the TOF distribution is acquired can thus be increased by a factor $\Delta T/\Delta t$. The maximum gain factor is, however, limited by the requirement that no signal be transmitted through earlier or later time windows. $\Delta T/\Delta t$ depends on the crystal–chopper and crystal–detector distances d_{CC} and d_{CD}: $\Delta T/\Delta t = 2d_{\mathrm{CD}}/d_{\mathrm{CC}}$. For our setup the maximum gain factor amounted to $\Delta T/\Delta t = 6$. In addition, chopper 2 reduced the contribution to the background signal due to He gas diffusing from the scattering chamber through the flight path into the detector. For the measurements reported here, the phase difference between the two choppers was set such that the time windows were centered around the time of arrival of each quasielastic peak.

During the long measurements, of several hours up to 30 h, the incident He beam tended to increase slowly in average energy and, to some extent, also in energy width. The effects of these variations were dealt with by dividing up the prolonged measurements of quasielastic scattering into a series of shorter measurements, separated by short reference measurements of the actual resolution function, at $K = 0$ (specular scattering). Using the average energy drift, determined from the reference measurements, all quasielastic and reference measurements in the series were then shifted in energy and summed to yield the drift-corrected quasielastic peak and the effective instrumental resolution function during the measurement of that peak. In this way the effects of possible slow variations in beam energy and beam quality were eliminated.

The following procedure was adopted in order to determine the FWHM ΔE of the quasielastic energy distribution from the experimental quasielastic peak. As mentioned before, our quasielastic peaks were superimposed on a relatively high inelastic background due to multiphonon processes. A second- or third-order polynomial was therefore fitted to this smoothly varying background on both sides of the peak and subtracted from the data. The width of the background-subtracted peak was then corrected for the instrumental resolution. For this purpose, the quasielastic energy profile was assumed to have a Lorentzian shape (12.25, 32), and the instrumental response was approximated by a Gaussian. Several methods are available to extract the FWHM of a Lorentzian profile from the convolution of the Lorentzian with a Gaussian instrument function of known width [12.26]. The background subtraction, however, was found to have a minor effect on the resulting peak shape, changing it slightly from the usual Voigt shape (convolution of a Lorentzian with a Gaussian). In order to reliably extract the Lorentzian widths from the measured peaks the signal, after background subtraction, was first fitted to a Gaussian of variable FWHM. We calibrated this width analysis by applying it to numerically generated convolutions of our Gaussian response function with Lorentzians of various widths, superimposed on a large, smoothly varying background, such as encountered in the actual experiments.

12.3 Experimental Results

12.3.1 Heterodiffusion: S on Cu(111)

As a first experimental example of the quasielastic scattering of He atoms we present results for the lateral diffusion of S atoms on a Cu(111) surface. The surface was produced by spark-cutting and mechanical polishing. It was then further prepared in situ by repeated cycles of Ne^+ sputtering (1 keV, 1 μA, 15 min at room temperature) and annealing at 1000 K. Surface cleanliness was checked with Auger electron spectroscopy (AES). After more than one hundred cycles the AES intensities indicated surface contaminant levels (O, C, S) of less than 0.5%. For most experimental applications this may be considered a clean surface. However, the bulk of the crystal had not been leached of S. Thus, after prolonged heating at 800 K or above, surface segregation was still apparent.

The rate of segregation of S at 820 K was estimated from the measured S and Cu AES peaks at energies of 152 and 60 eV respectively, using the sensitivities from [12.27]. These AES measurements were performed at room temperature, which adds extra uncertainty to the determined surface coverages of S segregated at 820 K. The calibration indicated a segregation rate of S onto the Cu(111) surface in the order of 1 atomic percent per hour, with a weak dependence on coverage.

At lower temperatures step edges are known to orient preferentially along the close-packed $\langle 110 \rangle$ directions on the (111) surface of a fcc crystal. These steps strongly scatter He atoms diffusely along the $\langle 112 \rangle$ azimuths perpendicular to these defects [12.28]. The quasielastic He atom scattering measurements on Cu(111) were performed along the [$1\bar{1}0$] azimuth which was chosen to minimize the diffuse elastic scattering intensity from static defects such as the $\langle 110 \rangle$ step edges.

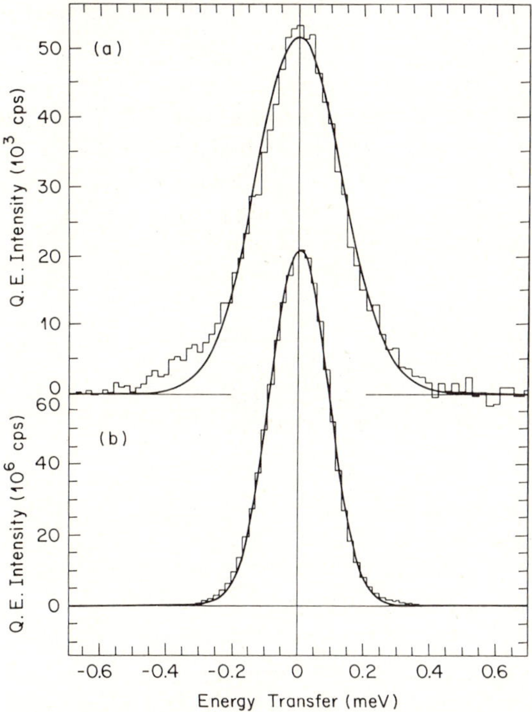

Fig.12.2. Energy distributions of He atoms scattered from a S-contaminated Cu(111) surface at 820 K. The measurement time of 3 h started $4\frac{1}{2}$ h after the freshly cleaned crystal was heated to 820 K. The central energy of the incident He beam was E_i = 7.49 meV: (a) K = 0.65 Å^{-1} along the [$1\bar{1}0$] azimuth; FWHM = 307 μeV; (b) Specular peak (K = 0), showing the experimental resolution function. The solid curve is a Gaussian fit with a FWHM of 208 μeV

Figure 12.2 shows the background-subtracted energy distributions of He atoms scattered along the [$1\bar{1}0$] azimuth from the S-contaminated Cu(111) surface at 820 K, for two incidence angles, corresponding to parallel momentum transfers K of 0.65 Å^{-1} and 0 Å^{-1}. The 208 μeV energy width of the specular distribution, Fig.12.2b, was determined solely by the experimental conditions of the incident He beam. The additional 99 μeV width observed under the same conditions (Sect.12.2) at nonzero K, Fig.12.2a, is evidence for the lateral diffusive motion of the dominant diffusely scattering species. We assume this mobile surface species to be adatoms, which, in this case, are the segregated S atoms.

Quasielastic energy widths ΔE obtained from measurements at 820 K, such as the one in Fig.12.2, are shown in Fig.12.3 as a function of the magnitude of the parallel momentum transfer K. These widths have already been corrected for the instrumental response function, as explained in Sect.12.2. The solid curve in Fig.12.3 indicates that the quasielastic energy widths measured along the [1$\bar{1}$0] azimuth can be described rather well by the simple parabolic $\Delta E(K)$ relation for random continuous diffusion (12.26), with a lateral diffusion coefficient for the S adatoms at 820 K of $D = 2.87 \times 10^{-5}$ cm^{2}s^{-1}. The other two curves in Fig.12.3 show the expected $\Delta E(K)$ relation for two nearest-neighbor jump diffusion models, both with the same macroscopic diffusion coefficient of 2.87×10^{-5} cm^{2}s^{-1}. Although the adsorption site of S on Cu(111) has not been investigated, S is known to adsorb on (111) surfaces of other fcc metals, at low coverages, in the 3-fold hollow 'fcc-type' sites [12.29–35]. The dashed curve in Fig.12.3 assumes only the fcc-sites to be available for the diffusing S adatoms, whereas the dashed-dotted curve allows the S adatoms to reside also in the hcp-sites. As derived in (12.33) both jump diffusion models produce periodic $\Delta E(K)$ relations [12.36]. The repeat distance in reciprocal space along the [1$\bar{1}$0] azimuth is identical for the two jump models and amounts to 4.92 Å^{-1}. The data in Fig.12.3 extend no further than ~ 0.9Å^{-1}. At larger parallel momentum transfers the low quasielastic intensities proved prohibitively small, and no reliable ΔE-values could be determined. It is clear from Fig.12.3 that this experimental K-range is too limited to really decide between the three diffusion models, although the first jump model, with fcc-sites only (dashed curve), does not seem to describe the data as well as the other two models.

During the course of these measurements at elevated crystal temperature the S coverage increased continuously. Any coverage dependence of the diffusion coefficient would have directly affected the quasielastic energy width. Figure 12.4 displays the experimental diffusion coefficient as a function of the time after raising the temperature of the freshly cleaned surface to 820 K. The diffusion coefficients in Fig.12.4 have been obtained from single measurements at K-values larger than 0.5 Å^{-1}, in order to optimize the experimental precision (vertical error bars). The parabolic form of $\Delta E(K)$ was assumed to convert energy widths into diffusion coefficients. The horizontal bars do not represent experimental errors but indicate the total acquisition time for each measurement. The upper scale of Fig.12.4 shows the S coverage estimated from AES measurements after cooling the crystal to room temperature. Apart from two deviating points the data in Fig.12.4 are described well by a constant value of the diffusion coefficient of $(2.86 \pm 0.10) \times 10^{-5}$ cm^{2}s^{-1}. Both lower data points are believed to be associated with irreproducibilities of surface preparation. Unusually rough, e.g. poorly annealed or contaminated surfaces probably yield a

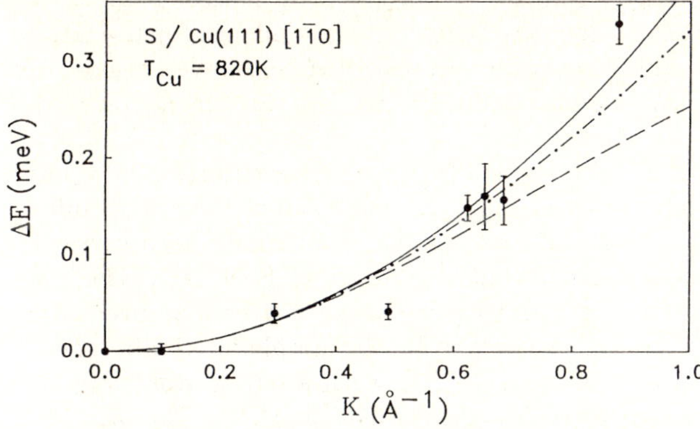

Fig.12.3. Dependence of the quasielastic energy width ΔE on the parallel momentum transfer K for 7.49 meV He atoms scattered from a S-contaminated Cu(111) surface along the [1$\bar{1}$0] azimuth, at 820 K. Solid curve: random continuous diffusion. Dashed curve: jump diffusion with S atoms in only fcc sites. Dashed-dotted curve: jump-diffusion with S atoms in both fcc and hcp sites. Each of the three curves was calculated for a diffusion coefficient of $D = 2.87 \times 10^{-5} \mathrm{cm^2 s^{-1}}$

lowered estimate for the diffusion constant. The first of the two lower points was measured, starting at zero time, with initially no S adatoms present. Indeed, this measurement was most strongly affected by the presence of other (static) defects. At times later than 20 h the quasielastic He scattering measurements proved harder as the diffuse intensity along the [1$\bar{1}$0] azimuth decreased significantly. It is not clear whether this was due to changes in the form factor or the structure factor at higher coverages, as discussed in Sect.12.1.3. Alternatively, adsorbate–induced changes in the dynamic properties of the substrate, and a consequent increase in the proportion of inelastic (multiphonon) scattering might also have reduced the quasielastic intensities.

The data in Fig.12.4 show no evidence for a change in the measured diffusion coefficient or any strongly changing diffusion mechanism in the first 20 h of S segregation. The insensitivity of our results to the S coverage suggests that coverages up to 0.16 monolayer are sufficiently low to keep the *transport diffusion coefficient* constant. As discussed in Sect.12.1.3, this indicates independent diffusive motion of the S adatoms on the Cu(111) surface, implying the lack of noticeable S–S interactions [12.17–21].

12.3.2 Self-Diffusion: Surface Melting of Pb(110)

Our second example of quasielastic atom scattering concerns the two-dimensional self-diffusion of Pb adatoms on the Pb(110) surface at temperatures close to the bulk melting point of Pb, $T_{\mathrm{m}}^{\mathrm{Pb}} = 600.7$ K [12.37–39]. At temperatures above $\sim \frac{3}{4} T_{\mathrm{m}}^{\mathrm{Pb}}$ the (110) face of Pb is known to undergo a continuous and reversible order–disorder transition, called surface melting [12.40]. A variety of experimental probes has been used over the past five years to gather information on the degradation of crystalline order at surfaces of crystals close to

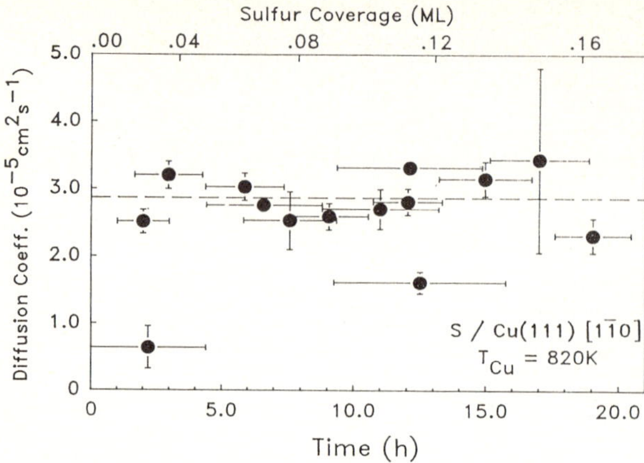

Fig.12.4. Diffusion coefficient for S adatoms on Cu(111) as a function of the time after raising the temperature of the freshly cleaned surface to 820 K. Horizontal bars denote the measurement period. The upper scale shows the coverages of segregated S, as estimated from AES measurements at room temperature. The dashed horizontal line indicates the average diffusion coefficient $D = 2.86 \times 10^{-5}\,\mathrm{cm^2 s^{-1}}$ of all measured diffusion coefficients with the exception of the two lower data points

melting [12.41, 42]. At surfaces such as Pb(110) the thickness of the disordered surface region diverges as the temperature approaches T_m.

During the initial stages of the surface-melting transition the spontaneous generation of surface adatoms and vacancies is believed to play an important role in destabilizing the surface [12.43, 44]. Molecular dynamics simulations suggest that the atoms of a melting surface exhibit liquid–like diffusivities parallel to the surface plane [12.45, 46]. Such high mobilities cannot be measured with the conventional methods used for atomic-scale studies of surface self-diffusion, such as field-ion microscopy [12.47] or the field-emission current-fluctuation technique [12.48]. At high temperatures, self-diffusion on the surfaces of three-dimensional solids has been investigated only with macroscopic-scale techniques, such as mass-transfer and tracer-diffusion measurements [12.49, 50]. Atomic-scale information on surface diffusivities just below T_m has been obtained only for thin methane films adsorbed on MgO or graphite powder, using the quasielastic scattering of thermal-energy neutrons [12.4–6].

We have used quasielastic He-atom scattering to measure the temperature dependence of the lateral diffusion coefficients of Pb adatoms on Pb(110) and to determine the actual diffusion mechanism. The Pb(110) surface was produced by spark-cutting and chemical polishing. The surface was cleaned in situ by cycles of Ar^+ or Xe^+ sputtering and annealing, or by sputtering at elevated temperature. AES and He diffraction were used to ensure that the surface was clean and ordered. Owing to the extremely low surface free energy of Pb, contamination of the surface remained below the detection limit of AES even after prolonged measurements at high temperatures. It was checked that long heat treatments also did not lead to reduction of the He-diffraction intensities.

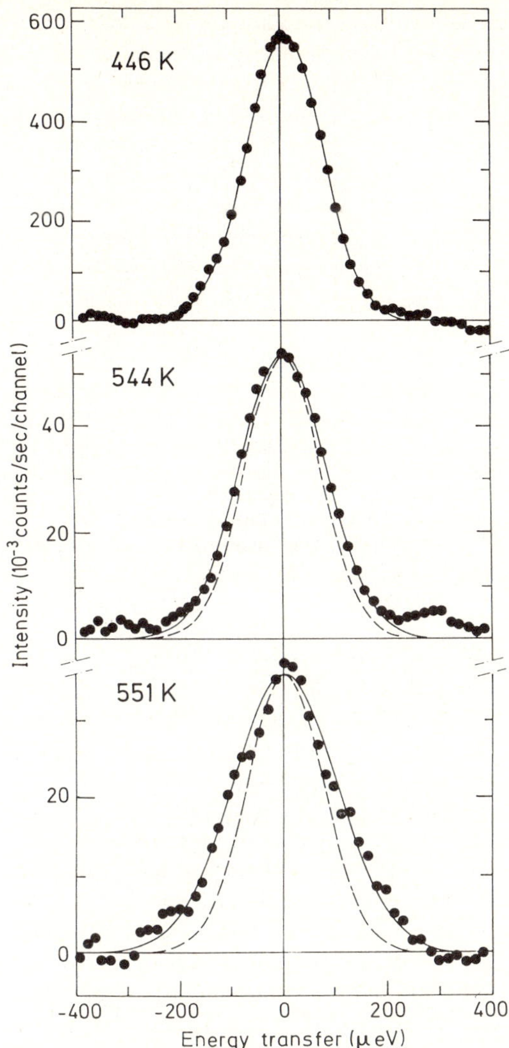

Fig.12.5. Energy distributions of He atoms scattered from a Pb(110) surface at three crystal temperatures. The central energy of the incident He beam was $E_\text{i} = 6.5$ meV, and the parallel momentum transfer was $K = 0.64\text{Å}^{-1}$ along the [001] azimuth: (a) 446 K, (b) 544 K and (c) 551 K. The dashed curves are Gaussians with a FWHM of 163 μeV, corresponding to the experimental resolution function

Figure 12.5 shows a selection of measured energy distributions of He atoms scattered from Pb(110) at crystal temperatures of 446, 544 and 551 K, with scattering conditions corresponding to a parallel momentum transfer of 0.64 Å^{-1} along the [001] azimuth. Like the data in Fig.12.2 the measurements in Fig.12.5 have been corrected by subtraction of a smoothly varying inelastic background. The dashed Gaussian curves indicate the instrumental energy resolution of 163 μeV, determined at room temperature and, independently, in the specular direction ($K = 0$) at each of the elevated temperatures. The energy profiles in Fig.12.5 clearly demonstrate the quasielastic energy broadening with increasing temperature. Also apparent from Fig.12.5 is the dramatic decrease in quasielastic intensity at temperatures above $\sim$500 K. A comparable loss of intensity is also observed for the specular and the diffraction peaks. This phenomenon is most probably caused by the pronounced anharmonicity of the surface vibrations at these high temperatures [12.51, 52]. At temperatures above $\sim$570 K, the quasielastic intensities were too small to allow a determination of the energy width ΔE.

Detailed information on the microscopic self-diffusion mechanism was obtained from measurements of ΔE as a function of parallel momentum transfer $\boldsymbol{K}$, along the two high-symmetry directions [001] and [1$\bar{1}$0], as well as the intermediate [1$\bar{1}$1] direction. One fixed crystal temperature of 521 K was selected for these measurements since it provided an optimal compromise between the quasielastic intensity and the diffusional energy broadening. The results are shown in Fig.12.6. From the data in Fig.12.6 we immediately draw several important qualitative conclusions. Firstly, in none of the three directions is the energy width ΔE proportional to K^2 over the entire K range. This clearly shows that, at this temperature, the self-diffusion on Pb(110) cannot be described as continuous random diffusion (12.23–26). This is different from the situation encountered in the case of S-diffusion on Cu(111), where the data do not allow a distinction between continuous diffusion and other diffusion models. Secondly, for K values smaller than $\sim 0.5\text{Å}^{-1}$, where the parabolic (macroscopic) description of (12.26) seems appropriate, the ΔE values are different along each of the three azimuths. This demonstrates that the diffusion coefficient depends strongly on the surface azimuthal direction. Thirdly, for K values larger than 0.5 Å^{-1}, the three data sets have different shapes, which indicates that the microscopic diffusion mechanisms are also different for the three directions.

The full and dashed-dotted curves in Fig.12.6 are the $\Delta E(\boldsymbol{K})$ relations calculated for two jump-diffusion models with independent jumps along the [001] and [1$\bar{1}$0] directions. Since the Pb(110) surface has rectangular symmetry, we know from (12.36) that energy widths measured along each of the two principal directions depend only on the jumps along that particular direction. The data obtained along the [001] azimuth show one broad maximum in ΔE centered around the Brillouin-zone boundary, $K = \pi/L_{001}$, and a minimum at the reciprocal-lattice point, $K = 2\pi/L_{001} = 1.27\text{Å}^{-1}$. This simple shape is described well by the first one-dimensional diffusion model in Table 12.1, for jumps over single lattice spacings $\pm L_{001}$ along the [001] direction. The solid curve in Fig.12.6a was calculated for an average time between successive jumps of $\tau_{001} = 9.5 \times 10^{-11}$ s, corresponding to a diffusion coefficient along [001] of $D_{001} = 1.3 \times 10^{-5}$ cm^{2}s^{-1}. This is an appreciable fraction of the bulk-liquid value of 2.2×10^{-5} cm^{2}s^{-1} for Pb at the melting point [12.53], and more than

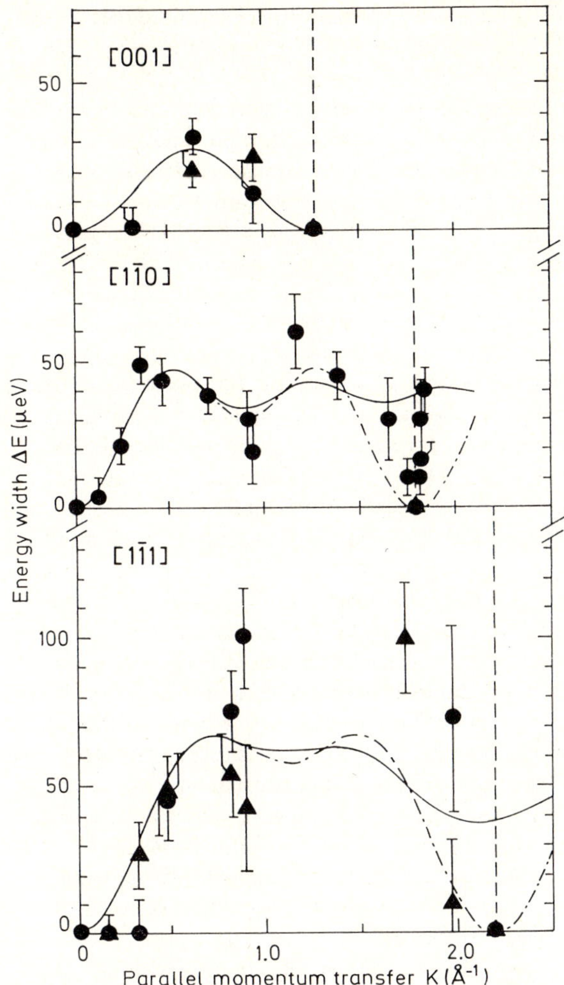

Fig.12.6. Dependence of the quasielastic energy width ΔE on parallel momentum transfer K for He atoms scattered from a Pb(110) surface at 521 K. The incident beam energy ranged from 4.9 to 6.1 meV: (a) K along the [001] direction, (b) K along [1$\bar{1}$0] and (c) K along [1$\bar{1}$1]. Dashed vertical lines denote the reciprocal-lattice points. The solid and dashed-dotted curves are fits for the two jump-diffusion models discussed in the text

four orders of magnitude higher than the value of $4.5 \times 10^{-10} \mathrm{cm}^2\mathrm{s}^{-1}$ for bulk-solid Pb at T_m [12.54].

The data in Fig.12.6b for the [1$\bar{1}$0] azimuth cannot be described by the simple sort of jump model appropriate along [001]. The energy width appears to go through a local minimum at the Brillouin-zone boundary, $K = \pi/L_{1\bar{1}0}$. The dashed-dotted curve in Fig.12.6b was calculated according to the second diffusion model of Table 12.1, for equally probable jumps over single and double lattice distances along the $[1\bar{1}0]$ direction, $\pm L_{1\bar{1}0}$ and $\pm 2L_{1\bar{1}0}$. The best-fit value for the average time between successive jumps within this model amounts to $\tau_{1\bar{1}0} = 4.3 \times 10^{-11}$ s, which corresponds to a diffusion coefficient of $D_{1\bar{1}0} = 3.5 \times 10^{-5} \mathrm{cm}^2\mathrm{s}^{-1}$. At the reciprocal-lattice point, $K = 2\pi/L_{1\bar{1}0} = 1.80\text{Å}^{-1}$, the experimental ΔE shows a narrow minimum, beyond which it sharply returns to a value around 40 μeV. The intensities suggest that the scattering around $K = 1.80$ Å^{-1} is dominated by a purely *elastic* diffraction contribution from the Pb(110) substrate, which is not yet disordered completely at 521 K. Ignoring the few low data points in the direct vicinity of 1.80Å^{-1}, we see that the experimental energy width does not appear to exhibit the global minimum around the reciprocal-lattice point, expected according to the single- and double-jump model, or any other jump model with lattice periodicity. Better agreement between theory and experiment was obtained by removing the periodicity from the model and allowing a continuous distribution of equally probable jump lengths between zero and a maximum jump length $j_{\max}$ (see third entry of Table 12.1). The best fit with this model (solid curve in Fig.12.6b) was reached for $\tau_{1\bar{1}0} = 3.4 \times 10^{-11}$s and $j_{\max} = 8.7$Å, corresponding to a diffusion coefficient of $D_{1\bar{1}0} = 3.8 \times 10^{-5}\mathrm{cm}^2\mathrm{s}^{-1}$, close to the results for the single- and double-jump model.

As mentioned before, we describe the self-diffusion on Pb(110) as a combination of independent diffusional jumps along the orthogonal $[001]$ and $[1\bar{1}0]$ directions. Following (12.35) and (12.36) we obtain the diffusion coefficient and the quasielastic energy width for the intermediate $[1\bar{1}1]$ azimuth as linear combinations of the results along $[001]$ and $[1\bar{1}0]$. The two curves in Fig.12.6c represent the appropriate linear combinations of the two fits in Fig.12.6b for the $[1\bar{1}0]$ direction with the fit in Fig.12.6a for the $[001]$ azimuth. As for the $[1\bar{1}0]$ direction, the energy distributions along the $[1\bar{1}1]$ azimuth are not broadened at the reciprocal-lattice point (2.20 Å^{-1}), due to a dominant diffraction contribution from the substrate. Within the substantial error margins the other data are described well by the solid curve. The corresponding diffusion coefficient along $[1\bar{1}1]$ amounts to $D_{1\bar{1}1} = 2.9 \times 10^{-5}\mathrm{cm}^2\mathrm{s}^{-1}$.

Figure 12.7 shows the energy widths ΔE measured at the Brillouin-zone-boundary positions $K = \pm 0.64\text{Å}^{-1}$ along $[001]$ and at $K = -0.90\text{Å}^{-1}$ along $[1\bar{1}0]$, as a function of the crystal temperature. Assuming that the diffusion mechanisms along these two directions do not change over the temperature range of Fig.12.7, the results from Fig.12.6, at 521 K, were used to convert the energy widths into the diffusion coefficients on the right-hand vertical scale of Fig.12.7. The dashed lines in Fig.12.7 represent the bulk-diffusion coefficient of liquid Pb at T_{m} [12.53]. At the surface this value is already reached at temperatures appreciably below T_{m}. The solid curves in Fig.12.7 show an attempt to describe the surface diffusion coefficients with the *Arrhenius* expression $D_{\mathrm{s}}(T) = D_0 \exp(-Q_{\mathrm{s}}/k_{\mathrm{B}}T)$, where k_{B} is *Boltzmann*'s constant. Within the large statistical scatter, the data for the $[001]$ direction can be fitted by this expression with an activation energy $Q_{\mathrm{s}} = 1.0 \pm 0.3$ eV and a preexpo-

$D_0 = 6.2 \times 10^4 \mathrm{cm^2 s^{-1}}$. The solid curve for the [1$\bar{1}$0] azimuth was calculated for the same activation energy and a preexponential factor of $1.8 \times 10^5 \mathrm{cm^2 s^{-1}}$. These results are to be compared with the activation energies for self-diffusion in bulk solid and liquid Pb, which are 1.11 and 0.19 eV respectively [12.53, 54]. The value of 1.0 eV found here for the [001] direction on Pb(110) is much higher than the bulk-liquid value and close to the activation energy for the bulk solid. A similar observation has been reported for the activation energy for self-diffusion in melting methane films [12.4–6]. These observations suggest that in both cases the lateral diffusion at a melting surface is noticeably affected by the presence of residual crystalline order at the surface.

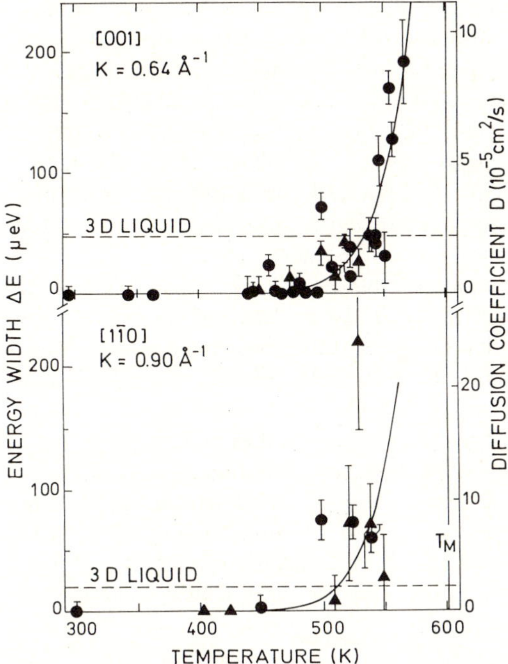

Fig.12.7. Temperature dependence of the quasielastic energy width ΔE for He atoms scattered from Pb(110): (a) $\boldsymbol{K} = \pm 0.64$ Å^{-1} along the [001] azimuth and (b) $\boldsymbol{K} = -0.90$ Å^{-1} along [1$\bar{1}$0]. The incident beam energies were 2.2 meV (triangles) and 6.5 meV (circles). The right-hand vertical axis shows the diffusion coefficients calculated from the energy widths assuming the diffusion mechanisms derived at 521 K from Fig.12.6 to remain unchanged over the whole temperature range. The dashed horizontal lines denote the diffusion coefficient of bulk liquid Pb at T_m [12.53]. The solid Arrhenius curves are discussed in the text.

12.4 Conclusion

In summary, the quasielastic scattering of He atoms is used to investigate diffusion phenomena on surfaces on a microscopic scale. The two examples presented in this chapter illustrate how this technique can be employed to determine the lateral diffusion coefficients of adatoms and to obtain insight in their microscopic diffusion mechanisms.

The diffuse elastic or quasielastic intensity in He-atom scattering always arises from defects on the surface. In this chapter we have assumed these defects to be single adatoms. Generally, not only single adatoms, but also adatom clusters, vacancies, step edges and other surface imperfections give rise to diffuse scattering. The quasielastic energy width is of course governed by the lateral diffusive motion of the dominant diffuse scatterer present on the surface. In view of our limited energy resolution, steps and other extended surface defects may be safely regarded as *static* defects. The fact that we observe quasielastic energy broadening clearly shows that static defects do not dominate the diffuse scattering in our experiments. Among adatoms and surface vacancies the former probably have the higher cross section for diffuse scattering [12.55] as well as the higher lateral mobilities. In principle, one could discriminate between scattering from single adatoms and adatom clusters [12.56] on the basis of the form factor of the scattering species, i.e. the K-dependence of the quasielastic intensity (12.20). Such an analysis has not been attempted here, but instead, single adatoms were simply assumed to outnumber di-, tri- and other multiatom clusters.

A major advantage of quasielastic He-atom scattering over quasielastic neutron scattering in the study of diffusion on surfaces lies in the extreme surface sensitivity of low-energy He atoms and their large cross section for diffuse scattering from surface adatoms. This allows experiments on the surface of one single crystal rather than on powders. Moreover, quasielastic He-atom scattering can be used in principle for any adsorbate-substrate combination rather than the limited set of combinations of highly incoherent neutron scatterers adsorbed on highly coherently scattering powder substrates [12.4–6].

A disadvantage of quasielastic He-atom scattering with respect to quasielastic neutron scattering is that the He atoms cannot distinguish between the *self* and *distinct* contributions to the total correlation function $G(\boldsymbol{R}, t)$. As discussed in Sect.12.1, this limits the validity of the simple theory to low adatom concentrations. An additional problem is that measurements such as those on jump diffusion of Pb adatoms on the Pb(110) surface (Sect.12.3.2) do not permit one to discriminate between diffusion mechanisms in which a single adatom jumps from one site to another, and exchange mechanisms in which the adatom actually changes places with a substrate atom, which, in turn, is displaced to a new adatom position. In particular, the jump diffusion across the close-packed [1$\bar{1}$0] rows could take place in this way. In fact, an exchange mechanism has been proposed for diffusion of adatoms across the close-packed rows on W(221), Pt(110) and Ir(110) [12.48, 57–59] .

The velocity resolution of the present He-scattering setups, which corresponds to an energy resolution of $\sim 80~\mu$eV or more, restricts the application of the quasielastic-scattering technique to surfaces with highly mobile adatoms, i.e. with $D \gtrsim 5 \times 10^{-6} \mathrm{cm}^2 \mathrm{s}^{-1}$. Such diffusivities are more than three orders of magnitude higher than those attainable with other microscopic techniques and fall in the range where otherwise only macroscopic methods are available which yield, at best, the transport diffusion coefficient [12.3]. Improvements of the instrumentation will prove necessary in order to measure diffusion coefficients below $5 \times 10^{-6} \mathrm{cm}^2 \mathrm{s}^{-1}$ and to extend the measurements to higher K values. Such improvements will involve the stability, the energy spread and the intensity of the primary He beam as well as the time resolution and the efficiency of the detection system.

Acknowledgements

Both authors would like to express their gratitude for the hospitality and the support of Prof. J.P. Toennies and his group in Göttingen (FRG), which enabled them to perform the work described in this chapter. The Alexander-von-Humboldt Foundation (Bonn, FRG) is gratefully acknowledged for financial support.

References

12.1   M. Bée: *Quasielastic Neutron Scattering* (Adam Hilger, Bristol 1988)

12.2   T. Springer: *Quasielastic Neutron Scattering for the Investigation of Diffusive Motions in Solids and Liquids* (Springer, Berlin, Heidelberg 1972)

12.3   R. Gomer: Rep. Prog. Phys. to be published

12.4   M. Bienfait: Europhys. Lett. **4**, 79 (1987)

12.5   M. Bienfait, J.P. Coulomb, J.P. Palmari: Surf. Sci. **182**, 557 (1987)

12.6   M. Bienfait, P. Zeppenfeld, J.M. Gay, J.P. Palmari: Surf. Sci. **226**, 327 (1990)

12.7   A.C. Levi, R. Spadacini, G.E. Tommei: Surf. Sci. **121**, 504 (1982)

12.8   L. Van Hove: Phys. Rev. **95**, 249 (1954)

12.9   The introduction of a periodic corrugation of the substrate would, of course, redistribute the elastic specular intensity in (12.5) and (12.6) over a set of diffraction peaks at the two-dimensional reciprocal-lattice points of the substrate.

12.10 Note that not only the eikonal approximation but also other, more exact approaches (e.g. the RR′-method and the sudden approximation) would lead to a factorization of the nonspecular term in (12.6) into a form factor and a factor which contains the scattering function $S(\boldsymbol{K},\omega)$.

12.11 B. Chakraborty, S. Holloway, J.K. Nørskov: Surf. Sci. **152/153**, 660 (1985)

12.12 In some cases, such as quantum diffusion of H on metal surfaces, the simple description of $G_s(\boldsymbol{R},t)$ in equation (12.12) in terms of vibrational and diffusive motion may be completely inappropriate: M.J. Puska, R.M. Nieminen, M. Manninen, B. Chakraborty, S. Holloway, J.K. Nørskov: Phys. Rev. Lett. **51**, 1081 (1983)

12.13 The 'perpendicular' Debye-Waller factor does not include the effect of the attractive potential well of the surface on the perpendicular momentum of the He atom.

12.14 G.H. Vineyard: Phys. Rev. **110**, 999 (1958)

12.15 Perpendicular vibrational motion of the *substrate* would result in an additional 'perpendicular' substrate Debye-Waller factor in front of the $F(\boldsymbol{K},0)$-term in (12.5), (12.6) and (12.20), and in front of the term proportional to $(1 - 2\rho a_{\mathrm{at}})$ in the specular scattering contribution in (12.5) and (12.6). The proper treatment of the Debye-Waller factor in the case of atom-surface scattering can be found in: A.C. Levi, H. Suhl: Surf. Sci. **88**, 221 (1979)

12.16 C.T. Chudley, R.J. Elliott: Proc. Phys. Soc. **77**, 353 (1961)

12.17 C.H. Mak, H.C. Andersen, S.M. George: J. Chem. Phys. **88**, 4052 (1988)

12.18 R. Kutner: Phys. Lett. **81A**, 239 (1981)

12.19 K.W. Kehr, R. Kutner, K. Binder: Phys. Rev. **B23**, 4931 (1981)

12.20 R. Kutner, K. Binder, K.W. Kehr: Phys. Rev. **B28**, 1846 (1983)

12.21 R. Kutner, K. Binder, K.W. Kehr: Phys. Rev. **B26**, 2967 (1982)

12.22 G. Lilienkamp, J.P. Toennies: J. Chem. Phys. **78**, 5210 (1983)

12.23 D.M. Smilgies, J.P. Toennies: Rev. Sci. Instrum. **59**, 2185 (1988)

12.24 J.P. Toennies: In *Surface Phonons*, ed. by W. Kress, F. DeWette, Springer Ser. Surf. Sci. Vol. 21 (Springer, Berlin, Heidelberg 1991)

12.25 H.H. Sawin, D.D. Wilkinson, W.M. Chan, S. Smiriga, R.P. Merrill: J. Vac. Sci. Technol. **14**, 1205 (1977)

12.26 G.K. Weitheim, M.A. Butler, K.W. West, D.N.E. Buchanan: Rev. Sci. Instrum. **45**, 1369 (1974)

12.27 L.E. Davis, N.C. MacDonald, P.W. Palmberg, G.E. Riach, R.E. Weber: *Handbook of Auger Electron Spectroscopy* (Physical Electronics Industries, Eden Prairie 1976)

12.28 B.J. Hinch, A. Lock, J.P. Toennies, G. Zhang: J. Vac. Sci. Technol. **B7**, 1260 (1989)

12.29 J.E. Demuth, D.W. Jepsen, P.M. Marcus: Phys. Rev. Lett. **32**, 1182 (1974)

12.30 C.-M. Chan, W.H. Weinberg: J. Chem. Phys. **71**, 3988 (1979)

12.31 F. Maca, M. Scheffer, W. Berndt: Surf. Sci. **160**, 467 (1985)

12.32 K. Hayek, H. Glassl, A. Gutmann, H. Leonhard, M. Prutton, S.P. Tear, M.R. Welton-Cook: Surf. Sci. **152/153**, 419 (1985)

12.33 P.C. Wong, M.Y. Zhou, K.C. Hui, K.A.R. Mitchell: Surf. Sci. **163**, 172 (1985)

12.34 Th. Fauster, H. Dürr, D. Hartwig: Surf. Sci. **178**, 657 (1986)

12.35 D.R. Warburton, P.L. Wincott, G. Thornton, F.M. Quinn, D. Norman: Surf. Sci. **211/212**, 71 (1989)

12.36 In the case of jump diffusion on both fcc- and hcp-sites $\Delta E(\boldsymbol{K})$ cannot be calculated using equation (12.33), since together these sites do not form a Bravais lattice. Equation (12.27) has to be rewritten as a set of two coupled differential equations for the intermediate functions for diffusion starting from a fcc-site and diffusion starting from a hcp-site. The solution remains a single Lorentzian (12.32), but with a slightly more complicated expression for the FWHM than (12.33) [12.1,2].

12.37 J.W.M. Frenken, J.P. Toennies, Ch. Wöll: Phys. Rev. Lett. **60**, 1727 (1988)

12.38 J.W.M. Frenken. B.J. Hinch, J.P. Toennies: Surf. Sci. **211/212**, 21 (1989)

12.39 J.W.M. Frenken. B.J. Hinch, J.P. Toennies, Ch. Wöll: Phys. Rev. **B41**, 938 (1990)

12.40 J.W.M. Frenken, P.M.J. Marée, J.F. van der Veen: Phys. Rev. **B34**, 7506 (1986)

12.41 J.F. van der Veen, B. Pluis, A.W. Denier van der Gon: In R. Vanselow, R.F. Howe (eds.): *Chemistry and Physics of Solid Surfaces* Vol. VII, Springer Ser. Surf. Sci., Vol. 10 (Springer, Berlin, Heidelberg 1988) p. 455

12.42 J.G. Dash: Contemp. Phys. **30**, 89 (1989)

12.43 P. Stoltze, J.K. Nørskov, U. Landman: Surf. Sci. **220**, L693 (1989)

12.44 A.W. Denier van der Gon, R.J. Smith, P. Stoltze, D. Frenkel, J.W.M. Frenken: To be published

12.45 J.Q. Broughton, G.H. Gilmer: J. Chem Phys. **79**, 5119 (1983)

12.46 V. Rosato, G. Cicotti, V. Pontikis: Phys. Rev. **B33**, 1860 (1986)

12.47 D.W. Bassett: In Vu Thien Binh (ed.): *Surface Mobilities on Solid Materials* (Plenum, New York 1983) pp. 63 and 83

12.48 M. Tringides, R. Gomer: J. Chem. Phys. **84**, 4049 (1986)

12.49 H.P. Bonzel: In Vu Thien Binh (ed.): *Surface Mobilities on Solid Materials* (Plenum, New York 1983) p. 195
12.50 G.E. Rhead: Surf. Sci. **47**, 207 (1975)
12.51 G. Armand, D. Gorse, J. Lapujoulade, J.R. Manson: Europhys. Lett. **3**, 1113 (1987)
12.52 P. Zeppenfeld, K. Kern, R. David, G. Comsa: Phys. Rev. Lett. **62**, 63 (1989)
12.53 N.H. Nachtrieb: Ber. Bunsenges. **80**, 678 (1976)
12.54 J.W. Miller: Phys. Rev. **181**, 1095 (1969)
12.55 E. Zaremba: Surf. Sci. **151**, 91 (1985)
12.56 P.J. Feibelman: Phys. Rev. Lett. **58**, 2766 (1987)
12.57 Q.J. Gao, T.T. Tsong: Phys. Rev. Lett. **57**, 452 (1986)
12.58 T.T. Tsong, Q.J. Gao: Surf. Sci. **182**, L257 (1987)
12.59 Q.J. Gao, T.T. Tsong: Phys. Rev. **B36**, 2547 (1987)

Index